Lecture Notes in Mathematics

Edited by A. Dold and B. Eckmann

527

Manfred Denker
Christian Grillenberger
Karl Sigmund

Ergodic Theory on Compact Spaces

Springer-Verlag
Berlin · Heidelberg · New York 1976

Authors
Manfred Denker
Christian Grillenberger
Institut für Mathematische Statistik
Lotzestraße 13
D-3400 Göttingen

Karl Sigmund
Mathematisches Institut
Strudlhofgasse 4
A-1090 Wien

Library of Congress Cataloging in Publication Data
Denker, Manfred, 1944-
Ergodic theory on compact spaces.

(Lecture notes in mathematics ; vol. 527)
Bibliography: p.
Includes index.
1. Topological dynamics. 2. Ergodic theory.
3. Metric spaces. 4. Locally compact spaces.
I. Grillenberger, Christian, 1941- joint author.
II. Sigmund, Karl, 1945- joint author. III. Title. IV. Series: Lecture notes in mathematics (Berlin) ; vol. 527.
QA3.L28 vol. 527 [QA611.5] 510'.8s [514'.3]
76-19105

AMS Subject Classifications (1970): 28A50, 28A65, 54H20

ISBN 3-540-07797-9 Springer-Verlag Berlin · Heidelberg · New York
ISBN 0-387-07797-9 Springer-Verlag New York · Heidelberg · Berlin

Printed in Germany
Printing and binding: Beltz Offsetdruck, Hemsbach/Bergstr.

Contents

Introduction

The initial problems of ergodic theory arose in a differentiable framework: smooth flows on energy manifolds preserving Liouville measure. It was Poincaré who first introduced purely measure theoretic considerations. Both, differentiable and measure theoretic ergodic theory, developped tremendously, especially since Kolmogorov's definition of the entropy invariant. Topological ergodic theory grew as an intermediary between the two fields. It deals with continuous transformations and invariant Borel measures and can be traced back to a memoir of Krylov-Bogoliubov [127] in 1937.

In this volume we try to give a survey of the topological ergodic theory on compact spaces. For convenience, the underlying space is assumed to be metric and the transformation to be a homeomorphism.

Each of the authors having his own predilection in the field and his own style, the attempt towards a unifying approach was somehow handicapped. Hopefully, the 'essence' will - however - be distinguished.

Krylov and Bogoliubov proved that there always exist invariant measures. If there is only one invariant measure, the system is called uniquely ergodic. The prevalence of these systems was shown by Jewett and Krieger. In the general case it might be possible to single out one invariant measure by some variational principle. If this is so, one has an intrinsically ergodic system. Important examples of this kind are found by Parry and Bowen. Both, uniquely and intrinsically ergodic, systems will be treated extensively in these notes.

In spite of our efforts to cover the recent publications and to give the right persons credit for the right theorems we are aware of possible errors and wish to apologize for any such mistake . Also we wish to apologize if somebody's work closely connected with the aim of this survey is not emphasized as it should be; in particular, many concrete examples of dynamical systems bearing algebraic, combinatorial, differentiable or other structures are not discussed here.

We would like to single out two references: The Lecture Notes of Walters [185] for an excellent introduction to ergodic theory and of Bowen [31] for a deep study of equilibrium states and ergodic theory of important classes of differentiable dynamical systems.

We have borrowed proofs from many sources and are particularly indebted to Bowen and Misiurewics. Also it should be noted here that chapter 2 from proposition 2.17 onwards and the chapters 9, 13, 26-31 are contained in Grillenberger's "Habilitationsschrift" and some theorems are due to him.

Finally we would like to thank U.Krengel for his encouragement and his support. Likewise, we are grateful to Mrs. M.Powell for her excellent skill and cooperation in preparation of the manuscript and to A.Achilles for his tremendous help in proof-reading.

The authors.

1. Measure-Theoretic Dynamical Systems

(1.1) Definition: A family Σ of subsets of the space $\mathfrak{X}$ which is closed under the operations of complementation and countable union and which contains $\mathfrak{X}$ and $\emptyset$ is called a σ-algebra. A pair $(\mathfrak{X}, \Sigma)$ consisting of a space $\mathfrak{X}$ and a σ-algebra Σ of subsets of $\mathfrak{X}$ is called a measurable space, the elements of Σ are the measurable sets.

Let $(\mathfrak{X}, \Sigma)$ and $(\mathfrak{X}', \Sigma')$ be two measurable spaces. A map $\Psi : \mathfrak{X} \to \mathfrak{X}'$ is said to be measurable (with respect to Σ and Σ') if $\Psi^{-1}(B') \in \Sigma$ for all $B' \in \Sigma'$.

(1.2) Definition: A set function m defined on some σ-algebra Σ of subsets of $\mathfrak{X}$ is called a measure if it is positive, countably additive and normalized ($m(\mathfrak{X}) = 1$). The triple $(\mathfrak{X}, \Sigma, m)$ is called a measure space. We shall often write $(\mathfrak{X}, m)$ instead of $(\mathfrak{X}, \Sigma, m)$.

(1.3) Definition: If $E \subset \Sigma$ with $m(E) > 0$, we denote by Σ_E the trace σ-algebra $\{A \cap E | A \in \Sigma\}$ and by m_E the conditional measure defined by $m_E(A) = \dfrac{m(A \cap E)}{m(E)}$.

(1.4) Definition: Let $(\mathfrak{X}, \Sigma, m)$ and $(\mathfrak{X}', \Sigma', m')$ be two measure spaces. A map $\Psi : \mathfrak{X} \to \mathfrak{X}'$ is said to be measure preserving if it is measurable with respect to Σ and Σ' and if $m(\Psi^{-1}(B')) = m'(B')$ for all $B' \in \Sigma'$. Ψ is said to be an isomorphism between $(\mathfrak{X}, \Sigma, m)$ and $(\mathfrak{X}', \Sigma', m')$ if Ψ is invertible and if both Ψ and Ψ^{-1} are measure preserving.

The following notion is a useful technicality:

(1.5) Definition: A measure space $(\mathfrak{X}, \Sigma, m)$ is said to be complete if Σ contains all subsets of sets of measure zero (i.e. if $A \subset B$ and $B \in \Sigma$, then $m(B) = 0$ implies $A \in \Sigma$). In this case one also says that the measure m is complete.

(1.6) Proposition: Let $(\mathfrak{X}, \Sigma, m)$ be a given measure space. $\Sigma_m := \{A \subset \mathfrak{X} | \exists B, C \in \Sigma : A \Delta B \subset C, m(C) = 0\}$ is the smallest σ-Algebra, containing Σ and all m-null-sets. There exists a unique extension of m to a measure $\overline{m}$ on Σ_m. Σ_m (resp. $\overline{m}$) is called the completion of Σ (resp. m)

One has only to define Σ_m as the family of all $A \subset \mathfrak{X}$ which can be written in the form $A = B \cup N$, where $B \in \Sigma$ and $N \subset M$ for some $M \in \Sigma$ with $m(M) = 0$. Then $\overline{m}$ defined on Σ_m by $\overline{m}(A) = m(B)$ is the completion of m.

In accordance with the probabilistic interpretation of measure spaces, one usually neglects sets of measure 0 in ergodic theory. For example, two sets $A, B \subset \mathfrak{X}$ will be considered to be equal if their symmetric difference $A \Delta B = A \cup B \setminus A \cap B$ has m-measure 0: one writes $A = B \bmod 0$. Similarly, two maps Ψ and Ψ' on $\mathfrak{X}$ will be identified if $m(\{x \in \mathfrak{X} | \Psi(x) \neq \Psi'(x)\}) = 0$: one writes $\Psi = \Psi' \bmod 0$. A transformation defined on a subset $M \subset \mathfrak{X}$ with $m(M) = 1$ will still be considered as a transformation of $\mathfrak{X}$: it is obvious what is meant by saying that such a transformation is measurable, or measure preserving. In particular:

(1.7) Definition: Two measure spaces $(\mathfrak{X}, \Sigma, m)$ and $(\mathfrak{X}', \Sigma', m')$ are said to be measure theoretically isomorphic (or isomorphic mod 0) if there exist sets $M \in \Sigma$, $M' \in \Sigma'$ and a map $\Phi : M \to M'$ such that:

(a) $m(M) = 1$ and $m'(M') = 1$;

(b) Φ is an isomorphism between (M, Σ_M, m_M) and $(M', \Sigma'_{M'}, m'_{M'})$ in the sense of (1.4).

In ergodic theory it is often convenient to exclude from consideration certain measure spaces which are rather pathological and of no importance in applications. In particular, one often restricts the attention to Lebesgue spaces. It should be accentuated that this is a very weak and natural restriction (cf. (2.17)).

(1.8) Definition: A measure space $(\mathfrak{X}, m)$ is said to be a Lebesgue space if it is measure theoretically isomorphic to a measure space of the form $([0,1], \lambda)$ where λ is a Lebesgue-Stieltjes measure on the unit interval $[0,1]$.

(1.9) Definition: A point $x \in \mathfrak{X}$ is said to be an atom of the measure m if $m(\{x\}) > 0$.

It is obvious that a measure can have at most a countable number of atoms. Every Lebesgue space is measure theoretically isomorphic to a measure space consisting of atoms $x_1, x_2, \ldots$ with measures $m(\{x_j\}) = p_j$, and an interval of length $1 - \sum_{j=1}^{\infty} p_j$ with the usual Lebesgue measure.

(1.10) Definition: Let h be a measurable map from $(\mathfrak{X},\Sigma)$ into $(\mathfrak{X}',\Sigma')$. If m is a measure on $(\mathfrak{X},\Sigma)$ then the set function defined by $E' \to m(h^{-1}(E'))$ for $E' \in \Sigma'$ is a measure on $(\mathfrak{X}',\Sigma')$ denoted by hm and called the measure transported by h.

(1.11) Definition: A triple $(\mathfrak{X}, m, \Psi)$ consisting of a measure space $(\mathfrak{X}, m)$ and an isomorphism Ψ from $(\mathfrak{X}, m)$ onto itself is called a measure theoretical dynamical system (m.t. dynamical system).

(1.12) Definition: Two m.t. dynamical systems $(\mathfrak{X}, m, \Psi)$ and $(\mathfrak{X}', m', \Psi')$ are said to be measure theoretically conjugate (or isomorphic mod 0) if there exist sets $M \in \Sigma$, $M' \in \Sigma'$ and a map $\Phi : M \to M'$ such that conditions (a) and (b) of (1.7) are valid and furthermore

(c) $\Phi \circ \Psi = \Psi' \circ \Phi$.

(Note that $\Psi(x) \in M$ for m - a.e. $x \in M$, and that, therefore, $\Phi \circ \Psi$ is defined m - a.e.)

Thus the following diagram - where the functions are defined almost everywhere - is commutative

$$\begin{array}{ccc} \mathfrak{X} & \xrightarrow{\Psi} & \mathfrak{X} \\ \Phi \downarrow & & \downarrow \Phi \\ \mathfrak{X}' & \xrightarrow{\Psi'} & \mathfrak{X}' \end{array}$$

A related notion is that of a factor:

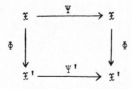

(1.13) Definition: An m.t. dynamical system $(\mathfrak{X}', m', \Psi')$ is said to be a factor of the m.t. dynamical system $(\mathfrak{X}, m, \Psi)$ if there exist sets $M \in \Sigma$, $M' \in \Sigma'$ and a map $\Phi : M \to M'$ such that conditions (a) of (1.7) and (c) of (1.12) are valid and condition (b) in (1.7) is replaced by:

(b') Φ is a measure preserving (but not necessarily invertible) map from (M, Σ_M, m_M) onto $(M', \Sigma'_{M'}, m'_{M'})$.

Φ is called a measure theoretic homomorphism.

(1.14) Definition: The product of two m.t. dynamical systems $(\mathfrak{X}, m, \Psi)$ and $(\mathfrak{X}', m', \Psi')$ is the m.t. dynamical system $(\mathfrak{X} \times \mathfrak{X}', m \times m', \Psi \times \Psi')$ where $m \times m'$ is the product measure and where the product transformation $\Psi \times \Psi' : \mathfrak{X} \times \mathfrak{X}' \to \mathfrak{X} \times \mathfrak{X}'$ is defined by

$$\Psi \times \Psi'(x,x') = (\Psi(x), \Psi'(x')) \quad \text{for} \quad x \in \mathfrak{X},\ x' \in \mathfrak{X}'.$$

Clearly $(\mathfrak{X}, m, \Psi)$ and $(\mathfrak{X}', m', \Psi')$ are factors of $(\mathfrak{X} \times \mathfrak{X}', m \times m', \Psi \times \Psi')$. The product $\Psi \times \Psi'$ is often denoted by $\Psi^{(2)}$. Similarly, one defines $\Psi^{(n)}$, for $n > 2$.

(1.15) Poincaré's recurrence theorem: Let $(\mathfrak{X}, m, \Psi)$ be an m.t. dynamical system and $E \subset \mathfrak{X}$ with $m(E) > 0$. Then for m - a.e. point $x \in E$ there exists an $n = n(x) > 0$ such that $\Psi^n(x) \in E$.

We refer to [80] or [57] for a proof.

(1.16) Definition: Let $(\mathfrak{X}, m, \Psi)$ be an m.t. dynamical system and $m(E) > 0$. Define $\Psi_E(x) = \Psi^{\bar{n}}(x)$, where $\bar{n}$ is the smallest integer $n > 0$ such that $\Psi^n(x) \in E$.

By (1.15) Ψ_E is defined m - a.e. on E.

It is easy to see that (E, m_E, Ψ_E) is an m.t. dynamical system (cf. [80] or [57]). This system is called the induced system on E. Note that if E is Ψ-invariant then Ψ_E is just the restriction of Ψ to E.

<u>(1.17) Definition:</u> $(\mathfrak{X}, m, \Psi)$ is called <u>aperiodic</u> if

$$m(\{x \in \mathfrak{X} \mid \Psi^n(x) = x \text{ for some } n \in \mathbb{N}\}) = 0$$

<u>(1.18) Lemma of Rohlin</u>: If $(\mathfrak{X}, m, \Psi)$ is aperiodic, then for any $\varepsilon > 0$ and any integer $n > 0$ there exists a set $E \subset \mathfrak{X}$ such that $E, \Psi E, \ldots, \Psi^{n-1} E$ are disjoint and $m(\bigcup_{j=0}^{n-1} \Psi^j E) > 1 - \varepsilon$. E is called a <u>$(\Psi, n, \varepsilon)$-Rohlin set.</u>

See e.g. [80], p.71 for a proof.

2. Measures on Compact Metric Spaces

Let X be a compact space with metric d.

(2.1) Definition: The smallest σ-algebra of subsets of X which contains all open (and, therefore, all closed) subsets of X is denoted by $\mathfrak{B}(X)$ or by $\mathfrak{B}$ and is called the Borel-σ-algebra of X. Its elements are called the Borel sets of X.

(2.2) Definition: By a measure on X we shall understand a measure defined on $\mathfrak{B}(X)$ (or, occassionally, the completion of such a measure). We denote by $\mathfrak{M}(X)$ the set of all measures on X.

(2.3) Proposition: All $\mu \in \mathfrak{M}(X)$ are regular, i.e. one has $\mu(B) = \inf \{\mu(U) | B \subset U,\ U \text{ open}\} = \sup \{\mu(C) | C \subset B,\ C \text{ closed}\}$ for all $B \in \mathfrak{B}(X)$.

Proof: Let $\mathfrak{R}$ denote the family of all $B \in \mathfrak{B}(X)$ satisfying these relations. It is obvious that $\mathfrak{R}$ contains X and $\emptyset$. Furthermore it is easy to see that $\mathfrak{R}$ is closed under complementation and that it is a monotonic family in the sense that if $B_1 \subset B_2 \subset \ldots$ is an increasing sequence of elements of $\mathfrak{R}$, then $\bigcup_{i=1}^{\infty} \mathfrak{B}_i \in \mathfrak{R}$. A standard argument shows that $\mathfrak{R}$, therefore, is a σ-algebra.

If $C \subset X$ is closed, then the sets $U_n = \{x \in X | d(x,C) < \frac{1}{n}\}$ form a nested sequence of open sets such that $C = \bigcap_{n=1}^{\infty} U_n$. Since $\mu(U_n) \to \mu(C)$, this implies that $C \in \mathfrak{R}$. Thus, $\mathfrak{R}$ contains all closed sets and, therefore, is equal to $\mathfrak{B}(X)$. □

(2.4) Definition: C(X) denotes the Banach space of all continuous complex valued functions on X with the norm $f \to \|f\| = \sup \{|f(x)|\ |x \in X\}$.

Every measure on X induces a normalized nonnegative linear functional on C(X) by $f \to \int f\, d\mu$. It is well known that the converse is also true.

<u>(2.5) Riesz representation theorem:</u> To any positive linear functional L on C(X) with L(1) = 1, there corresponds a unique $\mu \in \mathfrak{M}(X)$ such that $L(f) = \int f d\mu$ for all $f \in C(X)$.

For a proof we refer to [152].

It is obvious that $\mathfrak{M}(X)$ is a convex space: if $\mu_1, \mu_2 \in \mathfrak{M}(X)$ and $0 \leq \alpha \leq 1$ then $\alpha\mu_1 + (1 - \alpha)\mu_2 \in \mathfrak{M}(X)$.

One obtains a topology on $\mathfrak{M}(X)$ by viewing it as a space of linear functionals endowed with the weak star topology, usually called weak topology in this context.

<u>(2.6) Definition:</u> The <u>weak topology</u> on $\mathfrak{M}(X)$ is defined by taking as a basis of open neighborhoods for $\mu \in \mathfrak{M}(X)$ the sets

$$V_\mu(f_1,\ldots,f_k;\ \epsilon_1,\ldots,\epsilon_k) = \{\nu \in \mathfrak{M}(X) \mid |\int f_j d\mu - \int f_j d\nu| < \epsilon_j, j=1,\ldots,k\}$$

with $\epsilon_j > 0$ and $f_j \in C(X)$.

<u>(2.7) Proposition:</u> Let $\mu_n, \mu \in \mathfrak{M}(X)$. Then the following conditions are equivalent:

(a) $\lim \int f\, d\mu_n = \int f\, d\mu$ for all $f \in C(X)$;

(b) $\limsup \mu_n(C) \leq \mu(C)$ for all closed $C \subset X$;

(c) $\liminf \mu_n(U) \geq \mu(U)$ for all open $U \subset X$;

(d) $\lim \mu_n(A) = \mu(A)$ for all $A \in \mathfrak{B}(X)$ with $\mu(\mathrm{bd}A) = 0$

(e) μ_n converges weakly to μ ($\mu_n \to \mu$).

<u>Proof:</u> (a)$\Leftrightarrow$(e) is trivial, as well as (b)$\Leftrightarrow$(c).

(a)$\Rightarrow$(b). Let $C \subset X$ be closed, and write $U_k = \{x \mid d(x,C) < \frac{1}{k}\}$

Clearly $U_k \downarrow C$ and thus $\mu(U_k) \downarrow \mu(C)$. Let f_k denote a continuous function such that $0 \leq f_k(x) \leq 1$ for all $x \in X$, $f_k(x) = 1$ for $x \in C$ and $f_k(x) = 0$ for $x \notin U_k$. Then:

$$\limsup \mu_n(C) \leq \limsup_{n\to\infty} \int f_k\, d\mu_n = \int f_k\, d\mu \leq \mu(U_k)$$

for all k and hence $\limsup \mu_n(C) \leq \mu(C)$.

(b) and (c) $\Rightarrow$(d). Let $A \in \mathfrak{B}(X)$ and denote by $\bar{A}$, int A and bdA closure, interior and boundary of A, respectively.

If $\mu(\text{bd}A) = 0$ then $\mu(\text{int } A) = \mu(A) = \mu(\bar{A})$. Hence

$$\limsup \mu_n(A) \leq \limsup \mu_n(\bar{A}) \leq \mu(\bar{A}) = \mu(A)$$

$$\liminf \mu_n(A) \geq \liminf \mu_n(\text{int } A) \geq \mu(\text{int } A) = \mu(A)$$

and so $\lim \mu_n(A) = \mu(A)$.

(d)$\Rightarrow$(a). Let $f \in C(X)$ be real, and write $m = \min_{x\in X} f(x)-1$, $M = \max_{x\in X} f(x) + 1$. f transports the measure μ to the measure $f\mu$ on $[m, M]$. It is easy to see that for every $\varepsilon > 0$ there exists a partition $m = t_o < t_1 < \ldots < t_k = M$ such that $t_j - t_{j-1} < \varepsilon$ and $f\mu(\{t_j\}) = 0$ for all j. Set $A_j = \{x | t_{j-1} \leq f(x) < t_j\}$ for $j = 1, \ldots, k$. $A_1,\ldots,A_k$ form a partition of X into Borel sets, and since

$$\bar{A}_j \setminus \text{int } A_j \subset \{x \in X | f(x) = t_{j-1} \text{ or } f(x) = t_j\}$$

one has $\mu\ (\text{bd}A_j) = 0$ for $j = 1,\ldots,k$. Hence $|\mu_n(A_j) - \mu(A_j)| < \varepsilon$ for n large enough, and since f can be approximated uniformly up to ε by $\sum t_j\ 1_{A_j}$ (where 1_A is the characteristic function of A), a simple estimate shows that $|\int f d\mu_n - \int f d\mu\ | < 3\varepsilon$ for n large enough. □

<u>(2.8) Proposition:</u> $\mathfrak{M}(X)$ is a compact metrizable space in the weak topology.

<u>Proof</u>: Since X is a compact metric space, $C(X)$ is separable. Let (f_k) be a dense sequence in the unit ball of $C(X)$. Let P denote the map $\mu \rightarrow (\int f_k\ d\mu)$ from $\mathfrak{M}(X)$ into the product space $[-1, 1]^{\mathbb{N}}$. Clearly P is an injective and continuous map. It is easy to see that P^{-1} is also continuous. Indeed, if $P(\mu_n)$ converges to $P(\mu)$, i.e. if for all k one has $\int f_k\ d\mu_n \rightarrow \int f_k\ d\mu$, then the density of f_k implies that $\int f\ d\mu_n \rightarrow \int f d\mu$ for every f in the unit ball of $C(X)$, and hence that $\mu_n \rightarrow \mu$. Thus P is a homeomorphism onto its image $P(\mathfrak{M}(X))$, and it is enough to show that this image is a closed subset of the compact metric space $[-1, 1]^{\mathbb{N}}$.

Let $P(\mu_n)$ converge to some element $\omega = (\omega_k)$ in $[-1, 1]^{\mathbb{N}}$. For every f in the unit ball of $C(X)$ there exists a subsequence f_{k_j} such that $\|f_{k_j} - f\| \rightarrow 0$. It follows easily that $\int f\, d\mu_n$ converges for every $f \in C(X)$. The map $L : f \longrightarrow \lim \int f\, d\mu_n$ clearly is linear, positive and such that $L(1) = 1$, and hence corresponds by (2.5) to some $\mu \in \mathfrak{M}(X)$. Since $L(f_k) = \omega_k$, one has $P(\mu) = (\omega_k)$; thus $P(\mu)$ is the limit of $P(\mu_n)$, and $P(\mathfrak{M}(X))$ is compact. □

We mention without proof that a distance on $\mathfrak{M}(X)$ which is compatible with the weak topology is given by the Prohorov metric

$$\bar{d}(\mu,\nu) = \inf\{\epsilon \mid \mu(B^\epsilon) \le \nu(B)+\epsilon \text{ and } \nu(B^\epsilon) \le \mu(B)+\epsilon \text{ for all } B \in \mathfrak{B}(X)\}$$

where B^ϵ denotes the set $\{x \in X \mid d(x,B) \le \epsilon\}$. (cf. [194])

(2.9) Definition: For $\mu \in \mathfrak{M}(X)$, the set of all $x \in X$ with the property that $\mu(U) > 0$ for any open U containing x is called the support of μ and denoted by Supp μ.

Alternatively, Supp μ is the (well defined) smallest closed set C with $\mu(C) = 1$.

(2.10) Definition: A measure μ whose support reduces to a single point is called a point-measure.

To any $x \in X$ corresponds a point measure $\delta(x)$ defined by

$$\delta(x)(B) = \begin{cases} 1 & \text{if } x \in B \\ 0 & \text{otherwise,} \end{cases}$$

for $B \in \mathfrak{B}(X)$. One has obviously $\int f d\delta(x) = f(x)$ for $f \in C(X)$.

(2.11) Proposition: The map $x \longrightarrow \delta(x)$ is a homeomorphism from X onto a subset of $\mathfrak{M}(X)$.

Proof: It is obvious that $x \longrightarrow \delta(x)$ is continuous and one-to-one. Since X is compact, the inverse is also continuous. □

(2.12) Proposition: The set of convex combinations of point measures (i.e. the set of measures with finite support) is dense in $\mathfrak{M}(X)$.

We refer to (5.8) for the proof.

(2.13) Definition: For $n \in \mathbb{N}$ let $\mathfrak{M}_n(X)$ denote the set of measures on X of the form

$$\frac{1}{n} \sum_{i=1}^{n} \delta(x_i),$$

where the x_i are (not necessarily distinct) elements of X.

The following is easy to check:

(2.14) Proposition: $\mathfrak{M}_n(X)$ is closed in $\mathfrak{M}(X)$ and $\bigcup_{n \in \mathbb{N}} \mathfrak{M}_n(X)$ is dense in $\mathfrak{M}(X)$.

(2.15) Definition: A measure μ on X is said to be nonatomic if $\mu(\{x\}) = 0$ for all $x \in X$.

If X is uncountable, then nonatomic measures exist and are even prevalent in a certain sense. We first recall a few definitions:

Let Y be a topological space. A set $F \subset Y$ is said to be nowhere dense in Y if the complement of the closure of F is dense in Y. F is said to be of first category if it is the countable union of nowhere dense sets. $G \subset Y$ is said to be residual if it is the complement of a set of first category. $H \subset Y$ is said to be a G_δ-set if it is the countable intersection of open sets in Y. The theorem of Baire says that if Y is a complete metric space then any residual set is dense in Y (see [128]).

Let Y be a complete metric space. Clearly a set $H \subset Y$ is a countable intersection of open dense sets iff H is a dense G_δ in Y, and a set is residual iff it contains a dense G_δ. From the topological point of view the residual sets are the large sets, and sets of first category are the small ones.

(2.16) Proposition: If the compact metric space X is uncountable, then $\mathfrak{M}(X)$ contains a nonatomic measure. If X, in particular, has no isolated points, then the set of nonatomic measures is a residual set in $\mathfrak{M}(X)$.

Proof: Assume first that X has no isolated points. For $r \in \mathbb{N}$ write

$$K_r = \{\mu \in \mathfrak{M}(X) \mid \exists x \in X \text{ with } \mu(\{x\}) \geq r^{-1}\}.$$

Each K_r is closed. Indeed, let μ_n be a sequence in K_r converging to some $\mu \in \mathfrak{M}(X)$. To each μ_n corresponds an $x_n \in X$ with $\mu_n(\{x_n\}) \geq r^{-1}$. Let x be a limit point of the x_n. For every closed neighborhood V of x one has $\mu(V) \geq r^{-1}$ by (2.7.b) It follows that $\mu(\{x\}) \geq r^{-1}$ and thus that $\mu \in K_r$. On the other hand, K_r is nowhere dense. Indeed, it is easy to see that every neighborhood $\mu \in K_r$ contains measures obtained by "splitting" every μ-atom of measure $\leq r^{-1}$ into several atoms lying very near by and having each measure $< r^{-1}$. Thus, the set $\bigcup_{r=1}^{\infty} K_r$ of measures having atoms is of first category in the compact metric space $\mathfrak{M}(X)$.

Now consider the general case. Let $N \subset X$ be the set of all $x \in X$ having a neighborhood N_x of at most countable cardinality. Clearly $N = \bigcup \{N_x \mid x \in N\}$. Since X is compact metric, X (and, therefore, N) satisfy the second axiom of countability. This implies that N has the Lindelöf-property. Thus there exist $x_1, x_2, \ldots \in N$ such that $N = \bigcup_k N_{x_k}$. Thus N is countable. Clearly $Y = X \setminus N \neq \emptyset$ is compact and has no isolated points. Thus there exists a nonatomic measure on Y and hence on X. □

So far, most of the material of this section is taken from [152]. We refer to [152] and [194] for more details on $\mathfrak{M}(X)$. In the remainder of the section we want to show that a large class of measure spaces are Lebesgue spaces.

(2.17) Proposition: If (X, μ) is a compact metric space with a normalized Borel measure (or completed Borel measure) μ, then it is m.t. isomorphic to a Borel measure (resp. Lebesgue-Stieltjes measure) on $[0,1]$.

We remark that the proposition also holds for complete separable metric spaces X; but for this case we did not show the regularity of μ which is essential for the proof.

As a preparation for the proof we consider the mappings in question.

<u>(2.18) Lemma:</u> Let (X, μ) be a compact metric space with a Borel (or completed Borel) measure μ, $(X', \mathfrak{B}')$ a separable metric space with its Borel σ-algebra $\mathfrak{B}'$, $N \subset X$ a set with $\mu(X \setminus N) = 1$, and $\Phi: X \setminus N \to X'$ a measurable and injective map. Then Φ is an m.t. isomorphism between $(X, \mathfrak{B}, \mu)$ and $(X', \mathfrak{B}', \Phi\mu)$ (resp. $(X, \mathfrak{B}_\mu, \mu)$ and $(X', \mathfrak{B}'_{\Phi\mu}, \Phi\mu)$).

<u>Proof:</u> Assume μ is a Borel measure. Since μ is regular, Lusin's theorem gives us a sequence $(K_i)_{i \in \mathbb{N}}$ of compact sets $K_i \subset X \setminus N$ such that $\Phi | K_i$ is continuous and $\mu(\bigcup_i K_i) = 1$. We put $K = \bigcup_i K_i$. On the compact set $\Phi(K_i)$ the inverse mapping Φ^{-1} is continuous and therefore Borel measurable. Putting these piecewise measurable mappings together, we see that $\Phi^{-1}: \Phi(K) \to K$ is Borel measurable. It is obvious that $\Phi(K) \in \mathfrak{B}'$ and $\Phi\mu\ (\Phi K) = 1$.

If μ is completed, it is still regular so that we can do the same as before and see that Φ is an m.t. isomorphism from $(X, \mathfrak{B}, \mu)$ to $(X', \mathfrak{B}', \mu')$. Now, null sets being carried over to null sets, we deduce easily that Φ is also an m.t. isomorphism between the completions of the two spaces. □

The following corollary shows that for our purposes (and indeed in the whole of this book) it does not matter if we deal with Borel measures or their completions.

<u>(2.19) Corollary:</u> Let (X, μ) and (X', μ') be compact metric spaces with measures μ, μ' (considered on the Borel algebras and their respective completions), and $\Phi: X \to X'$ a mapping. Equivalent are:

a) $\Phi : (X, \mathfrak{B}, \mu) \to (X', \mathfrak{B}', \mu')$ is an m.t. isomorphism

b) $\Phi : (X, \mathfrak{B}_\mu, \mu) \to (X', \mathfrak{B}'_{\mu'}, \mu')$ is an m.t. isomorphism.

<u>Proof:</u> a$\Rightarrow$b is immediate since null sets are transported into null sets. b$\Rightarrow$a: There are subsets $K \subset X$, $K' \subset X'$ such that $\Phi: K \to K'$ is bijective and $\mu(K) = 1$. The proof of (2.18) shows that K and K' may be assumed to be countable unions of compact sets K_i resp. ΦK_i on which Φ is continuous. So $\Phi: K \to K'$ is Borel measurable. □

The next step is the reduction to the auxiliary space $M = \{0,1\}^{\mathbb{N}}$ with the product topology of the discrete topology on $\{0,1\}$. It will then be easy to come to $[0,1]$. Because of (2.19), μ is a Borel measure from now on.

(2.20) Lemma: If (X, μ) is a compact metric space with a Borel measure μ, then it is m.t. isomorphic to $(\{0,1\}^{\mathbb{N}},\nu)$, where ν is some Borel measure on $\{0,1\}^{\mathbb{N}}$.

Proof: We take a sequence $(A_1,A_2,A_3,\ldots)$ of Borel sets in X which separates all points in X in the sense that for any $x \neq y \in X$, there is an $i \in \mathbb{N}$ with $x \in A_i$, $y \notin A_i$ (or vice versa). Then we map x to the sequence

$$\Phi x = (1_{A_i}(x))_{i \in \mathbb{N}} \quad \in \quad \{0,1\}^{\mathbb{N}}.$$

Φ is injective by the point separating property of $(A_i)_i$; therefore lemma (2.18) says that it is an m.t. isomorphism from (X,μ) to $(\{0,1\}^{\mathbb{N}}, \Phi\mu)$. □

Proof of proposition (2.17): We may now assume that $X = M = \{0,1\}^{\mathbb{N}}$ and the measure ν on M is a Borel measure without point masses. For later use (proof of (13.1)) we discuss the space M in some detail. We denote the elements of M by $p = (p_1,p_2,\ldots)$, $s = (s_1,s_2,\ldots)$ and so on. We write $p \to 0$ (resp. 1) if $p_i = 0$ (resp. 1) eventually, and $p \rightsquigarrow 0$ (resp. 1) if $p \to 0$ (resp. 1), but p is not the constant sequence $(0,0,0,\ldots)$ or $(1,1,1,\ldots)$. A finite sequence $P = (p_1,\ldots,p_n)$ of zeros and ones is a block, $n = l(P)$ its length, and the n-cylinder $[P]$ is the set $[P] = \{s \in M \mid (s_1,\ldots,s_n) = P\}$.

The set of blocks of constant length n as well as M carry the lexicographical order. This order gives us intervals $[a,b]$, $[a,b[$ etc. in M similarly as in $\mathbb{R}$, and the product topology on M is the same as the order topology (having as closed sets the order intervals $[a,b]$, $a,b \in M$). Cylinders are special intervals.

For any block $P \neq (0,0,\ldots,0)$ of length n we denote by $\tilde{P}$ the block immediately before it, i.e. if $P = (p_1,\ldots,p_k,1,0,\ldots,0)$, then $\tilde{P} = (p_1,\ldots,p_k,0,1,\ldots,1)$; if $P = (0,0,\ldots,0)$, then $P = \tilde{P}$.

For the moment we call a saturated interval an interval $[p,p']$ where neither $p \rightsquigarrow 0$ nor $p' \rightsquigarrow 1$. Such an interval has either one point, or two points (and then $p = (p_1,\dots,p_n\ 0\ 1\ 1\ 1\ \dots)$, $p' = (p_1,\dots,p_n\ 1\ 0\ 0\ 0\ \dots)$ for some $n \geq 0$), or it has nonempty interior. In the last case

$$\mathrm{int}[p,p'] = \bigcup\big([P] \mid [P] \subset [p,p']\big) = \lim_n \bigcup\big([P] \mid l(P) = n, [P] \subset [p,p']\big)$$

$$\{p\} \cup \mathrm{int}[p,p'] = \lim_n \bigcup\big([P] \cup [\tilde{P}] \mid l(P) = n,\ [P] \subset [p,p']\big)$$

(Here the "lim" means that the lim inf and lim sup coincide) For the measure ν, the order on M gives a distribution function F:

$$F(p) = \nu(\{s \in M \mid s \leq p\})$$

F is continuous and monotone, and the maximal (closed) intervals on which F is constant are saturated. (For if $p = (p_1,\dots,p_k,1000\ \dots)$, then $F(p) = F(p_1,\dots,p_k,011,\dots)$, since ν has no mass in p.)

We want to exclude a suitable ν-null set $N_F \subset M$ such that $F|M\setminus N_F$ is injective. As N_F we can take

$$\{(1,1,\dots)\} \cup \{p \mid p \rightsquigarrow 1\} \cup \bigcup\big(\{p\} \cup \mathrm{int}\,[p,p'] \mid [p,p'] \text{ is maximal with } \nu[p,p'] = 0\big);$$

from what we said above on saturated intervals, and from the fact that an n-cylinder always contains only two n+1-cylinders, it is not hard to see that

$$N_F = \{p \mid p \rightsquigarrow 1\} \cup \liminf_n \bigcup\big([P] \cup [\tilde{P}] \mid l(P) = n,\ \nu([P]) = 0\big).$$

N_F is a Borel set, and it is clear from lemma (2.18) that F itself is an m.t. isomorphism from (M, ν) to $([0,1],F\nu)$. Note also that $F\nu$ is the uniform distribution on $[0,1]$. □

We avoid here the abstract theory of Lebesgue spaces; (2.17) - (2.19) show that we can do this in our context. However, we refer to the standard exhibition of this theory in [158]. A basic theorem on Lebesgue spaces, in a form adapted to our applications, will also be given in (13.1).

All spaces considered from now on will be either Lebesgue spaces or such that their completions are Lebesgue spaces.

3. Invariant Measures for Continuous Transformations

(3.1.) Definition: Let X,X' be compact metric spaces and φ a continuous map from X into X'. φ transports every measure μ on X into a measure $\varphi\mu$ in $\mathfrak{M}(X')$. This map from $\mathfrak{M}(X)$ into $\mathfrak{M}(X')$, which is again denoted by φ, is called the extension of φ.

By (2.11) X can be embedded into $\mathfrak{M}(X)$. Clearly one has $\varphi(\delta(x) = \delta(\varphi(x))$ for all $x \in X$.

Note that for $f' \in C(X')$

$$\int f' d\varphi\mu = \int f'\circ\varphi \, d\mu$$

(3.2) Proposition: The extended transformation is a continuous affine map from $\mathfrak{M}(X)$ into $\mathfrak{M}(X')$. $\varphi: X \to X'$ is onto iff its extension is onto.

Proof: If $\mu_n \to \mu$ in $\mathfrak{M}(X)$ then $\int f d\mu_n \to \int f d\mu$ for all $f \in C(X)$ and therefore $\int f'\circ\varphi \, d\mu_n \to \int f'\circ\varphi \, d\mu$ for all $f' \in C(X')$. This implies $\varphi\mu_n \to \varphi\mu$. It is clear that

$$\varphi(\alpha\mu + (1-\alpha)\nu) = \alpha\cdot\varphi\mu + (1 - \alpha)\varphi\nu$$

for $\mu,\nu \in \mathfrak{M}(X)$ and $0 \leq \alpha \leq 1$. Since the convex combinations of point measures are dense in $\mathfrak{M}(X)$ (see 2.12) it follows that the extension $\varphi : \mathfrak{M}(X) \to \mathfrak{M}(X')$ is onto if $\varphi : X \to X'$ is onto. The converse is obvious, since X' is embedded in $\mathfrak{M}(X')$. $\square$

Now let $T : X \to X$ be a homeomorphism. Clearly:

(3.3) Proposition: The set of point measures and the sets $\mathfrak{M}_n(X)$ defined in (2.13) are invariant under the extension of T.

(3.4) Definition: The set of all T-invariant measures in X, i.e. the set of all $\mu \in \mathfrak{M}(X)$ with $T\mu = \mu$, is denoted by $\mathfrak{M}_T(X)$ (or by $\mathfrak{M}(T)$ if no confusion about X is possible).

(3.5) Proposition: $\mathfrak{M}_T(X)$ is a convex subset of $\mathfrak{M}(X)$, closed in the weak topology. If $T \neq Id$, then $\mathfrak{M}_T(X)$ is nowhere dense in $\mathfrak{M}(X)$.

Proof : Convexity and closeness are obvious. It remains to show that $\mathfrak{M}_T(X)$ contains no neighborhood if $T \neq \mathrm{Id}$. If $Tx \neq x$ and $\mu \in \mathfrak{M}_T(X)$ then $\epsilon \cdot \delta(x) + (1-\epsilon)\mu$ is not invariant, but converges to μ for $\epsilon \downarrow 0$. □

(3.6) Theorem of Krylov-Bogoliubov [127]: $\mathfrak{M}_T(X)$ is nonempty.

We shall obtain this as a corollary of proposition (3.8) below. It is a simple consequence of the fact that $\mathfrak{M}(X)$ is compact (see (2.8)).

(3.7) Definition: For $\mu \in \mathfrak{M}(X)$ and N a positive integer write

$$\mu^N = \frac{1}{N}(\mu + T\mu + \ldots + T^{N-1}\mu).$$

Clearly $\mu^N \in \mathfrak{M}(X)$. Let $V_T(\mu)$ be the set of all accumulation points of the sequence μ^N, i.e. the set

$$\{\nu \in \mathfrak{M}(X) \mid \exists\, N_k \to \infty \text{ with } \mu^{N_k} \to \nu\}$$

For $x \in X$, we shall write $V_T(x)$ instead of $V_T(\delta(x))$.

(3.8) Proposition: For $\mu \in \mathfrak{M}(X)$ the set $V_T(\mu)$ is a nonempty closed connected subset of $\mathfrak{M}_T(X)$.

Proof: $V_T(\mu) \neq \emptyset$ follows from (2.8). Obviously $V_T(\mu)$ is closed, and each accumulation point of μ^N is T-invariant. Finally it is clear that

$$\left| \int f d\mu^N - \int f d\mu^{N+1} \right| \to 0$$

for $f \in C(X)$ and $N \uparrow \infty$. This implies that the set of accumulation points is connected. □

(3.9) Definition: By a topological dynamical system we shall understand a compact metric space X together with a homeomorphism $T : X \to X$.

Note that T is measurable with respect to the Borel sets. If a measure $\mu \in \mathfrak{M}(X)$ is T-invariant, then we say that the m.t. dynamical system (X, μ, T) is supported by the top. dynamical system (X, T). (3.6) says that every top. dynamical system supports at least one m.t. dynamical system.

(3.10) Definition: Let (X, T) and (X', T') be two top. dynamical systems. (X',T') is said to be a factor of (X,T) if there exists a continuous surjective map $\varphi : X \to X'$ such that $\varphi \circ T = T' \circ \varphi$, i.e. such that the following diagram commutes:

$$\begin{array}{ccc} X & \xrightarrow{T} & X \\ \varphi \downarrow & & \downarrow \varphi \\ X' & \xrightarrow{T'} & X' \end{array}$$

φ is called a topological homomorphism. If φ is one-to-one, i.e. a homeomorphism, then (X,T) and (X',T') are said to be topologically conjugate, and φ is said to be an isomorphism between the two systems.

It is clear that φ sends $\mathfrak{M}_T(X)$ into $\mathfrak{M}_{T'}(X')$.

(3.11) Proposition: φ sends $\mathfrak{M}_T(X)$ onto $\mathfrak{M}_{T'}(X')$.

Proof: Let $\mu' \in \mathfrak{M}_{T'}(X')$ be given. By (3.2) there exists a $\nu \in \mathfrak{M}(X)$ such that $\varphi\nu = \mu'$. Since $T'\mu' = \mu'$ and $\varphi \circ T' = T \circ \varphi$, one has $\varphi(T\nu) = \mu'$ and more generally $\varphi(T^n\nu) = \mu'$. Since φ is affine on $\mathfrak{M}(X)$ (by (3.2)) this implies $\varphi(\nu^N) = \mu'$. Let $\mu \in V_T(\nu)$. φ is continuous, and therefore $\varphi\mu = \mu'$. But $\mu \in \mathfrak{M}_T(X)$. □

If (X,T) and (X',T') are top. conjugate, then the m.t. dynamical systems (X,μ,T) and $(X', \varphi\mu, T')$ are m.t. conjugate, for any $\mu \in \mathfrak{M}_T(X)$.

On the other hand, m.t. conjugacy does not say anything about top. conjugacy. Thus a top. dynamical system can support several m.t. dynamical systems which are not m.t. conjugate. Also, there are m.t. conjugate m.t. dynamical systems (X,μ,T) and (X',μ',T') supported by top. dynamical systems (X,T) and (X',T') which are not top. conjugate. This follows for example from the fact that any m.t. dynamical system is conjugate to a system (X,μ,T) where T is a homeomorphism of a totally disconnected space X (see (9.8)).

4. Time Averages

(4.1) Definition: Let f be a real or complex valued function defined on the space $\mathfrak{X}$ and Ψ a transformation from $\mathfrak{X}$ onto itself. If $\frac{1}{N}\sum_{j=0}^{N-1} f(\Psi^j(x))$ converges for some $x \in X$, we shall denote the limit by $f^*(x)$ and call it the time average of f at the point x.

(4.2) Statistical ergodic theorem (v.Neumann): Let $(\mathfrak{X},m,\Psi)$ be an m.t. dynamical system and $f \in L^2(\mathfrak{X},m)$. Then

$$\frac{1}{N}\sum_{j=0}^{N-1} f\circ\Psi^j$$

converges in the L^2-norm. The limit is the projection of f onto the subspace of Ψ-invariant functions in $L^2(\mathfrak{X},m)$.

(4.3) Individual ergodic theorem (Birkhoff): Let $(\mathfrak{X},m,\Psi)$ be an m.t. dynamical system and f an m-integrable function on $\mathfrak{X}$. Then

(a) $f^*(x)$ exists m-a.e.;

(b) $\lim \frac{1}{N}\sum_{j=0}^{N-1} f\circ\Psi^j = f^*$ holds in the L^1-norm;

(c) $f^*\circ\Psi = f^*$ m-a.e;

(d) $\int f^*\,dm = \int f\,dm$.

For the proofs we refer to [80] or to [57]. In [57] one can also find some far-reaching generalisations of this theorem.

Now let (X,T) be a top. dynamic system.

(4.4) Definition: A point $x \in X$ is said to be quasigeneric for the measure $\mu \in \mathfrak{M}(X)$ if there exists an increasing sequence of integers $N_k > 0$ such that

$$\lim \frac{1}{N_k}\sum_{j=0}^{N_k-1} f(T^j(x)) = \int f d\mu$$

for all $f \in C(X)$.

x is said to be generic for μ if $f^*(x) = \int f d\mu$ holds for all $f \in C(X)$.

It is obvious that if $\mu \in \mathfrak{M}(X)$ admits a quasigeneric point then μ is T-invariant. Remark that every point is quasigeneric for some measure, but not necessarily generic. In the notation of (3.7), x is quasigeneric for μ iff $\mu \in V_T(x)$.

(4.5) Definition: The set of all points which are generic for μ will be denoted by G_μ.

Thus $V_T(x)$ is a nonempty closed connected subset of $\mathfrak{M}_T(X)$ which reduces to a single point if x is generic. It may happen that for some $\mu \in \mathfrak{M}_T(X)$, the set G_μ is empty: for example, if T is the identity and μ not a point measure, then $G_\mu = \emptyset$ (see, however, (5.9)).

(4.6) Definition: A point $x \in X$ is said to be quasiregular with respect to T if it is generic with respect to some measure. We denote by $Q_T(X)$ the set of all quasiregular points.

Thus $x \in Q_T(X)$ iff $f^*(x)$ exists for all $f \in C(X)$. In this case the linear functional $f \to f^*(x)$ on $C(X)$ corresponds by (2.5) to a measure which we shall denote by μ_x. x is generic for μ_x. In general, the map $x \to \mu_x$ from $Q_T(X)$ into $\mathfrak{M}_T(X)$ is neither injective nor surjective.

Obviously $Q_T(X)$ is a disjoint union of the G_μ's, $\mu \in \mathfrak{M}_T(X)$.

(4.7) Proposition: $Q_T(X)$ and the G_μ's are Borel sets. $\mu(Q_T(X)) = 1$ holds for every $\mu \in \mathfrak{M}_T(X)$, in particular $Q_T(X) = \emptyset$.

Proof: For $f \in C(X)$ and $N \in \mathbb{N}$ write $f^N = \frac{1}{N}\sum_{j=0}^{N-1} f \circ T^j$, and

$Q(f) = \{x \in X \mid f^*(x) \text{ exists}\}$. One has

$$Q(f) = \{x \in X \mid (f^N(x)) \text{ is a Cauchy sequence}\}$$
$$= \bigcap_{k \in \mathbb{N}} \bigcup_{l \in \mathbb{N}} \bigcap_{N > l} \bigcap_{M > l} \{x \in X \mid |f^N(x) - f^M(x)| < \tfrac{1}{k}\}.$$

Since $\{x \in X \mid |f^N(x) - f^M(x)| < \frac{1}{k}\}$ is open, $Q(f)$ is a Borel set. By Birkhoff's ergodic theorem (4.3), $\mu(Q(f)) = 1$ for all $\mu \in \mathfrak{M}_T(X)$.

Clearly $Q_T(X)$ may be written as $\bigcap_{n=1}^{\infty} Q(f_n)$, where (f_n) is a dense sequence in $C(X)$. Hence $\mu(Q_T(X)) = 1$ for all $\mathfrak{M}_T(X)$ and $Q_T(X)$ is a Borel set. The proof that G_μ is a Borel set is similar.

□

(4.7) implies that $Q_T(X)$ is a large set, from the measure theoretic viewpoint. Another fact that indicates this large size is:

<u>(4.8) Proposition:</u> If X has infinitely many points, then $Q_T(X)$ has infinitely many points. If, in addition, X is locally connected, then $Q_T(X)$ has c points ($c = \text{card}\, \mathbb{R}$)

We refer to Dowker [54] for the proof of (4.8).

In spite of (4.7) and (4.8), it happens very often that $Q_T(X)$ is small from the topological point of view.

<u>(4.9) Proposition [54]:</u> If there exists an $x \in X$ which is not quasiregular, and whose positive orbit $\{T^n x\}_{n\in\mathbb{N}}$ is dense in X, then the set $Q_T(X)$ is of first category.

<u>Proof:</u> We use the same notations as in the proof of (4.7). If $x \notin Q_T(X)$, there exists an $f \in C(X)$ such that $x \notin Q(f)$. Let α and β be such that

$$\liminf f^N(x) < \alpha < \beta < \limsup f^N(x).$$

Write

$$E = \{y \in X \mid \liminf f^N(y) < \alpha \text{ and } \limsup f^N(y) > \beta\}$$

$$= [\bigcap_{N\geq 0} \bigcup_{i\geq 0} \{y \in X \mid f^{N+i}(y) < \alpha\}] \cap [\bigcap_{N\geq 0} \bigcup_{j\geq 0} \{y\in X \mid f^{N+j}(y)>\beta\}].$$

It is clear that E is a G_δ-set, E contains x and even the dense positive orbit of x. Henxe $X \setminus E$ is a set of first category. But obviously $Q_T(X)$ is a subset of $X\setminus E$.

□

In later sections we shall find many examples of transformations satisfying the conditions of (4.9).

5. Ergodicity

Let $(\mathfrak{X},m,\Psi)$ be an m.t. dynamical system. A function f on $\mathfrak{X}$ is said to be <u>Ψ-invariant</u> if $f\circ\Psi = f$ m-a.e., and a set $B \subset \mathfrak{X}$ is called Ψ-invariant if 1_B is invariant, i.e. if $\Psi^{-1}B = B \bmod 0$.

<u>(5.1) Definition:</u> $(\mathfrak{X},m,\Psi)$ is said to be <u>ergodic</u> if for every measurable Ψ-invariant $B \subset \mathfrak{X}$ one has either $m(B) = 0$ or $m(B) = 1$. One also says that the measure m is ergodic (with respect to Ψ) or that the transformation Ψ is ergodic (with respect to m).

Ergodicity means indecomposability: $\mathfrak{X}$ is not a disjoint union of two invariant sets having both positive measure. There exist many characterizations of ergodic systems. We list some of them, referring to one of the standard texts for a proof.

<u>(5.2) Proposition:</u> The m.t. dynamical system $(\mathfrak{X},m,\Psi)$ is ergodic iff one of the following equivalent properties holds

(a) if $f \in L^1(\mathfrak{X},m)$ is Ψ-invariant then $f = \text{const.}$ m-a.e.

(b) for any $f \in L^1(\mathfrak{X},m)$ one has $f^*(x) = \int f\,dm$ m-a.e.

(c) for any measurable $B \subset \mathfrak{X}$ with $m(B) > 0$ one has

$$m\left(\bigcup_{j=0}^{\infty} \Psi^{-j} B\right) = 1$$

(d) for any two measurable $A,B \subset \mathfrak{X}$ one has

$$\lim \frac{1}{N}\sum_{j=0}^{N-1} m\,(A \cap \Psi^{-j}B) = m(A)\,m(B)$$

Property (b) means that the time average is equal to the space average, a property very useful to physicists.

It may happen that the system $(\mathfrak{X},m,\Psi)$ is ergodic but the system $(\mathfrak{X},m,\Psi^k)$ is not, for some $k \in \mathbb{N}$. For example if $\mathfrak{X} = \{x_1,x_2\}$, and $m(\{x_1\}) = m(\{x_2\}) = \frac{1}{2}$, then the map Ψ permuting x_1 and x_2 is ergodic, but the map $\Psi^2 = \mathrm{Id}$ certainly is not.

<u>(5.3) Definition:</u> A measure theoretic dynamical system $(\mathfrak{X},m,\Psi)$ is said to be <u>totally ergodic</u> if the systems $(\mathfrak{X},m,\Psi^k)$ are ergodic, for $k = 1,2,\ldots$.

If $(\mathfrak{X},m,\Psi)$ is ergodic, then clearly $(\mathfrak{X},m,\Psi^{-1})$ is ergodic as well. It is easy to see that every factor of an ergodic system is ergodic.

Let $(\mathfrak{X},\Sigma,m)$ be a measure space and Ψ a measure preserving transformation. It often happens that there exists another measure $\tilde{m}$ on Σ which is preserved by Ψ. If the systems $(\mathfrak{X},m,\Psi)$ and $(\mathfrak{X},\tilde{m},\Psi)$ are both ergodic, however, then m and $\tilde{m}$ are either equal or singular. This is a corollary of the following proposition:

(5.4) Proposition: Let $(\mathfrak{X},m,\Psi)$ be an ergodic m.t. dynamical system. Let $\tilde{m}$ be another Ψ-invariant measure defined on the same σ-algebra as m. If $\tilde{m}$ is absolutely continuous with respect to m (i.e. if $\tilde{m}(B) = 0$ whenever $m(B) = 0$) then $\tilde{m} = m$.

Proof: By the theorem of Radon-Nikodym there exists a function $f \in L^1(\mathfrak{X},m)$, such that $\tilde{m}(B) = \int_B f\,dm$ for all measurable sets $B \subset \mathfrak{X}$. Since both m and $\tilde{m}$ are Ψ-invariant, the function f is Ψ-invariant and therefore m-almost everywhere equal to some constant. Since $\tilde{m}(\mathfrak{X}) = 1$, this constant must be 1. Hence $m = \tilde{m}$. □

(5.5) Definition: Let K be a convex set. A point $k \in K$ is said to be extremal if it cannot be written in the form

$$k = \alpha k_1 + (1 - \alpha)k_2$$

with k_1,k_2 two distinct points in K and $0 < \alpha < 1$.

Thus a point is extremal if it cannot lie in the interior of a segment contained in K.

Now let (X,T) be a top. dynamical systèm.

(5.6) Proposition: The ergodic T-invariant measures are the extremal points of the convex set $\mathfrak{M}_T(X)$.

Proof: (a) Let $\mu \in \mathfrak{M}_T(X)$ be non-ergodic. Then there exists a T-invariant Borel set $Y \subset X$ with $0 < \mu(Y) < 1$. Let μ_1 (resp. μ_2) be the conditional measure μ_Y (resp. $\mu_{X\setminus Y}$). It is easy to see that μ_1 and μ_2 belong to $\mathfrak{M}_T(X)$ and that

$$\mu = \mu(Y)\mu_1 + (1 - \mu(Y))\mu_2.$$

Thus μ is not an extremal point of $\mathfrak{M}_T(X)$.

(b) Let μ be non-extremal in $\mathfrak{M}_T(X)$. Then there exist two distinct measures $\mu_1, \mu_2 \in \mathfrak{M}_T(X)$ and an α with $0 < \alpha < 1$ such that $\mu = \alpha\mu_1 + (1-\alpha)\mu_2$. Clearly μ_1 is absolutely continuous with respect to μ. This implies that μ is non-ergodic, for otherwise (5.4) would imply $\mu_1 = \mu$, a contradiction.

(5.7) Proposition: The set of ergodic measures is a (non-empty) G_δ-subset of $\mathfrak{M}_T(X)$.

Proof [144]: Nonemptiness follows immediately from (5.6) and (3.6).

Now let (K,d) be a compact convex metric space, and let φ be the continuous map from $K \times K \times [0,1]$ into K defined by

$$\varphi(k_1,k_2,\alpha) = \alpha k_1 + (1 - \alpha)k_2.$$

For $n \in \mathbb{Z}$ define

$$D_n = \{(k_1,k_2,\alpha) \mid d(k_1,k_2) \geq \tfrac{1}{n},\ \alpha \in [\tfrac{1}{n}, \tfrac{n-1}{n}]\}.$$

D_n is a closed subset of the compact space $K \times K \times [0,1]$ and $F_n = \varphi(D_n)$ is a compact subset of K. It is easy to see that x is extremal iff x does not belong to $\bigcup_{n=1}^{\infty} F_n$. Hence the sets of extremal points of K is a G_δ. Taking $K = \mathfrak{M}_T(X)$ one obtains the result. □

(5.8) Proposition: The convex combinations of ergodic measures are dense in $\mathfrak{M}_T(X)$.

Proof: The theorem of Krein-Milman states that every compact convex set in a locally convex space is equal to the closed convex hull of its extremal points. Recall that the weak topology is locally convex. □

It is easy to see that if $T = Id$, the identity on X, then the ergodic measures are just the point-measures on X. Thus (5.8) implies (2.12).

(5.9) Proposition: If $\mu \in \mathfrak{M}_T(X)$ is ergodic, then μ-almost all points of X are generic for μ, i.e. $\mu(G_\mu) = 1$.

Proof: Let (f_n) be a dense sequence in the separable space $C(X)$. Denote by $G(f_n)$ the set $\{x \in X \mid f_n^*(x) = \int f_n \, d\mu\}$.

If μ is ergodic, parts (b), (d) of (4.3) and (b) of (5.2) imply that $\mu(G(f_n)) = 1$. Clearly $G_\mu = \bigcap_{n=1}^{\infty} G(f_n)$ and hence $\mu(G_\mu) = 1$. □

<u>(5.10) Proposition:</u> **If $\mu \in \mathfrak{M}_T(X)$ is non-ergodic, then $\mu(G_\mu) = 0$.**

<u>Proof:</u> **This follows immediately from the next proposition (5.12)** □

<u>(5.11) Definition:</u> Set $E_T(X) = \bigcup[G_\mu | \ \mu$ ergodic in $\mathfrak{M}_T(X)]$.
Clearly $x \in E_T(X)$ iff $x \in Q_T(X)$ and μ_x is ergodic.

<u>(5.12) Proposition:</u> $E_T(X)$ is a Borel set and $\mu(E_T(X)) = 1$ for all $\mu \in \mathfrak{M}_T(X)$.

<u>Proof:</u> Let $f \in C(X)$ be real valued and write f^N for $\frac{1}{N}\sum_{i=1}^{N-1} f\circ T^i$. Write

$$E(f) = \{x \in Q_T(X) | f^*(y) = \int f \, d\mu_x \text{ for } \mu_x \text{ - a.e. } y\}$$

For $x \in Q_T(X)$, $\int f(y)\, d\mu_x(y) = f^*(x)$ and therefore

$$E(f) = \{x \in Q_T(X) | \ \int |f^*(y) - f^*(x)|^2 \, d\mu_x(y) = 0\}.$$

Now $\int |f^*(y) - f^*(x)|^2 \, d\mu_x(y) =$

$$= \int | \lim_N f^N(y) - \lim_M f^M(x) \ |^2 \, d\mu_x(y) =$$

$$= \lim_M \lim_N \int | \ f^N(y) - f^M(x)|^2 d\mu_x(y) =$$

$$= \lim_M \lim_N \ [|f^N - f^M(x)|^2]^*(x) =$$

$$= \lim_M \lim_N \lim_K [|f^N - f^M(x)|^2]^K(x) =$$

$$= \lim_M \lim_N \lim_K \frac{1}{K}\sum_{j=0}^{K-1} | \ f^N \circ T^j(x) - f^M(x) \ |^2.$$

This shows that $E(f)$ is a Borel set.

Now let $\mu \in \mathfrak{M}_T(X)$. We claim that $\mu(E(f)) = 1$, i.e. that

$$\int | \ f^*(y) - f^*(x) \ |^2 \, d\mu_x(y) = 0$$

holds for μ-a.e. $x \in Q_T(X)$. (Recall that $\mu(Q_T(X)) = 1$).

Since this term is positive, one has to show that

$$(*) \qquad \int \left(\int |f^*(y) - f^*(x)|^2 \, d\mu_x(y) \right) d\mu(x) = 0$$

Writing $\|\cdot\|$ and $\langle \cdot , \cdot \rangle$ for norm and scalar product in $L^2(X,\mu)$ and using the fact that μ is T-invariant, one obtains

$$\left| \frac{1}{K} \sum_{j=0}^{K-1} \|f^N \circ T^j - f^M\|^2 - \|f^N - f^M\|^2 \right|$$

$$= \left| \|f^N\|^2 + \|f^M\|^2 - 2 \left\langle \frac{1}{K} \sum_{j=0}^{K-1} f^N \circ T^j, f^M \right\rangle \right.$$

$$\left. - \|f^N\|^2 - \|f^M\|^2 + 2 \cdot \langle f^N, f^M \rangle \right|$$

$$\leq 2 \|f^M\| \cdot \left\| \frac{1}{K} \sum_{j=0}^{K-1} f^N \circ T^j - f^N \right\| .$$

By von Neumann's ergodic theorem (4.2),

$$\lim_K \frac{1}{K} \sum_{j=0}^{K-1} f^N \circ T^j = (f^N)^* = f^* \quad \text{in } L^2(X,\mu)\text{-norm}$$

so that the last expression tends to $2\|f^M\| \cdot \|f^* - f^N\|$.

Now (*) is equivalent to

$$\lim_M \lim_N \lim_K \frac{1}{K} \sum_{j=0}^{K-1} \|f^N \circ T^j - f^M\|^2 = 0$$

and hence to $\lim_N \lim_M \|f^N - f^M\|^2 = 0$. This last relation is valid by (4.2).

Now let (f_n) be a dense sequence in the real C(X) (and hence in $L^2(X,\mu)$, for any μ). Clearly $E_T(X) = \bigcap_{n=1}^{\infty} E(f_n)$. Hence $E_T(X)$ is a Borel set and $\mu(E_T(X)) = 1$. $\square$

<u>(5.13) Proposition:</u> Let $\mu \in \mathfrak{M}_T(X)$ be ergodic and $\nu \in \mathfrak{M}(X)$ be absolutely continuous with respect to μ. Then $\nu^N \longrightarrow \mu$.

<u>Proof:</u> Let $f \in C(X)$. Then $f^N(x) \longrightarrow \int f \, d\mu$ μ-a.e. and hence also ν-a.e. Therefore $\int f \, d\nu^N = \int f^N \, d\nu \longrightarrow \int f \, d\mu$. Thus μ is the only weak limit point of the sequence $(\nu^N)_N$. $\square$

<u>(5.14) Definition:</u> A top. dynamical system (X,T) is said to be <u>uniquely ergodic</u> if $\mathfrak{M}_T(X)$ consists of exactly one element μ_o.

Of course this measure must be ergodic, and $\nu^N \longrightarrow \mu_o$ for every $\nu \in \mathfrak{M}(X)$. Also $Q_T(X) = X$.

(5.15) Proposition [58 ,142]: The following conditions are equivalent

(a) (X,T) is uniquely ergodic;

(b) For each $f \in C(X)$ the sequence $\frac{1}{N}\sum_{j=0}^{N-1} f \circ T^j$ converges uniformly on X to a constant;

(c) For some $\mu \in \mathfrak{M}_T(X)$, every $x \in X$ is generic.

Proof: (a)$\Rightarrow$(b). Suppose that for some $g \in C(X)$ the sequence $\frac{1}{N}\sum_{j=o}^{N-1} g \circ T^j$ does not converge uniformly to $\int g \, d\mu_o$. Then there exist an $\alpha \neq \int g \, d\mu_o$, a sequence (x_i) in X and a sequence $N_i \longrightarrow \infty$ of positive integers such that

$$\frac{1}{N_i} \sum_{j=o}^{N_i-1} g(T^j x_i) \longrightarrow \alpha \, .$$

Extracting subsequences of (x_i) and (N_i), we may assume that $\lim \frac{1}{N_i}\sum_{j=o}^{N_i-1} f(T^j x_i)$ exists for all f in a dense subset of $C(X)$. But then this limit exists for all $f \in C(X)$ and defines a positive linear functional L on $C(X)$ and hence a measure μ, which obviously is invariant and thus equal to μ_o. But $L(g) = \alpha \neq \int g \, d\mu_o$, a contradiction.

(b)$\Rightarrow$(c) is obvious, since the constant referred to in the statement of (b) is $\int f d\mu$ for some $\mu \in \mathfrak{M}_T(X)$.

(c)$\Rightarrow$(a) also is obvious, since the existence of two invariant measures would imply the existence of two ergodic measures. But by (5.9), every ergodic measure has generic points. □

(5.16) Proposition: Let G be a compact group and $g \in G$ generating a dense subgroup. Let $L_g : x \longrightarrow gx$ be left translation on G. Then (G, L_g) is uniquely ergodic.

Proof: Any measure invariant under L_g is invariant under $L_g{}^n$, and hence, by density, with respect to all left translations. But the Haar measure is the unique measure with this property. □

In particular, irrational rotations of the circle are uniquely ergodic.

6. Mixing and Transitivity

(6.1) Definition: An m.t. dynamical system $(\mathfrak{X},m,\Psi)$ is called strongly mixing if

$$\lim_{n\to\infty} m\,(A \cap \Psi^{-n} B) = m(A)\cdot m(B)$$

and weakly mixing if

$$\lim_{N\to\infty} \frac{1}{N}\sum_{n=0}^{N-1} |m(A \cap \Psi^{-n}B) - m(A)\cdot m(B)| = 0$$

for all measurable sets $A,B \subset \mathfrak{X}$. One also says that the measure m is strongly (resp. weakly) mixing with respect to Ψ, or that the transformation Ψ is strongly (resp. weakly) mixing with respect to m.

(6.2) Proposition: A strongly mixing system is weakly mixing; a weakly mixing system is ergodic; the converse statements are not true in general.

The first assertion is obvious, the second one follows from (5.2.b). A simple example of an ergodic but not weakly mixing system is given by the permutation on the two point space $\{x_1,x_2\}$ with $m(\{x_1\}) = m(\{x_2\}) = 2^{-1}$.

Category arguments show that there exist many weakly mixing systems which are not strongly mixing (see [80] and [170]) but it is rather difficult to find examples of this kind: The first one has been constructed by Chacon in [195].

We now mention several characterizations of strongly mixing systems. For the proofs we refer to [57]:

(6.3) Proposition: $(\mathfrak{X},m,\Psi)$ is strongly mixing iff one of the following equivalent conditions holds:

(a) $\lim_{n\to\infty} m(A \cap \Psi^{-n} A) = [m(A)]^2$

for all measurable $A \subset \mathfrak{X}$;

(b) There exists a constant K such that

$$\limsup m\,(A \cap \Psi^{-n} B) \leq K\, m\,(A)\, m(B)$$

for all measurable $A,B \subset \mathfrak{X}$;

(c) for any increasing sequence of integers $k_j > 0$ and any $f \in L^1(\mathfrak{X},m)$ one has

$$\lim \frac{1}{N}\sum_{j=0}^{N-1} f \circ \Psi^{k_j} = \int f \, dm$$

in the L^1-norm.

It should be noted that this last limit relation does not hold m-a.e., in general (Cf. (5.2.b)).

(6.4) Proposition: $(\mathfrak{X},m,\Psi)$ is weakly mixing iff the product of $(\mathfrak{X},m,\Psi)$ with itself is ergodic.

We refer to [80] for a proof.

To the measure theoretic notions of mixing and ergodicity correspond similar concepts in the topological framework. Let (X,T) be a top. dynamical system.

(6.5) Definition: (X,T) is said to be topologically transitive if for all nonempty open $U,V \subset X$ one has $U \cap T^{-n}V \neq \emptyset$ for some $n \in \mathbb{Z}$. (X,T) is said to be topologically weakly mixing if the product $(X \times X, T \times T)$ is topologically transitive. (X,T) is said to be topologically strongly mixing if for all nonempty open $U,V \subset X$ one has $U \cap T^{-n}V \neq \emptyset$ for all n large enough. We shall also say that T is top. transitive (resp. weakly or strongly mixing).

It is easy to see that top. strong mixing implies top. weak mixing, which implies top. transitivity, and that if (X,T) has one of these properties, then any factor of (X,T) has the same property.

The permutation on a two-point space is an example of a transformation which is top. transitive but not top. weakly-mixing. Hence the product of two transformations which are top. transitive need not be top. transitive.

(6.6) Proposition: (X,T) is top. transitive iff the orbit $\bigcup_{n\in\mathbb{Z}} T^n U$ of every nonempty open $U \subset X$ is dense in X.

Thus top. transitivity is analogous to ergodicity (cf. (5.2.c.)).

(6.7) Proposition: Let $\mu \in \mathfrak{M}_T(X)$ have support X. If (X,μ,T) is ergodic, then (X,T) is top. transitive. If (X,μ,T) is strongly (resp. weakly) mixing, then (X,T) is top. strongly (resp. weakly) mixing.

One can see by simple examples that the converse is not true, in general. It has been conjectured, however, that if (X,T) admits a measure $\mu \in \mathfrak{M}_T(X)$ with support X, and if it is top. transitive, then it admits an ergodic measure $\mu \in \mathfrak{M}_T(X)$ with support X. B.Weiss has given a counterexample to this conjecture in [189].

If $T : X \longrightarrow X$ is top. transitive, then its extension $T : \mathfrak{M}(X) \longrightarrow \mathfrak{M}(X)$ need not be so. Consider for example the permutation on the space $X = \{x_1, x_2\}$. If one chooses sufficiently small neighborhoods U of $\delta(x_1)$ and V of $2^{-1}(\delta(x_1) + \delta(x_2))$ in $\mathfrak{M}(X)$, then the orbit of U will never intersect V.

(6.8) Proposition: If $T : X \longrightarrow X$ is top. strongly (resp. weakly) mixing, then its extension $T : \mathfrak{M}(X) \longrightarrow \mathfrak{M}(X)$ is also top. strongly (resp. weakly) mixing.

Proof: (1) Let T be top. strongly mixing. The n-fold product $T^{(n)}$ is top. strongly mixing. Let V_1 and V_2 be nonempty open sets in $\mathfrak{M}(X)$. By (2.14), for some n sufficiently large, there exist $x_j^1, x_j^2 \in X$ $(1 \leq j \leq n)$ such that

$$\mu_1 = \frac{1}{n} \sum_{j=1}^{n} \delta(x_j^1) \in \mathfrak{M}_n(X) \cap V_1$$

and

$$\mu_2 = \frac{1}{n} \sum_{j=1}^{n} \delta(x_j^2) \in \mathfrak{M}_n(X) \cap V_2 \ .$$

For $i = 1,2$ and $1 \leq j \leq n$ choose neighborhoods U_j^i of x_j^i such that if $y_j^i \in U_j^i$ then

$$\frac{1}{n} \sum_{j=1}^{n} \delta(y_j^i) \in V_i.$$

There exists an N_o such that for $N \geq N_o$

$$(T^{(n)})^N (U_1^1 \times \ldots \times U_n^1) \cap (U_1^2 \times \ldots \times U_n^2) \neq \emptyset$$

Pick $y_j^1 \in U_j^1 \cap T^{-N}(U_j^2)$. One has $\nu_1 = \frac{1}{n} \sum_{j=1}^{n} \delta(y_j^1) \in V_1$ and

$T^N \nu_2 = \frac{1}{n}\sum_{j=1}^{n} \delta(T^N y_j^2) \in V_2$. Hence $T^{-N} V_2 \cap V_1 \neq \emptyset$ for all $N \geq N_o$, and so the extended transformation is strongly mixing.

(2) The weakly mixing case is analogous. □

(6.9) Proposition: $T : X \longrightarrow X$ is top. transitive (resp. top. weakly mixing, resp. top strongly mixing) if its extension $T : \mathfrak{M}(X) \longrightarrow \mathfrak{M}(X)$ is top. transitive (resp. top. weakly mixing, resp. top. strongly mixing).

Proof: Assume that $T : \mathfrak{M}(X) \longrightarrow \mathfrak{M}(X)$ is top. transitive. Let U_1, U_2 be nonempty open in X. Write $V_i = \{\mu \in \mathfrak{M}(X) | \mu(U_i) > 0{,}9\}$ $(i = 1,2)$. It is easy to see that the V_i's are nonempty open in $\mathfrak{M}(X)$, and thus that $V_1 \cap T^{-n} V_2 \neq \emptyset$ for some $n \in \mathbb{Z}$. This obviously implies $U_1 \cap T^{-n} U_2 \neq \emptyset$. Hence $T : X \longrightarrow X$ is top. transitive.

The other cases are analogous. □

(6.10) Definition: A point $x \in X$ is said to be topologically transitive with respect to T if its orbit $\{T^n x\}$ is dense in X.

(6.11) Proposition: The following three conditions are equivalent:

(a) (X,T) is top. transitive;

(b) some point $x \in X$ is top. transitive;

(c) the set of top. transitive points is a dense G_δ in X.

Proof: (b)$\Longrightarrow$(a): If $x \in X$ is top. transitive, and if $U, V \neq \emptyset$ are open in X, there exist integers m,n with $T^n x \in U$, $T^m x \in V$. Thus $T^{m-n} U \cap V \neq \emptyset$.

(c)$\Longrightarrow$(b) is trivial.

(a)$\Longrightarrow$(c): Let $O_1, O_2, \ldots$ be a countable basis of open sets for X, and $E \subset X$ the set of points which are not top. transitive. If $x \in E$, there is an O_j such that $T^n x$ never hit O_j. Thus $E = \bigcup_{j \in \mathbb{N}} \bigcap_{n \in \mathbb{Z}} T^{-n} (X \setminus O_j)$.

The sets $\bigcup_{n \in \mathbb{Z}} T^{-n} O_j$ are open and dense in X, their complements $\bigcap_{n\in\mathbb{Z}} T^{-n}(X \setminus O_j)$ are nowhere dense, and hence $X \setminus E$ is a dense G_δ-set. □

(6.12) Definition: A nonempty closed T-invariant subset A of X is said to be minimal (with respect to T) if there exists no proper closed T-invariant subset of A. If X itself is minimal, one says that the top. dynamical system (X,T) (or the transformation T) is minimal.

With the help of Zorn's lemma, it is easy to see that X always contains some minimal subset. But the union of all minimal subsets of X is not necessarily equal to X. For example, if T is the transformation $x \longrightarrow x^2$ from $[0,1]$ onto itself, then the only minimal subsets are the endpoints.

The following propositions are obvious:

(6.13) Proposition: (X,T) is minimal iff every $x \in X$ is top. transitive.

(6.14) Proposition: If (X,T) is uniquely ergodic, then X contains only one minimal set. (X,T) is minimal iff the unique T-invariant measure has support X.

We also mention the following result due to Dowker and Lederer [55]:

(6.15) Proposition: If (X,T) satisfies $Q_T(X) = X$ and if X contains a single minimal set properly, then either (X,T) is uniquely ergodic, or $\mathfrak{M}_T(X)$ contains infinitely many ergodic measures.

(6.16) **Definition:** (X,T) is said to be strictly ergodic if it is uniquely ergodic and minimal.

(6.17) Definition: A point $x \in X$ is said to be nonwandering (with respect to T) if for every neighborhood U of x, there exists an $n > 0$ with $U \cap T^{-n} U \neq \emptyset$. The set of all nonwandering points is called the nonwandering set and denoted by $\Omega_T(X)$.

The following proposition is easy to check:

(6.18) Proposition: $\Omega_T(X)$ is a nonempty closed T-invariant subset of X. It contains all minimal sets, and in particular all periodic orbits.

Note that in general the nonwandering set of the restriction of T to $\Omega_T(X)$ need not coincide with $\Omega_T(X)$.

(6.19) Proposition: If $\mu \in \mathfrak{M}_T(X)$ then Supp $\mu \subset \Omega_T(X)$.

Proof: If $x \notin \Omega_T(X)$, there exists a neighborhood U such that $U \cap T^{-n} U = \emptyset$ for all $n \geq 1$, and hence such that all $T^{-n} U$, $n \geq 0$, are disjoint. If x were in Supp μ, one would have $\mu(T^{-n} U) = \mu(U) > 0$, a contradiction to $\mu(X) = 1$. □

7. Shifts and Subshifts

(7.1) Definition: Let X be compact metric. By $X^{\mathbb{Z}}$ we shall denote the infinite product space $\prod_{i=-\infty}^{+\infty} X_i$, where $X_i = X$ for all i, endowed with the product topology. X is called the state space.

Thus an element $x \in X^{\mathbb{Z}}$ is a bilateral sequence

$$x = (\dots x_{-1}, x_o, x_1, \dots)$$

where $x_i \in X$. x_i is called the i-th coordinate of x.

$X^{\mathbb{Z}}$ is a compact metric space. If $\bar{d}$ is a metric on X, one obtains a metric d on $X^{\mathbb{Z}}$ by

$$d(x,y) = \sum_{i=-\infty}^{+\infty} 2^{-|i|} \bar{d}(x_i, y_i)$$

for $x, y \in X^{\mathbb{Z}}$.

(7.2) Definition: The transformation $\sigma : X^{\mathbb{Z}} \longrightarrow X^{\mathbb{Z}}$ given by

$$(\sigma(x))_i = x_{i+1} \qquad \text{for } i \in \mathbb{Z},\ x \in X^{\mathbb{Z}}$$

is called the shift on $X^{\mathbb{Z}}$.

Thus the shift acts on $x \in X^{\mathbb{Z}}$ by translating the corresponding sequence by one step to the left. Clearly

(7.3) Proposition: $(X^{\mathbb{Z}}, \sigma)$ is a top. dynamical system.

(7.4) Definition: Let Λ be a closed subset of $X^{\mathbb{Z}}$ which is shift-invariant, i.e. such that $\sigma^{-1}(\Lambda) = \Lambda$. The transformation $\varphi|\Lambda$ is a homeomorphism. If no confusion can occur it will again be denoted by σ. The top. dynamical system $(\Lambda, \sigma|\Lambda)$ (or for short Λ) is called a subshift.

The following proposition is easy to verify:

(7.5) Proposition: The shift $\sigma : X^{\mathbb{Z}} \longrightarrow X^{\mathbb{Z}}$ is topologically transitive, and even topologically mixing. The periodic points are dense in $X^{\mathbb{Z}}$.

Shifts on finite spaces are of particular importance. Let S be a finite space with the discrete topology. We shall always assume card $S = s \geq 2$. S is often called an alphabet, its elements letters or symbols. Usually it will be convenient to assume $S = \{1,2,\ldots,s\}$.

The following proposition is well-known (see [70]):

(7.6) Proposition: $S^{\mathbb{Z}}$ is a totally disconnected perfect space and hence is homeomorphic to the Cantor discontinuum. In particular, $S^{\mathbb{Z}}$ is zero-dimensional.

(7.7) Definition: A finite sequence $(a_1,\ldots,a_N)$ of elements $a_j \in S$ is called a block (or N-block) in $S^{\mathbb{Z}}$. N is called the length of the block. The block $A = (a_1,\ldots,a_N)$ is said to occur in $x \in S^{\mathbb{Z}}$ at the place m if $x_m = a_1,\ldots,x_{m+N-1} = a_N$. In this case one writes $A \prec x$. It is said to be a centered block of $x \in S^{\mathbb{Z}}$ if $N = 2M + 1$ and if it occurs at the place $-M$, i.e. if it is of the form $(x_{-M},\ldots,x_M)$. The block A is said to occur in the subshift $\Lambda \subset S^{\mathbb{Z}}$ if there exists an $x \in \Lambda$ such that $A \prec x$. In this case one writes $A \prec \Lambda$.

(7.8) Definition: For any $m \in \mathbb{Z}$ and any block $A = (a_1,\ldots,a_N)$ in $S^{\mathbb{Z}}$ let ${}_m[a_1,\ldots,a_N] = {}_m[A]$ denote the set of all $x \in S^{\mathbb{Z}}$ such that $(a_1,\ldots,a_N)$ occurs in x at the place m. The set ${}_m[a_1,\ldots,a_N]$ is called a cylinder of length N based on the block $(a_1,\ldots,a_N)$ at the place m. It is said to be a centered cylinder if it is of the form ${}_{-M}[a_1,\ldots,a_{2M+1}]$.

In the literature a block is sometimes called a word and a cylinder is called a thin cylinder.

Remark that if x is an element of some centered cylinder of length $2M + 1$, then the i-th coordinate of x is specified for all $|i| \leq M$. If $x^{(n)} \in S$, $x \in S^{\mathbb{Z}}$ and $x^{(n)} \longrightarrow x$, then for each $i \in \mathbb{Z}$ there is an $N(i)$ such that $(x^{(n)})_i = x_i$ for all $n \geq N(i)$. Furthermore, it follows easily from the definition of the product topology that:

(7.9) Proposition: A cylinder is an open and closed subset of $S^{\mathbb{Z}}$. For any $x \in S^{\mathbb{Z}}$ the centered cylinders

$$_{-M}[x_{-M},\dots,x_M] \qquad M = 0,1,2,\dots$$

form a basis for the neighborhoods of x.

The following propositions are easy consequences of (7.9):

<u>(7.10) Proposition:</u> The subshift $\Lambda' \subset S^{\mathbb{Z}}$ is topologically transitive (resp. topologically mixing) iff for any two blocks $(a_1,\dots,a_N)$ and $(b_1,\dots,b_M)$ occuring in Λ one has

$$_{o}[a_1,\dots,a_N] \cap {}_{n}[b_1,\dots,b_M] \neq \emptyset$$

for some $n \in \mathbb{Z}$ (resp. for all n large enough).

<u>(7.11) Proposition:</u> A point $x \in \Lambda$ is topologically transitive for the subshift $\sigma|\Lambda$ iff every block which occurs in Λ occurs at some place in x.

A simple example of a map from $S^{\mathbb{Z}}$ to $S^{\mathbb{Z}}$ commuting with σ is obtained by permuting the elements of S. A generalization of this goes as follows: let S^m denote the set of all m-blocks, and let F be a mapping $S^m \longrightarrow S$. F induces in a natural way a mapping $F_\infty : S^{\mathbb{Z}} \longrightarrow S^{\mathbb{Z}}$ by:

$$(F_\infty(x))_j = F((x_j,\dots,x_{j+m-1}))$$

for $x \in S^{\mathbb{Z}}$, $j \in \mathbb{Z}$. It is easy to see that F_∞ is continuous and that $F_\infty \circ \sigma = \sigma \circ F_\infty$. Conversely, all shiftcommuting continuous maps are of the form F_∞, or $F_\infty \circ \sigma^k$, for some suitable F and $k \in \mathbb{Z}$. We shall prove this in a slightly more general setting:

<u>(7.12) Theorem of Hedlund [86]:</u> Let $S = \{1,\dots,s\}$ and $S' = \{1,\dots,s'\}$, and let $\Lambda \subset S^{\mathbb{Z}}$, $\Lambda' \subset S'^{\mathbb{Z}}$ be two subshifts. Suppose φ is a continous map from Λ into Λ' such that the following diagram commutes:

$$\begin{array}{ccc} \Lambda & \xrightarrow{\sigma} & \Lambda \\ \varphi\downarrow & & \downarrow\varphi \\ \Lambda' & \xrightarrow{\sigma} & \Lambda' \end{array}$$

(This is the case, in particular, if Λ' is a factor of Λ). Then there exists a $k \in \mathbb{Z}$, an $m \in \mathbb{N}$ and a map $F : S^m \longrightarrow S$ such that φ is the restriction of $F_\infty \circ \sigma^k$ to Λ.

Proof: The cylinders ${}_o[i]$, $i \in S'$, are open and closed in $S'^{\mathbb{Z}}$ and the sets $U_i = {}_o[i] \cap \Lambda'$ are open and closed in Λ'. $\{U_i \mid i \in S'\}$ is a partition of Λ'. Write $V_i = \varphi^{-1}(U_i)$. Then $\{V_i \mid i \in S'\}$ is a partition of Λ into open and closed sets. By (7.9) there exists a $k \in \mathbb{N}$ such that $x \in V_i$, $y \in V_j$ and $i \neq j$ implies that the centered $2k+1$-blocks of x and y are distinct.

Write $\mathfrak{B}_i$ for the set of $2k+1$-blocks which occur as centered $2k+1$-blocks of some $x \in V_i$. One clearly has $\mathfrak{B}_i \cap \mathfrak{B}_j = \emptyset$ for $i \neq j$. Since any $2k+1$-block occurring in Λ occurs as a centered block, $\bigcup_{i \in S'} \mathfrak{B}_i$ is the set of all $2k+1$-blocks occuring in Λ.

Define a map $F : S^{2k+1} \longrightarrow S'$ such that $F(\beta) = i$ if $\beta \in \mathfrak{B}_i$. We shall show that for any $x \in \Lambda$ one has $\varphi(x) = F_\infty(\sigma^k(x))$, i.e. that $y_m = z_{m-k}$ for all $m \in \mathbb{Z}$, where $y = \varphi(x)$ and $z = F_\infty(x)$.

It is easy to see that $y_o = z_{-k}$. Indeed one has $x \in V_i$ for some well-defined $i \in S'$. Thus on one hand $(\varphi(x))_o = y_o = i$, on the other hand $(x_{-k},\dots,x_k) \in \mathfrak{B}_i$, hence $F((x_{-k},\dots,x_k)) = i$ and, therefore, $z_{-k} = (F_\infty(x))_{-k} = i$. Since $y_m = (\varphi(x))_m =$ $= (\sigma^m(\varphi(x)))_o = (\varphi(\sigma^m(x)))_o$ and $z_{m-k} = (F_\infty(x))_{m-k} = (F_\infty(\sigma^m(x)))_{-k}$, the result follows. □

This theorem allowed Hedlund and others to obtain a wealth of information about shift-commuting mappings (see [86]). We shall only quote a few samples. Let $\Phi(S)$ denote the set of all continuous maps from $S^{\mathbb{Z}}$ into itself such that $\varphi \circ \sigma = \sigma \circ \varphi$.

(7.13) Proposition: There exist infinitely many $\varphi \in \Phi(S)$ which are not onto. If $\varphi \in \Phi(S)$ is not onto, there exists a periodic point $x \in S^{\mathbb{Z}}$ such that $\varphi^{-1}(x)$ is uncountable.

(7.14) Proposition: If $\varphi \in \Phi(S)$ is onto, there exists a constant M such that card $\varphi^{-1}(x) \leq M$ for all $x \in S^{\mathbb{Z}}$.

Let N be the maximal integer such that card $\varphi^{-1}(x) \geq N$ for all $x \in S^{\mathbb{Z}}$. If x is topologically transitive then card $\varphi^{-1}(x) = N$. Thus φ is N-to-one on a residual subset of $S^{\mathbb{Z}}$. φ is N-to-one on all of S^Z iff φ is open.

<u>(7.15) Proposition:</u> If $\varphi \in \Phi(S)$ and if $x \in S^Z$ is periodic, then $\varphi^{-1}(x)$ contains a periodic point. If φ is onto, and if x is periodic (resp. topologically transitive), then all points in $\varphi^{-1}(x)$ are periodic (resp. topologically transitive).

<u>(7.16) Proposition:</u> For $\varphi \in \Phi(S)$ and $\psi \in \Phi(S)$ the composition $\varphi \cdot \psi \in \Phi(S)$ is onto iff both φ and ψ are onto.

<u>(7.17) Proposition:</u> If $\varphi \in \Phi(S)$ is one-to-one, then it is onto and thus a homeomorphism.

8. Measures on the Shift Space

Let $S = \{1,\dots,s\}$ be a finite state space and σ the shift on $S^{\mathbb{Z}}$.

(8.1) Proposition: For $\mu \in \mathfrak{M}_\sigma(S^{\mathbb{Z}})$ the following properties are valid:

(1) $$\sum_{a_o \in S} \mu({}_o[a_o]) = 1$$

and for any block $(a_o,\dots,a_k)$ and any $n \in \mathbb{Z}$,

(2) $$\mu\,({}_n[a_o,\dots,a_k]) \geq 0;$$

(3) $$\mu\,({}_n[a_o,\dots,a_k]) = \sum_{a_{k+1} \in S} \mu\,({}_n[a_o,\dots,a_k,a_{k+1}]);$$

(4) $$\mu\,({}_n[a_o,\dots,a_k]) = \sum_{a_{-1} \in S} \mu\,({}_n[a_{-1},a_o,\dots,a_k]).$$

This is trivial to check. By a special case of Kolmogoroff's consistency theorem (see [152, chapter V]), these properties are sufficient to define a measure:

(8.2) Proposition: Let μ be a function on the set of cylinders of $S^{\mathbb{Z}}$ satisfying conditions (1)-(4). Then there exists a uniquely determined measure in $\mathfrak{M}_\sigma(S^{\mathbb{Z}})$ which agrees with μ on those cylinders.

We refer to [17, pp.33-35] for a simple direct proof of (8.2). Remark that it is condition (4) which implies the shift-invariance.

(8.3) Proposition: $\mu \in \mathfrak{M}_\sigma(S^{\mathbb{Z}})$ is ergodic iff

(5) $$\lim_{N\to\infty} \frac{1}{N} \sum_{n=o}^{N-1} \mu({}_o[A] \cap {}_n[B]) = \mu({}_o[A])\,\mu({}_o[B]);$$

weakly mixing iff

(6) $$\lim_{N\to\infty} \frac{1}{N} \sum_{n=o}^{N-1} |\mu({}_o[A] \cap {}_n[B]) - \mu({}_o[A])\,\mu({}_o[B])| = 0;$$

strongly mixing iff

(7) $\lim_{n\to\infty} \mu({}_o[A] \cap {}_n[B]) = \mu({}_o[A])\,\mu({}_o[B])$

holds for all blocks A and B occuring in $S^{\mathbb{Z}}$.

Remark that the limit on the left hand side of (5) exists for all $\mu \in \mathfrak{M}_\sigma(S^{\mathbb{Z}})$ by virtue of Birkhoff's ergodic theorem (4.3).

Particularly important classes of measures in $\mathfrak{M}_\sigma(S^{\mathbb{Z}})$ are defined by probability vectors and stochastic matrices. Our treatment follows [17] closely.

<u>(8.4) Definition:</u> An s-tuple $\pi = (p_1,\dots,p_s)$ with $p_i \geq 0$, $\sum_{i=1}^{s} p_i = 1$ is called a <u>probability vector.</u>

A probability vector defines a set function μ_π on the cylinders of $S^{\mathbb{Z}}$ by

$$\mu_\pi({}_n[a_o,\dots,a_k]) = p_{a_o}\, p_{a_1} \dots p_{a_k}$$

It is easy to see that conditions (1)-(4) are satisfied.

<u>(8.5) Definition:</u> The corresponding shift-invariant measure will be denoted by μ_π and called a <u>Bernoulli measure.</u> The system $(S^{\mathbb{Z}}, \mu_\pi, \sigma)$ is called a <u>Bernoulli shift.</u>

Remark that π defines a measure on the state space S by assigning mass p_i to the point $i \in S$. Thus μ_π is just the corresponding product measure on the product space $S^{\mathbb{Z}}$.

In the sequel we shall have several occasions to use the Perron-Frobenius theorem. We first introduce some notation.

<u>(8.6) Definition:</u> An $s \times s$-matrix $R = (r_{ij})$ is said to be <u>positive</u> if $r_{ij} \geq 0$. R is said to be <u>irreducible</u> if for any pair of indices i,j with $1 \leq i,j \leq s$, the (i,j)-th coefficient of some power of R^n $(n > 0)$ is strictly positive.

<u>(8.7) Theorem of Perron-Frobenius:</u> The positive matrix $R = (r_{ij})$ has a positive eigenvalue r such that no eigenvalue of R has absolute value > r. One has

$$\min_i \sum_j r_{ij} \leq r \leq \max_i \sum_j r_{ij}.$$

To this dominant eigenvalue r there correspond positive left (row) and right (column) eigenvectors. If R is irreducible, the eigenvalue r is simple, and the corresponding eigenvectors are strictly positive.

For a proof of this theorem, we refer to [97 , chapter 13].

<u>(8.8) Definition:</u> An s×s-matrix $P = (p_{ij})$ is said to be a <u>stochastic matrix</u> if $p_{ij} \geq 0$ and $\sum_j p_{ij} = 1$ for all $i \in S$.

The column vector whose components are equal to 1 obviously is an eigenvector of P for the eigenvalue 1.

By the estimate in (8.7), 1 is a dominant eigenvalue, all other eigenvalues of P have absolute value ≤ 1. There exists a row eigenvector $\pi = (p_1,\dots,p_2)$ of P for the eigenvalue 1. We may assume that this eigenvector is positive (by (8.7)) and normalized. Thus, π is defined as a probability vector such that $\pi P = \pi$. If P is irreducible, π is uniquely defined.

A pair π, P where P is a stochastic matrix and π a probability vector with $\pi P = \pi$ defines a set function $\mu_{\pi P}$ on the cylinders of $S^{\mathbb{Z}}$ by

$$\mu_{\pi P}\ ({}_n[a_0,\dots,a_k]) = p_{a_0}\ p_{a_0 a_1} \cdots p_{a_{k-1}a_k}$$

It is easy to check that conditions (1)-(4) of (8.1) are satisfied.

<u>(8.9) Definition:</u> The corresponding shift-invariant measure will be denoted by $\mu_{\pi P}$ and called a <u>Markov measure.</u> The dynamical system $(S^{\mathbb{Z}}, \mu_{\pi P}, \sigma)$ is called a <u>Markov shift.</u>

Remark that if the columns of P are constant, i.e. $p_{ij} = p_{kj}$ for all $i,j,k \in S$, then $p_{ij} = p_j$ for all $j \in S$ and $\mu_{\pi P}$ is just μ_π. Thus Bernoulli measures are special cases of Markov measures.

The elements p_{ij} of the stochastic matrix P can be viewed as the transition probabilities from the state i to the state j, i.e. as conditional probabilities for the occurence of j, given the occurence of i one step earlier. The p_j can be viewed as "initial probabilities", or as stationary probability distribution on S. Thus $\mu\ ({}_n[a_0,\dots,a_k])$ is the probability of the occurence of a_o and the transitions from a_o to a_1 to a_2 ... to a_k.

From now on we shall always assume $p_j > 0$ for all $j \in S$. This is no real restriction, since otherwise no cylinder containing j would have positive measure: the state j might as well be deleted from the state space S.

Let $p_{ij}^{(n)}$ denote the elements of P^n, the n-th power of P $(n > 1)$. Since $p_{ij}^{(2)} = \sum_{k \in S} p_{ik}p_{kj}$, $p_{ij}^{(2)}$ can be viewed as probability for the transition from i to j in two steps (via all possible intermediary states k). Similarly $p_{ij}^{(n)}$ is the probability for the transition from i to j in n steps. The set of all stochastic s×s matrices forms a semi-group with identity. In particular, the matrices P^n are all stochastic ones. Every eigenvector of P for the eigenvalue 1 is also an eigenvector of P^n.

(8.10) Proposition: The stochastic matrix P is irreducible iff there is no permutation of the indices forming it into
$\begin{bmatrix} A & 0 \\ B & C \end{bmatrix}$, with A a square matrix of order < s.

Proof: If P is irreducible, then for any two states $i,j \in S$ there exists an n with $p_{ij}^{(n)} > 0$. Thus from any state i one can finally reach any state j. This is obviously equivalent to the condition that there is no proper subset S' of the state space S which is "closed" in the sense that $\sum_{j \in S'} p_{ij} = 1$ for all $i \in S$. Indeed, this would mean that there is no possible way from S' to a state outside of S'. If P is reducible, i.e. if there exists a proper "closed" set $S' \subset S$, let s' denote the cardinality of S'. A permutation of S mapping S' onto the set $\{1,\dots,s'\} \subset S$ would change the matrix P into a matrix of the form $\begin{bmatrix} A & 0 \\ B & C \end{bmatrix}$, where A is an s'×s'-matrix. □

It is obvious that (8.10) is valid for any positive - not necessarily stochastic - matrix R.

(8.11) Proposition: For any stochastic matrix P, the matrix

$$\bar{P} = \lim \frac{1}{N} \sum_{n=0}^{N-1} P^n$$

exists. $\bar{P} = (\bar{p}_{ij})$ is again a stochastic matrix, with $P\bar{P}=\bar{P}P=\bar{P}$ and $\bar{P}^2 = \bar{P}$. Any eigenvector of P for the eigenvalue 1 is an eigenvector of $\bar{P}$, too.

Proof: For $i,j \in S$ one has

(8) $$\mu_{\pi P}({}_o[i] \cap {}_n[j]) =$$
$$= \sum_{a_1 \in S} \cdots \sum_{a_{n-1} \in S} p_i p_{ia_1} \cdots p_{a_{n-1}j} = p_i p_{ij}^{(n)}$$

Since $\lim \frac{1}{N} \sum_{n=0}^{N-1} \mu_{\pi P} ({}_o[i] \cap {}_n[j])$

exists (see remark after (8.3)) and $p_i \neq 0$, it follows that

(9) $$\lim_{N \to \infty} \frac{1}{N} \sum_{n=0}^{N-1} p_{ij}^{(n)} = \bar{p}_{ij}$$

exists for all $i,j \in S$. (Here we define $p_{ij}^{(1)} = p_{ij}$ and $p_{ij}^{(o)} = \delta_{ij}$, the Kronecker symbol). Thus $\bar{P}$ exists. The other assertions are obvious. □

(8.12) Proposition (see [17]): The following conditions are equivalent:

(a) The measure $\mu_{\pi P}$ is ergodic;

(b) $\bar{p}_{ij}$ is independent of i, for all $j \in S$;

(c) $\bar{p}_{ij} > 0$ for all $i,j \in S$,

(d) P is irreducible;

(e) 1 is a simple eigenvalue of P.

Proof: One has

$$\lim_{N \to \infty} \frac{1}{N} \sum_{n=0}^{N-1} \mu_{\pi P} ({}_o[i] \cap {}_n[j]) = p_i \cdot \bar{p}_{ij}$$

by (8) and (9). If $\mu_{\pi P}$ is ergodic, this limit is equal to $\mu_{\pi P}({}_o[i])\cdot\mu_{\pi P}({}_o[j]) = p_i p_j$ and therefore $\bar{p}_{ij} = p_j$ is independent of i. Thus (a)$\Longrightarrow$(b).

If, conversely, $\bar{p}_{ij}$ is independent of i, then it follows from $\pi\bar{P} = \pi$ that $\bar{p}_{ij} = p_j$. For any two blocks $(a_o,\ldots,a_k)$ and $(b_o,\ldots,b_l)$ one has

$$\lim \frac{1}{N}\sum_{n=o}^{N-1} \mu_{\pi P}\left({}_o[a_o,\ldots,a_k] \cap {}_n[b_o,\ldots,b_l]\right) =$$

$$= p_{a_o} p_{a_o a_1} \cdots p_{a_{k-1} a_k} \left(\lim \frac{1}{N}\sum_{n=k}^{N-1} p_{a_k b_o}^{(n-k)}\right) p_{b_o b_1} \cdots p_{b_{l-1} b_l}$$

$$= \left(p_{a_o} p_{a_o a_1} \cdots p_{a_{k-1} a_k}\right)\left(p_{b_o} p_{b_o b_1} \cdots p_{b_{l-1} b_l}\right)$$

$$= \mu_{\pi P}({}_o[a_o,\ldots,a_k])\, \mu_{\pi P}({}_o[b_o,\ldots,b_l]).$$

It follows from (8.3) that $\mu_{\pi P}$ is ergodic. Thus (b)$\Longrightarrow$(a).

(b)$\Longrightarrow$(c) is trivial and (c)$\Longrightarrow$(d) is clear since $\bar{p}_{ij} > 0$ implies $p_{ij}^{(n)} > 0$ for some n. Also (b)$\Longrightarrow$(e) is obvious.

In order to prove (c)$\Longrightarrow$(b), fix $j \in S$ and write $\bar{p}_j = \max_i \bar{p}_{ij}$. Since $\bar{P}^2 = \bar{P}$, the j-th column of $\bar{P}$ is an eigenvector of $\bar{P}$ for the eigenvalue 1. If $\bar{p}_{ij} < \bar{p}_j$ for some $i \in S$, then

$$\bar{p}_{kj} = \sum_{i\in S} \bar{p}_{ki}\, \bar{p}_{ij} < \sum_{i\in S} \bar{p}_{ki}\, \bar{p}_j = \bar{p}_j \quad \text{for all } k \in S,$$

a contradiction. Hence $\bar{p}_{ij} = \bar{p}_j$ for all i. As for (e)$\Longrightarrow$(b), if 1 is a simple eigenvalue of P, it follows from $\bar{P} = P\bar{P}$ that the columns of $\bar{P}$ are constants.

It remains to show that (d)$\Longrightarrow$(c). Denote by S_i the set of states j such that $\bar{p}_{ij} > 0$. Since $\bar{P} = \bar{P}P$, one has $\bar{p}_{ij} = \sum_{k\in S} \bar{p}_{ik}\, p_{kj} \geq \bar{p}_{il}\, p_{lj}$. One has $\sum_{j\in S_i} p_{lj} = 1$ for all $l \in S_i$, since $p_{lj} > 0$ for some $l \in S_i$ and some $j \notin S_i$ would imply $\bar{p}_{ij} > 0$, a contradiction. Thus S_i is a "closed" subset of S, and since P is irreducible, S_i must be equal to S, i.e. $\bar{p}_{ij} > 0$ for all $j \in S$. □

As a corollary one obtains

(8.13) Proposition: Every Bernoulli measure μ_π is ergodic.

A state $j \in S$ is said to have period $m > 1$ if $p_{jj}^{(n)} = 0$ whenever n is not divisible by m, and m is the largest integer with this property. (This implies that only after m, 2m, 3m,... steps there is possibly a positive probability for the return to j). A state $j \in S$ is said to be aperiodic if it has no such period.

If P is irreducible and j has period m, then so has every other $k \in S$. Indeed, there exist N, M > 0 with $p_{jk}^{(N)} > 0$, $p_{kj}^{(M)} > 0$ and thus $p_{jj}^{(n+N+M)} \geq p_{jk}^{(N)} p_{kj}^{(M)} p_{kk}^{(n)}$ and $p_{kk}^{(n+N+M)} \geq p_{jk}^{(N)} p_{kj}^{(M)} p_{jj}^{(n)}$.

(8.14) Definition: The stochastic matrix P is said to be periodic of period q (resp. aperiodic) if it is irreducible and there exists a state $j \in S$ which is periodic of period q (resp. aperiodic).

The following proposition is easy to check.

(8.15) Proposition: If P has period q, then the state space S can be decomposed into disjoint subsets $S_1,\ldots,S_q$ (not necessarily of same size) such that a one-step transition from a state in S_j leads to a state in S_{j+1} (from S_q to S_1). Each S_j will be "closed" with respect to the transition matrix P^q, and the restriction of P^q to the states of S_j will be an aperiodic stochastic matrix.

An example of a periodic s×s-matrix of period s is given by

$$\begin{bmatrix} 0 & 1 & 0 & . & . & . & 0 \\ 0 & 0 & 1 & . & . & . & 0 \\ . & . & . & . & . & . & . \\ . & . & . & . & . & . & . \\ . & . & . & . & . & . & . \\ . & . & . & . & . & . & . \\ 1 & 0 & 0 & . & . & . & 0 \end{bmatrix}$$

Proposition (8.15) says that this is (possibly after relabelling the states in S) the form of a general periodic matrix, with the 1's replaced by stochastic sub-matrices, and the 0's by submatrices consisting of 0's.

We shall need the main theorem for finite Markov chains:

(8.16) Proposition: Let P be an aperiodic stochastic matrix and $\pi = (p_1,\dots,p_s)$ the uniquely defined probability vector such that $\pi P = \pi$. Then $\lim_{n\to\infty} p_{ij}^{(n)}$ exists and is equal to $p_j > 0$ for all $i,j \in S$.

For a proof, we refer to the standard texts of probability theory. As a corollary, one obtains the following proposition:

(8.17) Proposition: P is aperiodic iff there exists an N such that $p_{ij}^{(n)} > 0$ for all $n > N$ and all $i,j \in X$.

(8.18) Proposition: The following conditions are equivalent:

(a) $\mu_{\pi P}$ is weakly mixing;

(b) $\mu_{\pi P}$ is strongly mixing;

(c) P is aperiodic.

Proof: If $\mu_{\pi P}$ is weakly mixing, it follows by (6) that

$$\lim \frac{1}{N}\sum_{n=o}^{N-1} | p_{jj}^{(n)} - p_j | = 0$$

and hence from a well-known fact on Cesàro-averages that there exists a subset $D_j \subset \mathbb{N}$ of density 0 such that $p_{jj}^{(n)} \longrightarrow p_j$ for $n \notin D_j$, $n \longrightarrow \infty$. This implies obviously that the state j, and hence the irreducible matrix P, are aperiodic.

If P is aperiodic, one has by (8.16) that $\lim p_{ij}^{(n)} = p_j$ for all $i,j \in S$. It is easy to see that this implies (7) and hence that $\mu_{\pi P}$ is strongly mixing. □

As a corollary, one obtains:

(8.19) Proposition: Every Bernoulli measure μ_π is strongly mixing.

Actually this follows directly from (8.3).

9. Partitions and Generators

In the following $(\mathfrak{X},\Sigma,m,\Psi)$ is an m.t. dynamical system. Some parts of the definitions also make sense without the transformation. All sets are supposed to be measurable.

(9.1) Definition: a) A partition (of $\mathfrak{X}$) is a family $\alpha = (A_i)_{i\in I}$ of disjoint sets with at most a countable index set I and such that $\sum_{i\in I} m(A_i) = 1$ (see also remark b).

b) $\mathfrak{Z}_I$ is the set of all partitions with index set I.

c) The sets A_i are called the atoms of α. $A \in \alpha$ means that A is an atom of α.

d) $\Sigma(\alpha)$ is the smallest σ-algebra containing all atoms of α.

e) If α,β are partitions, $\alpha \subset \beta$ means that each non-empty atom of α is a union of atoms of β. Then we say that β is finer than α. If α is a partition and $\Sigma' \subset \Sigma$ a sub-σ-algebra, $\alpha \subset \Sigma'$ means that each atom of α belongs to Σ'.

Remarks:

a) It could be that some of the atoms A_i are of measure 0 or even empty.

b) Often other concepts of a partition are useful.

b1) In a purely topological situation, the condition $\Sigma\, m\,(A_i) = 1$ must be replaced by $\bigcup A_i = \mathfrak{X}$. In this case additional conditions may be imposed upon the atoms. We shall need such partitions in sections 15, 16, 18, 19, 20, 25.

b2) Sometimes the index set is superfluous, however, we need it for the comparison of different partitions (see (9.3.a))

b3) Often it is sufficient to consider the atoms of α as equivalence classes mod 0; in this case α and $\beta \in \mathfrak{Z}_I$ are identified if $\|\alpha,\beta\| = 0$ (see (9.3.a)). We can not do this because then the name mapping Φ_α (see (9.4.b)) would make no sense (see also remark b) to (9.3).).

(9.2) Definition:

a) If $\alpha \in \mathfrak{Z}_I$ and $B \in \Sigma$, $B \cap \alpha = \alpha \cap B$ is the partition $((B \cap A_i)_{i\in I}, \mathfrak{X} \setminus B)$ of $\mathfrak{X}$. (To have a well defined index set, we might give $\mathfrak{X} \setminus B$ the index I, so that $B \cap \alpha \in \mathfrak{Z}_{I\cup\{I\}}$; but whenever this operation occurs, the index set will not be important.)

b) If $\alpha \in \mathfrak{Z}_I$ and $B \in \Sigma$ with $m(B) > 0$, then $\alpha|B = \alpha_B$ is the partition $(B \cap A_i)_{i\in I}$ of B (with the restricted σ-algebra $\Sigma|B$ and the conditioned measure $m_B(A) = m(B)^{-1}\, m(A\cap B)$)

c) If $\alpha = (A_i)_{i\in I} \in \mathfrak{Z}_I$ and $\beta = (B_j)_{j\in J} \in \mathfrak{Z}_J$, then $\alpha \vee \beta \in \mathfrak{Z}_{I\times J}$ is the partition $(A_i \cap B_j)_{(i,j)\in I\times J}$. In a similar way $\bigvee_{k=1}^{n} \alpha_k$ is defined. It is called the refinement of the α_k.

d) If $\alpha = (A_i)_{i\in I} \in \mathfrak{Z}_I$, then $\Psi^j\alpha = (\Psi^j A_i)_{i\in I}$, and for $s \leq t \in \mathbb{Z}$: $(\alpha)_s^t = \bigvee_{s\leq k\leq t} \Psi^{-k}\alpha$

e) $\Sigma_\Psi(\alpha) = \bigvee_{j\in \mathbb{Z}} \Sigma(\Psi^j\alpha)$ is the smallest Ψ-invariant σ-algebra containing α or $\Sigma(\alpha)$ ($\bigvee_t \Sigma_t$ denotes the smallest σ-algebra containing all the σ-algebras Σ_t).

Remark: With our definition, $\alpha \vee \alpha \neq \alpha$, but $\alpha \vee \alpha \subset \alpha$ and $\alpha \subset \alpha \vee \alpha$ (see (9.1.e)).

(9.3) Definition: a) On $\mathfrak{Z}_I$ we define the following pseudo-metric (we shall say metric for short):

$$\| \alpha,\beta \| = \frac{1}{2} \sum_{i\in I} m(A_i \,\Delta\, B_i) = \sum_{i\in I} m(A_i \setminus B_i) \qquad (\alpha,\beta \in \mathfrak{Z}_I)$$

b) For partitions $\alpha \in \mathfrak{Z}_I$, $\beta \in \mathfrak{Z}_J$ and $\epsilon \geq 0$ we write $\alpha \overset{\epsilon}{\subset} \beta$ if there exists a partition $\alpha' \in \mathfrak{Z}_I$ such that $\alpha' \subset \beta$ and $\| \alpha,\alpha' \| \leq \epsilon$. If α is a partition, $\Sigma' \subset \Sigma$ a sub-σ-algebra, and $\epsilon > 0$, then $\alpha \overset{\epsilon}{\subset} \Sigma'$ is defined similarly.

c) If $\alpha \overset{0}{\subset} \beta$ we say that "β is finer than α mod 0." If $\alpha \overset{0}{\subset} \beta$ and $\beta \overset{0}{\subset} \alpha$ we write $\alpha \overset{0}{=} \beta$.

Remarks: a) The metric $\| .,. \|$ measures how well corresponding atoms of α and β overlap. It satisfies

$\|\alpha \vee \alpha', \beta \vee \beta'\| \leq \|\alpha,\beta\| + \|\alpha',\beta'\|$.

It is not difficult to see that $\|.,.\|$ is a complete metric on $\mathfrak{Z}_I$. If we have a Cauchy sequence $(\alpha_j)_{j\in\mathbb{N}}$ in $\mathfrak{Z}_I$, we use the notation $\alpha = \lim \alpha_j$ and mean that $\alpha \in \mathfrak{Z}_I$ is a fixed partition satisfying $\lim_j \|\alpha,\alpha_j\| = 0$. The choice of the limit is not important in our applications.

b) Identifying α and β if $\alpha \overset{o}{=} \beta$ means that both of the identifications in remark b2) and b3) to (9.1) are made. This identification is natural for all entropy considerations (section 10).

<u>(9.4) Definition:</u> a) If $\alpha \in \mathfrak{Z}_I$ and $n \in \mathbb{N}$, the <u>(α,Ψ,n)-name</u> of $x \in \mathfrak{X}$ is the index of the atom of $(\alpha)_o^{n-1}$ to which x belongs (see (9.2.c)), i.e. the sequence $(i_o(x),\dots,i_{n-1}(x)) \in I^n$ such that

$$\Psi^k x \in A_{i_k(x)} \quad (0 \leq k < n).$$

This name is defined for almost all $x \in \mathfrak{X}$.

b) If $\alpha \in \mathfrak{Z}_I$, the <u>$(\alpha,\Psi)$-name</u> of $x \in \mathfrak{X}$ is the sequence $\Phi_\alpha(x) = (i_k(x))_{k\in\mathbb{Z}} \in I^{\mathbb{Z}}$ such that $\Psi^k x \in A_{i_k}(x)$ $(k \in \mathbb{Z})$. The mapping $\Phi_\alpha : \mathfrak{X} \longrightarrow I^{\mathbb{Z}}$ is a.e. defined and measurable and makes the diagram

$$\begin{array}{ccc} \mathfrak{X} & \xrightarrow{\ \Psi\ } & \mathfrak{X} \\ \downarrow{\scriptstyle \Phi_\alpha} & & \downarrow{\scriptstyle \Phi_\alpha} \\ I^{\mathbb{Z}} & \xrightarrow{\ \sigma\ } & I^{\mathbb{Z}} \end{array}$$

(σ is the shift transformation on $I^{\mathbb{Z}}$) commutative.

The transported σ-invariant measure $\Phi_\alpha m$ on $I^{\mathbb{Z}}$ is called μ_α.

c) The partition α <u>separates almost all points</u> (under Ψ), if there is a set $N \in \Sigma$ with $m(N) = 0$ and $\Phi_\alpha x \neq \Phi_\alpha y$ $(x \neq y \in \mathfrak{X} \setminus N)$

d) A sequence $(\alpha_k)_{k\in\mathbb{N}}$ of partitions <u>generates</u> Σ if $\alpha_k \subset \alpha_{k+1}$ and $\bigvee_{k\in\mathbb{N}} \Sigma(\alpha_k) = \Sigma$ mod 0, i.e. if for $B \in \Sigma$ there is $B' \in \bigvee \Sigma(\alpha_k)$ such that $m(B \,\Delta\, B') = 0$; a sequence (α_k) <u>generates Σ under Ψ</u> if

$$\bigvee_{k\in\mathbb{N}} \Sigma_\Psi(\alpha_1 \vee \dots \vee \alpha_k) = \Sigma \qquad \text{mod } 0.$$

e) The partition α is a <u>generator</u> if $(\alpha)_{-k}^{k}$ $(k \in \mathbb{N})$ is a generating sequence (or if $\Sigma_\Psi(\alpha) = \Sigma \mod 0$), and a <u>strong generator</u> if $(\alpha)_{0}^{k}$ $(k \in \mathbb{N})$ is a generating sequence.

<u>Remarks:</u> a) Any identification of partitions makes the definition of Φ_α impossible. However, $\mu_\alpha = \mu_\beta$ if $\|\alpha,\beta\| = 0$.
b) In Lebesgue spaces generating sequences of finite partitions α_k always exist, since this trivially holds in $[0,1]$.
c) The increasing sequence (α_k) generates if and only if for any $B \in \Sigma$ and $\varepsilon > 0$ there is an $n \in \mathbb{N}$ and $B_n \in \Sigma(\alpha_n)$ with $m(B \Delta B_n) < \varepsilon$.
d) If $\alpha \overset{o}{=} \beta$ and α is a generator, then β is also a generator.

A partition may be interpreted as a physical measurement in the system $(\mathfrak{X}, \Sigma, m)$. If $\alpha \in \mathfrak{Z}_I$ and some experiment yields the result $x \in A_i$, the value we measure is $i \in I$. In this sense the refined partition $\alpha \vee \beta$ represents the simultaneous measurement of α and β, and $(\alpha)_s^t$ represents the repeated evaluation at constant time intervals of one unit of the same experiment when the time evolution of the system is described by Ψ. A generator is an experiment which, repeated continually, yields full information about Σ.

In many cases there are finite generators, but their construction is difficult; this will be treated in sections 28 and 30. The proof of the existence of generators without the condition of finiteness is surprisingly easy. Proofs can be found in [149], [159]. In our proof we follow an oral communication by U. Krengel.

<u>(9.5) Proposition:</u> If $(\mathfrak{X}, \Sigma, m, \Psi)$ is aperiodic, then there exists a strong generator.

<u>Proof:</u> a) First we assume that Ψ is ergodic. Then every set A with positive measure is a "sweep out set", i.e. $m(\bigcup_{j=0}^{\infty} \Psi^j A) = 1$. We choose a sequence of disjoint sets A_i such that $m(A_i) > 0$ and numbers $n_i \in \mathbb{N}$ such that

$$m(\bigcup_{j=0}^{n_i} \Psi^{-j} A_i) > 1 - 2^{-i}.$$

Further we take a generating sequence (α_i) of partitions and set

$$\beta_i = (\alpha_i)^{o}_{-n_i} \cap A_i .$$

For $0 \leq j \leq n_i$, $\alpha_i \cap \Psi^{-j} A_i \subset \Psi^{-j} \beta_i$, so

$$\alpha_i \overset{2^{-i}}{\subset} \alpha_i \cap \bigcup_{j=o}^{n_i} \Psi^{-j} A_i \subset (\beta_i)_o^{n_i} .$$

Let α be a minimal partition refining all the β_i, i.e. a partition enumerating $\mathfrak{X} \setminus \bigcup_{i\in\mathbb{N}} A_i$ and all atoms of the β_i which are not of the form $\mathfrak{X} \setminus A_i$. Then $(\alpha)_o^{n_i} \supset (\beta_i)_o^{n_i} \overset{2^{-i}}{\supset} \alpha_i$, and α is a strong generator.

Note that by the choice of the A_i, the atom $\mathfrak{X} \setminus \cup A_i$ of α can be made arbitrarily large, and that even $\alpha \cap \bigcup_{j\geq i} A_i$ is a strong generator $(i \in \mathbb{N})$.

b) In the case of a non-ergodic Ψ the only additional difficulty is finding suitable sets A_i. We use Rohlin's lemma (1.18) and obtain first $(\Psi^{-1}, 2^{2^i}, 2^{-2i})$-Rohlin sets F_i $(i \in \mathbb{N})$, i.e. sets F_i satisfying

$$F_i \cap \Psi^{-j} F_i = \emptyset \quad (1 \leq j < 2^{2^j})$$

(this implies in particular that $m(F_i) \leq 2^{-2^i}$)
and

$$m\Big(\bigcup_{o\leq j\leq 2^{2^i}} \Psi^{-j} F_i\Big) > 1 - 2^{-2i},$$

and put

$$A_i = F_i \setminus \bigcup_{j>i} F_j \qquad (i \in \mathbb{N}).$$

Since $m\big(\bigcup_{j>i} F_j\big) \leq \sum_{j=i+1}^{\infty} 2^{-2^j} < 2 \cdot 2^{-2^{i+1}}$,
it follows that

$$m\Big(\bigcup_{o\leq j\leq 2^{2^i}} \Psi^{-j} A_i\Big) > 1 - 2^{-2i} - 2^{2^i} \cdot 2^{-2^{i+1}} = 1-2^{-2i}-2\cdot 2^{-2^i}$$

$$> 1 - 2^{-i} \quad \text{for } i \geq 3.$$

This seqeunce (A_i) has all the necessary properties for the proof. □

(9.6) Proposition: Equivalent properties of the partition $\alpha \in \mathfrak{Z}_I$ are

a) α is a generator

b) α separates almost all points under Ψ

c) $(\mathfrak{X}, m, \Psi)$ is m.t. conjugate to $(I^{\mathbb{Z}}, \mu_\alpha, \sigma)$ by the map Φ_α.

Proof: a)$\Longrightarrow$b) (R.Adler, B.Weiß): We may forget the point masses of m (which belong to periodic orbits) and assume that $\mathfrak{X} = [0,1]$ and m is the Lebesgue measure. Let J be the set of closed intervals $[0,b]$ in $\mathfrak{X}$ with rational b. J is countable and separates points, i.e. for $x \neq y \in \mathfrak{X}$ there is some $B \in J$ with $x \in B$, $y \notin B$ or vice versa. Since $\Sigma_\Psi(\alpha) = \Sigma \bmod 0$, we may take for each $B \in J$ some $B' \in \Sigma_\Psi(\alpha)$ such that $m(B \,\Delta\, B') = 0$. The set

$$N = \bigcup_{j\in\mathbb{Z}} \bigcup_{B\in J} \Psi^j(B\Delta B') \cup \bigcup_{j\in\mathbb{Z}} \Psi^j(\mathfrak{X} \setminus \bigcup_{i\in I} A_i)$$

has measure zero. On the Ψ-invariant set $\mathfrak{X} \setminus N$, Φ_α is defined, and if $x \neq y \in \mathfrak{X} \setminus N$ there is some B' which separates x and y. Suppose now that $x \neq y \in \mathfrak{X} \setminus N$, but $\Phi_\alpha x = \Phi_\alpha y$. Then for all $t \in \mathbb{Z}$, x and y lie in the same atom of $\Psi^t\alpha$. But the system of all sets which do not separate x and y is a σ-algebra, so it must contain $\Sigma_\Psi(\alpha)$, and we have a contradiction.

b)$\Longrightarrow$c): We note first that $I^{\mathbb{Z}}$ with the product topology of the discrete topology on I is separable and metric. That Φ_α is an m.t. isomorphism between $(\mathfrak{X}, m)$ and $(I^{\mathbb{Z}}, \mu_\alpha)$ was proved in (2.18), and the rest is clear since $\Phi_\alpha \circ \Psi = \sigma \circ \Phi_\alpha$ a.e.

c)$\Longrightarrow$a): This is trivial since the Borel field in the shift space $I^{\mathbb{Z}}$ is generated by the cylinder sets, and their inverse images are almost surely atoms of partition $(\alpha)_s^t$ of $\mathfrak{X}$. □

(9.7) Corollary: Two aperiodic m.t. dynamical systems $(\mathfrak{X}, m, \Psi)$ and $(\mathfrak{X}', m', \Psi')$ are m.t. conjugate if and only if there exist generators α for Ψ and α' for Ψ' such that

$$\mu_\alpha = \mu_{\alpha'}$$

(9.8) Corollary: Every m.t. dynamical system is m.t. conjugate to a system carried by a top-dynamical system.

Proof: If the system is aperiodic, we choose some generator α with index set I. Either I is finite, or we may assume that $I = \{0,1,\frac{1}{2},\frac{1}{3},\frac{1}{4},\ldots\}$. In any case we obtain a compact shift space $I^{\mathbb{Z}}$, and the corollary follows from (9.6.c). If all points have a common (smallest) period k, then there exists a set F such that $\Psi^j F \cap F = \emptyset$ $(1 \leq j < k)$ and $m(F) = k^{-1}$ (exercise!). Since (F, m_F) is a Lebesgue space, the corollary is clear.

In the general case we decompose the space into an aperiodic part and parts with constant period, and apply the arguments above on each part separately. The direct sum of the compact spaces obtained is locally compact; now we add the Alexandroff point as a fixed point of the transformation. □

10. Information and Entropy

Let $(\mathfrak{X},\Sigma,m)$ be a measure space.

(10.1) Definition: For $B \in \Sigma$ the quantity $-\log m(B)$ is called the information given by B and denoted by $I(B)$.

If α is a partition then the function

$$\bar{I}(\alpha) : x \longrightarrow \sum_{A\in\alpha} 1_A(x) \cdot I(A)$$

(or its equivalence class mod 0) is called the information of α.

Obviously:

(10.2): Proposition:

(a) $\bar{I}(\alpha) \geq 0$;

(b) $\bar{I}(\alpha) = 0$ a.e. iff α is trivial;

(c) $\bar{I}(\alpha) \leq \bar{I}(\alpha')$ a.e. if $\alpha \overset{o}{\subset} \alpha'$;

(d) $\bar{I}(\alpha \vee \alpha') = \bar{I}(\alpha) + \bar{I}(\alpha')$ a.e. iff α and α' are independent.

The entropy of a partition α will be defined as the expectation of the information of the corresponding experiment, i.e. as $\int \bar{I}(\alpha)\, dm$. Thus

(10.3) Definition: The quantity

$$H_m(\alpha) = -\sum_{A\in\alpha} m(A) \log m(A)$$

is called the entropy of the partition α. We shall often omit the subscript m. (We define $0 \log 0 = 0$).

It should be noted that if $\alpha \overset{o}{=} \alpha'$ then $H(\alpha) = H(\alpha')$. Thus in dealing with entropy we need not distinguish between partitions and equivalence classes of partitions as we shall do in the following two sections.

(10.4) Proposition:

(a) $0 \leq H(\alpha)$, with equality iff α is trivial;

(b) If α has k atoms then $H(\alpha) \leq \log k$, with equality iff all atoms have measure $\frac{1}{k}$. If α is infinite it may happen that $H(\alpha) = +\infty$.

(c) $\alpha \overset{\circ}{\subset} \alpha'$ implies $H(\alpha) \leq H(\alpha')$.

(d) $H(\alpha \vee \alpha') \leq H(\alpha) + H(\alpha')$. For partitions with finite entropy equality holds iff α and α' are independent.

Proof: (a) is trivial.

(b) Applying Jensen's inequality to the function $z : x \longrightarrow -x \log x$, which is strictly convex in $[0,1]$, we obtain

$$H(\alpha) = \sum_{A\in\alpha} z(m(A)) = k \sum \frac{1}{k} z(m(A)) \leq k\, z\left(\frac{1}{k} \sum m(A)\right) =$$

$$= k\, z\left(\frac{1}{k}\right) = \log k$$

with equality iff $m(A) = \frac{1}{k}$ for all $A \in \alpha$.

(c) is a trivial consequence of (10.2.c) and the interpretation of $H(\alpha)$ as the expectation of $\overline{I}(\alpha)$.

(d) We may assume that all the atoms of α and α' have strictly positive measure.

$$H(\alpha \vee \alpha') = -\sum_{\substack{A\in\alpha \\ A'\in\alpha'}} m(A \cap A') \log m(A \cap A')$$

$$= -\sum_{A,A'} m(A \cap A') \log m(A)$$

$$- \sum_{A,A'} m(A) \frac{m(A \cap A')}{m(A)} \log \frac{m(A \cap A')}{m(A)}$$

$$= (1) + (2) .$$

Since $\sum_{A'\in\alpha} m(A \cap A') = m(A)$, the term (1) is equal to $H(\alpha)$. The term (2) can be written as

$$\sum_{A'\in\alpha'} \sum_{A\in\alpha} m(A) \cdot z\left(\frac{m(A \cap A')}{m(A)}\right) .$$

Applying Jensen's inequality, one obtains

$$(2) \qquad \leq \sum_{A'\in\alpha'} z\left(\sum_{A\in\alpha} m(A) \frac{m(A \cap A')}{m(A)}\right) =$$

$$= \sum_{A'\in\alpha'} z(m(A')) = H(\alpha') .$$

Now assume α and α' have finite entropy. Equality occurs iff, for each fixed $A' \in \alpha'$, $\frac{m(A \cap A')}{m(A)}$ does not depend on $A \in \alpha$. From

$$m(A \cap A') = m(A) \frac{m(A \cap A')}{m(A)}$$

one obtains, by summation over the $A \in \alpha$, that

$$\frac{m(A \cap A')}{m(A)} = m(A'),$$

i.e. that $A \in \alpha$ and $A' \in \alpha'$ are independent. □

Now let $(\mathfrak{X}, m, \Psi)$ be an m.t. dynamical system. Clearly

<u>(10.5) Proposition:</u> $H(\alpha) = H(\Psi^{-1} \alpha)$

<u>(10.6) Proposition:</u>

$$\lim_{N\to\infty} \frac{1}{N} H(\alpha \vee \Psi^{-1} \alpha \vee \ldots \vee \Psi^{-N+1} \alpha)$$

exists and is equal to the infimum.

<u>Proof:</u> If $H(\alpha) = \infty$, this is trivial. By (10.5) and (10.4.d)

$$H\left((\alpha)_0^{N+M-1}\right) \leq H\left((\alpha)_0^{N-1}\right) + H\left(\Psi^{-N}(\alpha)_0^{M-1}\right)$$

$$= H\left((\alpha)_0^{N-1}\right) + H\left((\alpha)_0^{M-1}\right)$$

Hence the sequence $S_N = H\left((\alpha)_0^{N-1}\right)$ is positive and subadditive. Now apply the next proposition. □

<u>(10.7) Proposition:</u> Let s_N be a sequence which is positive and subadditive. Then $\lim \frac{1}{N} s_N$ exists and is equal to $\inf \frac{1}{N} s_N$.

<u>Proof:</u> Write $\theta = \inf \frac{1}{N} s_N$. Let $M \in \mathbb{N}$ be fixed.

For $N \in \mathbb{N}$ write $N = kM + q$, with $0 \leq q < M$. Clearly $s_{kM} \leq ks_M$ and therefore $\frac{s_{kM}}{kM} \leq \frac{s_M}{M}$. Hence

$$\frac{s_N}{N} \leq \frac{s_{kM}+s_q}{kM+q} \leq \frac{s_{kM}+s_q}{kM} \leq \frac{s_M}{M} + \frac{s_q}{kM} .$$

If $N \to \infty$ then $k \to \infty$. Hence

$$\limsup \frac{s_N}{N} \leq \frac{s_M}{M},$$

and therefore $\limsup \frac{s_N}{N} \leq \theta$. On the other hand $\theta \leq \liminf \frac{s_N}{N}$. □

(10.8) Definition: Let α be a partition with finite entropy. Then

$$h_m(\alpha,\Psi) = \lim \frac{1}{N} H_m((\alpha)_o^{N-1})$$

is called the entropy of α with respect to Ψ. One often writes $h(\alpha,\Psi)$ instead of $h_m(\alpha,\Psi)$.

$H_m((\alpha)_o^{N-1})$ may be interpreted as expectation of the amount of information obtained by performing the same experiment α on N consecutive days. $h_m(\alpha,\Psi)$ therefore measures the average gain of information per day.

(10.9) Proposition:

(a) $0 \leq h(\alpha,\Psi) \leq H(\alpha)$;

(b) $\alpha \overset{o}{<} \alpha'$ implies $h(\alpha,\Psi) \leq h(\alpha',\Psi)$;

(c) $h(\alpha \vee \alpha',\Psi) \leq h(\alpha,\Psi) + h(\alpha',\Psi)$;

(d) $h\left((\alpha)_M^{M+k},\Psi\right) = h(\alpha,\Psi)$ for $M \in \mathbb{Z}$, $k \in \mathbb{N}$.

The proofs are easy. Note again that $h(\alpha,\Psi)$ depends only on the equivalence class of α.

(10.10) Definition: The quantity

$$h_m(\Psi) = \sup \{h_m(\alpha,\Psi) \mid \alpha \text{ is a partition with } H(\alpha) < \infty\}$$

is called the entropy of $(\mathfrak{X},m,\Psi)$, or the entropy of Ψ (with respect to m). One often writes $h(\Psi)$ instead of $h_m(\Psi)$.

(10.11) Proposition: If two m.t. dynamical systems $(\mathfrak{X},m,\Psi)$ and $(\mathfrak{X}',m',\Psi')$ are m.t. conjugate, then

$$h_m(\Psi) = h_{m'}(\Psi').$$

Proof: This is trivial, since the m.t. conjugacy between $(\mathfrak{X},m,\Psi)$ and $(\mathfrak{X}',m',\Psi')$ sets up a one-to-one correspondence between equivalence classes of partitions α on $\mathfrak{X}$ and α' on $\mathfrak{X}'$. □

(10.12) Proposition:

(a) $0 \leq h(\Psi)$;

(b) $h(\mathrm{Id}) = 0$;

(c) $h(\Psi^k) = |k|\, h(\Psi)$ for every $k \in \mathbb{Z}$.

Proof: (a) and (b) are trivial. For (c), assume first that $k > 0$. One has

$$h\Big(\bigvee_{j=o}^{k-1} \Psi^{-j}\alpha,\ \Psi^k\Big) = \lim \frac{1}{N} H\big((\alpha)_o^{Nk-1}\big)$$

$$= \lim k \frac{1}{kN} H\big((\alpha)_o^{Nk-1}\big) =$$

$$= k\, h(\alpha,\ \Psi).$$

Hence

$$k\, h(\Psi) = k \sup h(\alpha,\Psi) = \sup h\Big(\bigvee_{j=o}^{k-1} \Psi^{-j}\alpha,\Psi^k\Big) \leq h(\Psi^k)$$

where the sup extends over all α with $H(\alpha) < \infty$. On the other hand

$$h(\alpha,\Psi^k) \leq h\Big(\bigvee_{j=o}^{k-1} \Psi^{-j}\alpha,\Psi^k\Big) = k\, h(\alpha,\Psi)$$

(the inequality follows from (10.9.b) and therefore $h(\Psi^k) \leq k\, h(\Psi)$.

For $k = 0$, the statement follows from part (b).

For $k < 0$, it is enough to show that $h(\Psi) = h(\Psi^{-1})$. But this follows easily from the fact that

$$\alpha \vee \Psi^{-1}\alpha \vee \quad \vee \Psi^{-N+1}\alpha = \Psi^{-N+1}(\alpha \vee \Psi\alpha \vee \quad \vee \Psi^{N-1}\alpha)$$

and thus that $h(\alpha,\Psi) = h(\alpha,\Psi^{-1})$. □

In particular, if there exists a $k \neq 0$ such that $\Psi^k = \mathrm{Id}$ (for example when $\mathfrak{X}$ is finite) then $h(\Psi) = 0$.

(10.13) Proposition: $h_m(\alpha,\Psi)$ and $h_m(\Psi)$ are affine functions of m.

Proof: Let m', m" be Ψ-invariant and

$$m = \lambda\, m' + (1-\lambda)m''$$

with $0 < \lambda < 1$. Using the convexity of $z : x \longrightarrow -x \log x$ we get for $A \in \Sigma$

$$0 \leq -m(A) \log m(A) + \lambda\, m'(A) \log m'(A) + (1-\lambda)\, m''(A) \log m''(A) =$$

$$= -\lambda\, m'(A) \big(\log m(A) - \log \lambda\, m'(A)\big) - (1-\lambda)\, m''(A) \big(\log m(A) - \log(1-\lambda)\, m''(A)\big)$$

$$-m'(A)\, \lambda \log \lambda \;-\; m''(A)\, (1-\lambda) \log (1-\lambda)$$

$$\leq -m'(A)\, \lambda \log \lambda \;-\; m''(A)\, (1-\lambda) \log (1-\lambda)$$

so that

$$0 \leq H_m\big((\alpha)_o^{N-1}\big) - \lambda\, H_{m'}\big((\alpha)_o^{N-1}\big) - (1-\lambda)\, H_{m''}\big((\alpha)_o^{N-1}\big) \leq \log 2.$$

Dividing by N and letting $N \to \infty$ shows that

$$(*) \qquad h_m(\alpha,\Psi) = \lambda\, h_{m'}(\alpha,\lambda) + (1-\lambda) h_{m''}(\alpha,\Psi).$$

Now suppose that $h_m(\Psi)$, $h_{m'}(\Psi)$ and $h_{m''}(\Psi)$ are finite. Choose partitions $\alpha, \alpha', \alpha''$ such that $h_m(\Psi) - h_m(\alpha,\Psi)$ etc. ... are very small. Applying (*) with $\alpha \vee \alpha' \vee \alpha''$ instead of α and using (10.9.b) one sees that $h_m(\Psi)$ can be approximated arbitrarly well by $\lambda\, h_{m'}(\Psi) + (1-\lambda)\, h_{m''}(\Psi)$ and hence that $m \longrightarrow h_m(\Psi)$ is affine. If some of the entropies considered are infinite, the proof is analogous. □

11. The Computation of Entropy

Let $(\mathfrak{X},\Sigma,m)$ be a measure space.

(11.1) Definition: Let α and β be two partitions. The quantity

$$H_m(\alpha|\beta) = -\sum_{B\in\beta} m(B) \sum_{A\in\alpha} m(A|B) \log m(A|B)$$

(where the summation is over all $B \in \beta$ with $m(B) > 0$) is called the conditional entropy of the partition α given β. We shall often omit the subscript m.

As in the previous sections, one can interpret conditional entropy in terms of experiments. $H(\alpha|\beta)$ is the expectation of the information gained by the experiment α, if one has already performed the experiment β. The following relations can easily be checked by using the same arguments as in the previous section:

(11.2) Proposition:

(a) $0 \leq H(\alpha|\beta)$. Equality holds iff $\alpha \overset{o}{\subset} \beta$;

(b) If each atom of β intersects at most k atoms of α, then $H(\alpha|\beta) \leq \log k$;

(c) $H(\alpha|\beta) \leq H(\alpha)$. If α and β are finite, then equality holds iff α and β are independent,

(d) $\alpha \overset{o}{\subset} \alpha'$ implies $H(\alpha|\beta) \leq H(\alpha'|\beta)$;

(e) $\beta \overset{o}{\subset} \beta'$ implies $H(\alpha|\beta) \geq H(\alpha|\beta')$;

(f) $H(\alpha \vee \alpha'|\beta) = H(\alpha|\beta) + H(\alpha'|\alpha \vee \beta)$,

and in particular

$H(\alpha|\beta) = H(\alpha \vee \beta) - H(\beta)$;

(g) $H(\alpha \vee \alpha'|\beta) \leq H(\alpha|\beta) + H(\alpha'|\beta)$.

Now let $(\mathfrak{X},m,\Psi)$ be an m.t. dynamical system. Clearly:

(11.3) Proposition: $H(\alpha|\beta) = H(\Psi^{-1}\alpha|\Psi^{-1}\beta)$

(11.4) Proposition: $h(\alpha,\Psi) = \lim_{n\to\infty} H(\Psi^{-n}\alpha|(\alpha)_o^{n-1})$

if $H(\alpha) < \infty$.

Proof: Write $h_n = H\big((\alpha)_o^{n-1}\big)$ and $s_n = h_n - h_{n-1}$. By (11.2.f) one has

$$s_n = H(\Psi^{-n}\alpha | (\alpha)_o^{n-1}) .$$

By (11.2.e) and (11.3) one obtains $s_{n+1} \leq s_n$. Hence s_n converges to some limit, say s. But

$$h_n = H(\alpha) + s_1 + s_2 + \ldots + s_n$$

and hence $\frac{h_n}{n}$ converges to s. On the other hand $\lim \frac{h_n}{n} = h(\alpha,\Psi)$ by (10.8). □

Thus we obtain another interpretation of $h(\alpha,\Psi)$: it is (asymptotically) the expectation of the information gained by repeating the experiment α in n-th time.

(11.5) Definition: For two partitions α,α' one defines

$$d(\alpha,\alpha') = H(\alpha|\alpha') + H(\alpha'|\alpha).$$

(11.6) Proposition: $d(\alpha,\alpha')$ is a metric on the space of equivalence classes of partitions.

Proof: $d(\alpha,\alpha') \geq 0$ and $d(\alpha',\alpha) = d(\alpha,\alpha')$ are obvious. If $d(\alpha,\alpha') = 0$ then $H(\alpha|\alpha') = H(\alpha'|\alpha) = 0$. Hence by (11.2.a) $\alpha \overset{o}{\subset} \alpha' \overset{o}{\subset} \alpha$ and thus $\alpha \overset{o}{=} \alpha'$. Let α'' be another partition. By (11.2.f)

$$\begin{aligned} H(\alpha|\alpha'') &= H(\alpha \vee \alpha'') - H(\alpha'') \\ &\leq H(\alpha \vee \alpha' \vee \alpha'') - H(\alpha' \vee \alpha'') + H(\alpha' \vee \alpha'') - H(\alpha'') \\ &= H(\alpha|\alpha' \vee \alpha'') + H(\alpha'|\alpha'') \leq H(\alpha|\alpha') + H(\alpha'|\alpha''). \end{aligned}$$

Similarly

$$H(\alpha''|\alpha) \leq H(\alpha'|\alpha) + H(\alpha''|\alpha')$$

and therefore $d(\alpha,\alpha'') \leq d(\alpha,\alpha') + d(\alpha',\alpha'')$. □

(11.7) Proposition: $h(\alpha,\Psi) - h(\alpha',\Psi) \leq H(\alpha|\alpha')$

and therefore

$$|h(\alpha,\Psi) - h(\alpha',\Psi)| \leq \max\{H(\alpha|\alpha'),H(\alpha'|\alpha)\} \leq d(\alpha,\alpha').$$

In particular, $\alpha \longrightarrow h(\alpha,\Psi)$ is continuous with respect to the metric d.

Proof: Applying (11.2.f) twice one obtains

$$H\left((\alpha)_o^{n-1}\right) - H\left((\alpha')_o^{n-1}\right) = H\left((\alpha)_o^{n-1}|(\alpha')_o^{n-1}\right) - H\left((\alpha')_o^{n-1}|(\alpha)_o^{n-1}\right)$$

$$\leq H\left((\alpha)_o^{n-1} \mid (\alpha')_o^{n-1}\right).$$

By (11.2.g)

$$H\left((\alpha)_o^{n-1} \mid (\alpha')_o^{n-1}\right) \leq H\left(\alpha \mid (\alpha')_o^{n-1}\right) + H\left(\Psi^{-1}\alpha \mid (\alpha')_o^{n-1}\right) + \dots + H\left(\Psi^{-n+1}\alpha \mid (\alpha')_o^{n-1}\right)$$

and, since $\Psi^{-j}\alpha' \subset (\alpha')_o^{n-1}$ for $0 \leq j < n$, (11.2.e) and (11.3) imply

$$H\left((\alpha)_o^{n-1} \mid (\alpha')_o^{n-1}\right) \leq H(\alpha|\alpha') + H(\Psi^{-1}\alpha|\Psi^{-1}\alpha') + \dots + H(\Psi^{-n+1}\alpha|\Psi^{-n+1}\alpha')$$

$$= n\, H(\alpha|\alpha').$$

Hence $\frac{1}{n} H\left((\alpha)_o^{n-1}\right) - \frac{1}{n} H\left((\alpha')_o^{n-1}\right) \leq H(\alpha|\alpha')$

Letting $n \longrightarrow \infty$, one obtains

$$h(\alpha,\Psi) - h(\alpha',\Psi) \leq H(\alpha|\alpha'). \qquad \square$$

(11.8) Definition: The set of equivalence classes of partitions of $(\mathfrak{X},m)$ with finite entropy, together with the metric d defined in (11.6), will be denoted by $\mathfrak{Z}$.

(11.9) Proposition: The finite partitions are dense in $\mathfrak{Z}$.

Proof: For $\alpha = (A_1, A_2, \dots) \in \mathfrak{Z}$ write $\alpha_n = (A_1, \dots, A_{n-1}, B_n)$ with $B_n = \mathfrak{X} \setminus (A_1 \cup \dots \cup A_{n-1})$. $\alpha_n \subset \alpha$ and hence $H(\alpha_n|\alpha) = 0$.

Thus

$$d(\alpha,\alpha_n) = H(\alpha|\alpha_n) = H(\alpha) - H(\alpha_n)$$
$$= m(B_n)\log m(B_n) - \sum_{j=n}^{\infty} m(A_j)\log m(A_j)$$

Since $H(\alpha)$ is finite, this expression converges to 0 for $n \longrightarrow \infty$. $\square$

(11.10) Proposition: Let $k \in \mathbb{N}$, $\varepsilon > 0$. Then there exists a $\delta = \delta(k,\varepsilon)$ such that for any two partitions α, α' with k elements such that $\|\alpha,\alpha'\| < \delta$ we have $H(\alpha|\alpha') < \varepsilon$ and hence also

$$d(\alpha,\alpha') < 2\varepsilon$$
$$|H(\alpha) - H(\alpha')| < \varepsilon$$
$$|h(\alpha,\Psi) - h(\alpha',\Psi)| < \varepsilon$$

Proof: If $\alpha = (A_1,\dots,A_k)$, $\alpha' = (A_1',\dots,A_k')$, set

$$\alpha'' = (A_i \cap A_j' \ (i \neq j); \ \bigcup_{1 \le s \le k} A_s \cap A_s').$$

Then $\alpha \vee \alpha' = \alpha' \vee \alpha''$, and α'' has $k(k-1)$ sets of total measure less than δ and one set of measure $> 1-\delta$. Hence $H(\alpha'') \geq -(1-\delta)\cdot\log(1-\delta) - \delta\cdot\log\frac{\delta}{k(k-1)} < \epsilon$ if δ is small enough. Thus
$H(\alpha') + H(\alpha|\alpha') = H(\alpha\vee\alpha') = H(\alpha'\vee\alpha'') < H(\alpha') + \epsilon$, and $H(\alpha|\alpha') < \epsilon$. By symmetry also $H(\alpha'|\alpha) < \epsilon$, hence $d(\alpha,\alpha') < 2\epsilon$, and the rest follows from $H(\alpha) - H(\alpha') = H(\alpha|\alpha') - H(\alpha'|\alpha)$ and by (11.7). □

(11.11) Proposition: Let $\alpha_1 \overset{o}{\subset} \alpha_2 \overset{o}{\subset} \dots$ be a sequence in $\mathfrak{Z}$ which generates Σ. Then $\lim_{n\to\infty} h(\alpha_n,\Psi) = h(\Psi)$.

Proof: Since $h(\Psi) = \sup\{h(\alpha,\Psi) \mid \alpha \in \mathfrak{Z}\}$, one obtains from (11.7) and (11.9) that

$$h(\Psi) = \sup\{h(\alpha,\Psi) \mid \alpha \text{ finite}\}.$$

If α has k elements, there exists for any $\epsilon > 0$, and n and a partition $\alpha' \subset \alpha_n$ with k elements such that $\|\alpha,\alpha'\| \leq 2\,\delta(k,\epsilon)$, where $\delta(k,\epsilon)$ is as in 11.10. Thus

$$h(\alpha_n,\Psi) \geq h(\alpha',\Psi) \geq h(\alpha,\Psi) - \epsilon.$$

This shows that $\lim_{n\to\infty} h(\alpha_n,\Psi) = h(\Psi)$. □

(11.12) Theorem of Kolmogoroff-Sinai:
Let $(\mathfrak{X},m,\Psi)$ be an m.t. dynamical system and α a generator with $H(\alpha) < \infty$. Then

$$h(\Psi) = h(\alpha,\Psi).$$

Proof: Write $\alpha_n = (\alpha)_{-n}^{n}$. Clearly α_n is an increasing sequence in $\mathfrak{Z}$ which generates Σ. Thus one has only to apply (11.11) and (10.9.d). □

(11.13) Proposition: Let $(\mathfrak{X},m,\Psi)$ and $(\mathfrak{X}',m',\Psi')$ be two m.t. dynamical systems. Then

$$h_{m\times m'}(\Psi\times\Psi') = h_m(\Psi) + h_{m'}(\Psi')$$

Proof: Let α_n (resp. α_n')be an increasing sequence of partitions of $\mathfrak{X}$ (resp. $\mathfrak{X}'$) which generates. Each α_n induces a partition β_n of $\mathfrak{X}\times\mathfrak{X}'$, the atoms of β_n being of the form $A\times\mathfrak{X}'$, where A runs through the atoms of α_n. Similarly α_n' induces a partition β_n' of $\mathfrak{X}\times\mathfrak{X}'$. It is easy to see that $\gamma_n = \beta_n \vee \beta_n'$ is an increasing sequence of partitions of $\mathfrak{X}\times\mathfrak{X}'$ which generates. Since β_n and β_n' are independent, one has

$$H_{m\times m'}((\gamma_n)_o^{N-1}) = H_{m\times m'}((\beta_n)_o^{N-1}) + H_{m\times m'}((\beta_n')_o^{N-1}).$$

But clearly

$$H_{m\times m'}((\beta_n)_o^{N-1}) = H_m((\alpha_n)_o^{N-1})$$

and

$$H_{m\times m'}((\beta_n')_o^{N-1}) = H_{m'}((\alpha_n')_o^{N-1}).$$

Therefore, dividing by N and letting $N\longrightarrow\infty$, one obtains

$$h_{m\times m'}(\gamma_n,\Psi\times\Psi') = h_m(\alpha_n,\Psi) + h_{m'}(\alpha_n',\Psi').$$

For $n\longrightarrow\infty$, one has therefore by (11.11)

$$h_{m\times m'}(\Psi\times\Psi') = h_m(\Psi) + h_{m'}(\Psi'). \qquad \square$$

(11.14) Proposition: Let $(\mathfrak{X}',m',\Psi')$ be a factor of $(\mathfrak{X},m,\Psi)$. Then

$$h_{m'}(\Psi') \leq h_m(\Psi).$$

Proof: Let Φ denote the homeomorphism $\mathfrak{X}\rightarrow\mathfrak{X}'$ as in (1.13). Let α' be a partition of $\mathfrak{X}'$. Then $\Phi^{-1}\alpha'$ is a partition of $\mathfrak{X}$, and $H_{m'}(\alpha') = H_m(\Phi^{-1}\alpha')$. It is easy to see that $\Phi^{-1}((\alpha')_o^n) = (\Phi^{-1}\alpha')_o^n$ and therefore that

$$h_{m'}(\alpha',\Psi') = h_m(\Phi^{-1}\alpha',\Psi).$$

Hence

$$h_m(\Psi) = \sup_{\alpha} h_m(\alpha,\Psi) \geq \sup_{\alpha'} h_m(\Phi^{-1}\alpha',\Psi) =$$

$$= \sup_{\alpha'} h_{m'}(\alpha',\Psi') = h_{m'}(\Psi'). \qquad \square$$

(11.15) Definition:

$(\mathfrak{X},m,\Psi)$ is said to be a K-system (Kolmogoroff-system) if there exists a generator α such that its "tailfield" $\bigcap_{n=1}^{\infty} \bigvee_{j=n}^{\infty} \Sigma(\Psi^j\alpha)$ is trivial in the sense that it contains only sets of measure 0 or 1.

(11.16) Theorem [8,160]: The following conditions are equivalent

(a) $(\mathfrak{X},m,\Psi)$ is a K-system;

(b) every nontrivial factor of $(\mathfrak{X},m,\Psi)$ has positive entropy;

(c) $h(\alpha,\Psi) > 0$ for any nontrivial partition α.

K-systems are strongly mixing.

12. Entropy for Bernoulli and Markov Shifts

Let $\sigma : S^{\mathbb{Z}} \longrightarrow S^{\mathbb{Z}}$ be the shift on a finite state space S.

(12.1) Proposition: For any $\mu \in \mathfrak{M}_\sigma(S^{\mathbb{Z}})$ one has

$$h_\mu(\sigma) = - \lim_{N\to\infty} \frac{1}{N} \sum_{A\in S^N} \mu({}_o[A]) \log \mu({}_o[A]).$$

Proof: Write $\mathfrak{U} = ({}_o[i] \mid i\in S)$. Thus $\mathfrak{U}$ is a partition of $S^{\mathbb{Z}}$; each of its atoms consists of all those $x \in S^{\mathbb{Z}}$ with given 0-th coordinate. It is obvious that $\mathfrak{U}$ is a generator for $(S^{\mathbb{Z}},\mu,\sigma)$. The atoms of $(\mathfrak{U})_o^{N-1}$ are the cylinders of the form ${}_o[A]$, $A \in S^N$. Hence the theorem of Kolmogoroff-Sinai (11.12) implies the result. □

One can also write

$$h_\mu(\sigma) = - \lim \frac{1}{2N+1} \sum_{a_{-N},\dots,a_N\in S} \mu({}_{-N}[a_{-N},\dots,a_N]) \cdot \log \mu({}_{-N}[a_{-N},\dots,a_N]).$$

At this point entropy comes closest to the notion of entropy used by physicists. Consider a one-dimensional physical lattice system - particles in a row, for example - indexed by $\mathbb{Z}$. Thus one has an infinite collection of objects, one at each point of $\mathbb{Z}$.

Suppose that each object can be in one of several possible states, - spin up or spin down, for example. If S denotes the set of possible states of those objects, then $S^{\mathbb{Z}}$ denotes the set of possible configurations of the lattice system. The measures μ on $S^{\mathbb{Z}}$, which determine the probabilities of those configurations, will be the statistical states. Now consider a finite collection of objects in the lattice system - those with index between -N and N, say. The different configurations are given by the blocks $(a_{-N},\dots,a_N)$, with $a_j \in S$. Then $-\mu({}_{-N}[a_{-N},\dots,a_N]) \log \mu({}_{-N}[a_{-N},\dots,a_N])$ is an expression for the information of the statement: "the finite collection is in the configuration $(a_{-N},\dots,a_N)$", and

$$- \sum_{a_{-N},\dots,a_N} \mu({}_{-N}[a_{-N},\dots,a_N]) \log \mu({}_{-N}[a_{-N},\dots,a_N])$$

is an expression for the information contained in the finite collection, or for its randomness, or disorder. If $\mu \in \mathfrak{M}_\sigma(S^{\mathbb{Z}})$

is σ-invariant, then $h_\mu(\sigma)$ is an expression for the disorder of the entire lattice system, obtained by taking the averages over the entropy of larger and larger finite subsystems.

Thus the main difference between the entropy in physics and the entropy in dynamical systems is that in the former case, $\mathbb{Z}$ is interpreted as one-dimensional space, while in the latter case, it is interpreted as time.

(12.2) Proposition: For any $\mu \in \mathfrak{M}_\sigma(S^{\mathbb{Z}})$ one has $h_\mu(\sigma) \leq \log s$, where s is the cardinality of S.

Proof: This is a trivial consequence of (1.1) and (10.9.a). □

(12.3) Proposition: Let $\mu_{\pi P}$ be a Markov measure given by the stochastic matrix $P = (p_{ij})$ and the probability vector $\pi = (p_i)$. Then

$$h_{\mu_{\pi P}}(\sigma) = - \sum_{i,j \in S} p_i \, p_{ij} \log p_{ij}.$$

Proof: By (12.1) one has $h_{\mu_{\pi P}}(\sigma) =$

$$= - \lim \frac{1}{N} \sum p_{a_o} p_{a_o a_1} \cdots p_{a_{N-2} a_{N-1}} \log p_{a_o} p_{a_o a_1} \cdots p_{a_{N-2} a_{N-1}}.$$

It is easy to see by induction that the sum on the right hand side is equal to

$$(N - 1) \sum_{i,j \in S} p_i p_{ij} \log p_{ij} + \sum_{i \in S} p_i \log p_i .$$

This implies the result. □

(12.4) Corollary: Let μ_π be a Bernoulli measure given by the probability vector $\pi = (p_i)$. Then

$$h_{\mu_\pi}(\sigma) = - \sum_{i \in S} p_i \log p_i.$$

If, in particular, $\pi = (\frac{1}{s},\ldots,\frac{1}{s})$, then $h_{\mu_\pi}(\sigma) = \log s$.

It follows in particular that the Bernoulli shift given by $\pi = (\frac{1}{2},\frac{1}{2})$ is not measure theoretically conjugate to the Bernoulli shift given by $\pi = (\frac{1}{3},\frac{1}{3},\frac{1}{3})$. In 1958 Kolmogorov introduced the notion of entropy in ergodic theory in order to show that these two shifts were different, thus solving a problem that had been open for more than 20 years.

(12.5) Corollary: Bernoulli shifts and Markov shifts have positive entropy.

Much more is true, in fact:

(12.6) Proposition: Strongly mixing Markov shifts (and in particular Bernoulli shifts) are Kolmogoroff systems.

Proof: Let $\mu_{\pi P}$ be a strongly mixing Markov measure. By (8.18) the stochastic matrix P is aperiodic and thus by (8.16) satisfies $p_{ij}^{(n)} \longrightarrow p_j$ for $i,j \in S$. Writing $\pi_n = \max_{i,j\in S} | p_{ij}^{(n)} - p_j |$ one, therefore, has $\pi_n \longrightarrow 0$.

Suppose that A is a cylinder at the place m, and B a cylinder belonging to $\bigvee_{j=m-n}^{\infty} \Sigma(\sigma^j \mathfrak{U})$, with $n \geq 0$. The sets in $\bigvee_{j=m-n}^{\infty} \Sigma(\sigma^j \mathfrak{U})$ depend only on coordinates $\leq n - m$. It is easy to check that

$$| \mu_{\pi P}(A \cap B) - \mu_{\pi P}(A)\, \mu_{\pi P}(B)| \leq \pi_n.$$

It follows that this relation is valid for all sets $B \in \bigvee_{j=m-n}^{\infty} \Sigma(\sigma^j \mathfrak{U})$ and thus for all $B \in \bigcap_{n=1}^{\infty} \bigvee_{j=n}^{\infty} \Sigma(\sigma^j \mathfrak{U})$.

Since this relation holds for all cylinders A of $S^{\mathbb{Z}}$, it holds for all Borel sets $A \subset S^{\mathbb{Z}}$ and in particular if $A = B \in \bigcap_{n=1}^{\infty} \bigvee_{j=n}^{\infty} \Sigma(\sigma^j \mathfrak{U})$. Hence $\mu_{\pi P}(B)$ is either 0 or 1, and, therefore, the tailfield consists of sets of measure 0 or 1. □

Since entropy is an invariant, two dynamical systems with different entropy cannot be measure theoretically conjugate. There exist simple examples of dynamical systems with the same entropy which are not conjugate. Thus entropy is not a complete invariant in the class of all m.t. dynamical systems. It was conjectured, however, that entropy is a complete invariant in the class of all Bernoulli shifts, and even in the class of all Kolmogoroff systems. (This latter statement would imply, of course, that every Kolmogoroff system is measure theoretically conjugate to a Bernoulli shift).

These conjectures remained open for more than ten years. During this time, a certain amount of positive evidence accumulated. Meshalkin [209] proved measure theoretical conjugacy for certain Bernoulli shifts of very particular type, as for example for those given by $\pi = (\frac{1}{2},\frac{1}{8},\frac{1}{8},\frac{1}{8},\frac{1}{8})$ and $\pi = (\frac{1}{4},\frac{1}{4},\frac{1}{4},\frac{1}{4})$ (entropy log 4 in both cases). Similar partial results for certain Markov shifts were obtained by Adler and Weiss [6], who used them to prove that entropy was a complete invariant for automorphisms of the two-torus (see section 24). A very significant progress was achieved in 1963 when Sinai proved the following:

<u>(12.7) Theorem:</u> Let $(\mathfrak{X},m,\Psi)$ be an m.t. dynamical system and $(S^{\mathbb{Z}}, \mu_\pi, \sigma)$ a Bernoulli shift such that

$$h_m(\Psi) \geq h_{\mu_\pi}(\sigma).$$

Then $(S^{\mathbb{Z}}, \mu_\pi, \sigma)$ is a factor of $(\mathfrak{X},m,\Psi)$.

<u>(12.8) Definition:</u> Two m.t. dynamical systems $(\mathfrak{X},m,\Psi)$ and $(\mathfrak{X}',m',\Psi')$ are said to be <u>weakly m.t. conjugate</u> if each is a factor of the other.

Thus (12.7) implies

<u>(12.9) Corollary:</u> Two Bernoulli shifts with the same entropy are weakly m.t. conjugate.

But Sinai's theorem did not yet solve the isomorphism problem for Bernoulli shifts, since nothing guaranteed that the factor maps between two weakly m.t. conjugate systems were inverse of each other. In fact, it raised a new problem, namely whether weak conjugacy implied conjugacy.

The real breakthrough occured in 1969, with Ornstein's isomorphism theorem, which was followed by a continuous flow of theorems by Ornstein and his co-workers. Here we shall quote only a few of these results. We refer to [141] for a more extensive treatment.

<u>(12.10) Theorem</u> of Ornstein: Two Bernoulli shifts with the same entropy are m.t. conjugate.

This result has been generalized in the following way: Let μ (resp. μ') be a measure on $\mathfrak{X}$ (resp. $\mathfrak{X}'$) and denote by $\mu^{\mathbb{Z}}$ (resp. $\mu'^{\mathbb{Z}}$) the corresponding product measure on $\mathfrak{X}^{\mathbb{Z}}$ (resp. $\mathfrak{X}'^{\mathbb{Z}}$).

Then $(\mathfrak{X}^{\mathbb{Z}},\mu^{\mathbb{Z}},\sigma)$ and $(\mathfrak{X}'^{\mathbb{Z}},\mu'^{\mathbb{Z}}\sigma)$ are measure theoretical conjugate iff they have the same entropy. (Note that the entropy of $(\mathfrak{X}^{\mathbb{Z}},\mu^{\mathbb{Z}},\sigma)$ is infinite if μ has a nonatomic part).

<u>(12.11) Theorem:</u> Any nontrivial factor of a Bernoulli shift is m.t. conjugate to a Bernoulli shift.

<u>(12.12) Theorem:</u> There exist K-systems which are not m.t. conjugate to Bernoulli shifts.

Thus entropy is a complete invariant for Bernoulli shifts but not for K-systems.

<u>(12.13) Theorem:</u> There exist dynamical systems which are weakly m.t. conjugate, but not m.t. conjugate.

<u>(12.14) Theorem:</u> Every mixing Markov shift is m.t. conjugate to a Bernoulli shift.

13. Ergodic Decompositions

This section will be required only for the development of generator theorems for aperiodic transformations in sections 30 and 31. It may be omitted in the first reading.

The Ergodic Decomposition of an Invariant Measure

Desintegration of measures is a topic in the general theory of Lebesgue spaces (see [158]). Since we avoid this theory, our approach to the ergodic decomposition will be through topological dynamical systems which by (9.8) can be used to treat the general case.

Let (X,T) be a topological dynamical system and $\nu \in \mathfrak{M}_T(X)$. By proposition (5.12), ν-almost every point $x \in X$ is quasi-regular and generic for an ergodic Borel measure μ_y, i.e. for $g \in C(X)$

$$\lim \frac{1}{n} \sum_{i=0}^{n-1} g(T^i y) = \int g \, d\mu_y \qquad \nu\text{ - a.e.}$$

This means, by the ergodic theorem, that $\int g \, d\mu_y$ is a version of the conditional expectation of g with respect to the σ-algebra of T-invariant sets in $\mathfrak{B}_\nu$. Therefore we also have

$$\int g \, d\nu = \int \left(\int g \, d\mu_y \right) d\nu(y) \qquad (g \in C(X)).$$

For all $y \in X$ such that μ_y is ergodic, the set

$$\Gamma_y = \{ z \in X \mid \mu_y = \mu_z \}$$

is Borel-measurable, T-invariant and $\mu_y(\Gamma_y) = 1$. We call Γ_y the <u>ergodic fibre</u> of y. Note that the completed σ-algebra $\mathfrak{B}_{\mu_y} | \Gamma_y$ coincides with $\mathfrak{B}_\nu | \Gamma_y$ only if $\nu(\Gamma_y) > 0$.

From the monotone convergence theorem for the integrals with respect to ν and μ_y and for the conditional expectation with respect to the σ-algebra of T-invariant sets and from the ergodic theorem one can deduce by standard approximation arguments that for every ν-integrable f

(1) f is μ_y-integrable for ν - a.e. y

(2) $y \longrightarrow \int f\, d\mu_y$ is measurable

(3) $\int f\, d\nu = \int (\int f\, d\mu_y)\, d\nu(y)$

(4) $\lim \frac{1}{n} \sum_{i=o}^{n-1} f(T^i y) = \int f\, d\mu_y$ ν - a.e.

(The class of functions f for which (1)-(4) hold is extended successively from C(X) to $\{1_K | K$ compact$\}$, $\{1_U | U$ open$\}$, $\{1_M | M \in \mathfrak{B}_\nu\}$, $\{f | f$ takes finitely many values$\}$, and $\{f | f$ is ν-integrable$\}$). Since (3) implies that for integrable f, f' with $f = f'$ a.e. the set $\{y | \mu_y(\{f \neq f'\}) > 0\}$ has ν-measure 0, (3) extends also to $f \in L_1(\nu)$, the space of equivalence classes, where $\int f\, d\mu_y$ has to be understood as an element of $L_1(\nu)$; (4) means that $\int f\, d\mu_y$ is the conditional expectation of f with respect to the σ-algebra of T-invariant subsets of $\mathfrak{B}_\nu$.

Let now $(\mathfrak{X}, m, \Psi)$ be an m.t. dynamical system. From corollary (9.8) it follows that $(\mathfrak{X}, m, \Psi)$ is conjugate to (X, ν, T), where (X, T) is a topological dynamical system, by a mapping $\Phi : \mathfrak{X} \longrightarrow X$. Hence we are able to pull back to $\mathfrak{X}$ all that was constructed on X:

$$C_x = \Phi^{-1}\ \Gamma_{\Phi x}$$

$$m_x = \Phi^{-1}\ \mu_{\Phi x}$$

m_x and C_x are defined for almost all $x \in \mathfrak{X}, \Psi$ -invariant, and

(3') $\int f\, dm = \int (\int f\, dm_x)\, dm(x)$

for m-integrable f resp. for $f \in L_1(m)$ in the same sense as above.

We consider m_x as a measure on $\mathfrak{X}$ or on C_x. We call Σ_x the completion of $\Sigma | C_x$ under m_x. The family of m.t. dynamical systems $(C_x, \Sigma_x, m_x, \Psi | C_x)_{x \in \mathfrak{X}}$ or of measures $(m_x)_{x \in \mathfrak{X}}$ is called the <u>ergodic decomposition</u> of $(\mathfrak{X}, m, \Psi)$ or of m.

This terminology is justified by a certain uniqueness: Let

$$\Phi : (\mathfrak{X}, m, \Psi) \longrightarrow (X, \nu, T)$$

$$\Phi' : (\mathfrak{X},m,\Psi) \longrightarrow (X',\nu',T')$$

be two conjugacies of $\mathfrak{X}$ with systems carried by top. dynamical systems, giving rise to two ergodic decompositions $(C_x,\Sigma_x,m_x)_{x\in\mathfrak{X}}$ and $(C'_x,\Sigma'_x,m'_x)_{x\in\mathfrak{X}}$. Then for almost all $x \in \mathfrak{X}$, $m'_x = m_x$ and $m(\bigcup_{x:m_x=m'_x} (C_x \cap C'_x)) = 1$.

To prove this, we assume (deleting a Ψ-invariant null set in $\mathfrak{X}$) that

$$\mathfrak{X} = \Phi^{-1}\{y \in X|\mu_y \text{ is ergodic}\} = \Phi'^{-1}\{y' \in X'|\mu_{y'} \text{ is ergodic}\},$$

and that $\Phi^{-1}\mathfrak{B} = \Phi'^{-1}\mathfrak{B}'$ (where $\mathfrak{B},\mathfrak{B}'$ are the Borel algebras on X,X'). Then we also have a countable set F of bounded, $\Phi^{-1}\mathfrak{B}$-measurable functions which are the pullbacks to $\mathfrak{X}$ of a certain dense subset of $C(X)$. We can easily see that a measure w on $\Phi^{-1}\mathfrak{B}$ (all m_x and m'_x are such measures) is determined by its values $\int f\,dw$ on F. Hence we have

$$\{x\in\mathfrak{X}|m'_x \neq m_x\} = \bigcup_{f\in F} \{x| \int f\,dm'_x \neq \int f\,dm_x = \lim \frac{1}{n}\sum_{i=0}^{n-1} f(\Psi^i x)\}.$$

But since for almost all x, $\int f\,d m'_x = \lim \frac{1}{n}\sum_{i=0}^{n-1} f(\Psi^i x)$,

this set is an m-null set, i.e. $m_x = m'_x$ a.e. For $x \in \mathfrak{X}$ with $m_x = m'_x$, we have of course $m_x(C_x \cap C'_x) = 1$. The set $\bigcup_{x:m_x=m'_x} (C_x \cap C'_x)$ is measurable, since it is the same as $\{x \in \mathfrak{X}|m_x = m'_x\}$. Now from (3') the assertion follows.

If Ψ is aperiodic, it is clear that for almost all $x \in \mathfrak{X}$ the m.t. dynamical system (C_x,m_x,Ψ) also is aperiodic, so that the measure m_x must be non-atomic and isomorphic to the Lebesgue measure on $[0,1]$. However, more is true:

<u>(13.1) Proposition:</u> If $(\mathfrak{X},m,\Psi)$ is aperiodic, then there is an m.t. isomorphism

$$\Phi : (\mathfrak{X},m) \longrightarrow (\mathfrak{X}',m') \times ([0,1],\lambda)$$

where $(\mathfrak{X}',m')$ is a Lebesgue space and λ the Lebesgue measure on $[0,1]$, and for almost all $x\in\mathfrak{X}$ Φ is also an m.t. isomorphism

$$\Phi : (C_x,\Sigma_x,m_x) \longrightarrow \Phi_1(x) \times ([0,1],\lambda)$$

($\Phi_i(x)$ is the i-th coordinate of $\Phi(x)$).

Proof: We assume that $\mathfrak{X} = \bigcup_{x\in\mathfrak{X}} C_x$ and that all systems (C_x, m_x, Ψ) are aperiodic. Further we assume that for all $x \in \mathfrak{X}$, $m(C_x) = 0$. This we may do because the countably many fibres with $m(C_x) > 0$ can be treated separately and offer no difficulties. Using the pullback from a conjugate measure on a top.dynamical system (the same which gave the ergodic decomposition) we see that there is a sequence $A_1, A_2, \ldots$ of Ψ-invariant sets which separates the fibres, i.e. if $x \in \mathfrak{X}$, $i \in \mathbb{N}$, then $C_x \subset A_i$ or $C_x \subset \mathfrak{X} \setminus A_i$, and if $C_x \neq C_y$ then for some $i \in \mathbb{N}$, $x \in A_i$, $y \notin A_i$ or vice versa.

In the following we use again the auxiliary space $M = \{0,1\}^{\mathbb{N}}$ as in the proof of proposition (2.17) and the special null sets introduced there.

The sequence $(A_i)_i$ produces the mapping $a : \mathfrak{X} \longrightarrow M$ by

$$ax = (1_{A_i}(x))_{i\in\mathbb{N}}.$$

a is an injective mapping of fibres, i.e. $ax = ay \Leftrightarrow C_x = C_y$. By our assumption that $m(C_x) = 0$, this implies that the image measure am has no point masses. Let F be the distribution function of am on M. It is clear that $m(a^{-1}N_F) = 0$ and $F \circ a : \mathfrak{X} \setminus a^{-1} N_F \longrightarrow [0,1]$ is an injective mapping of fibres, transporting m into the Lebesgue measure on $[0,1]$. We set $\Phi_1 = F \circ a$.

For the definition of Φ_2 we repeat the same procedure on the fibres. We take a sequence $Q_1, Q_2, \ldots$ in Σ which separates all points of $\mathfrak{X}$ and for which $Q_i \cap C_x \in \Sigma_x$ ($x \in$, $i \in \mathbb{N}$), and define the mapping $q : \mathfrak{X} \longrightarrow M$ by

$$qx = (1_{Q_i}(x))_{i\in\mathbb{N}}.$$

F_x is the distribution function of qm_x on M. To show the measurability of $x \longrightarrow (F_x \circ q)(x)$, we set for a block $P = (p_1, \ldots, p_n)$

$$Q^P = Q_1^{p_1} \cap \ldots \cap Q_n^{p_n}$$

where

$$Q_i^j = \begin{cases} Q_i & j=1 \\ \mathfrak{X}\setminus Q_i & j=0 \end{cases} \qquad (i \in \mathbb{N}).$$

By the continuity of each F_x,

$$F_x \circ q(x) = \lim_n F_x(1_{Q_1}(x),\ldots,1_{Q_n}(x),1,1,1,\ldots).$$

But $x \longrightarrow F_x(1_{Q_1}(x),\ldots,1_{Q_n}(x),1,1,1,\ldots)$

is measurable since for P with $l(P) = n$, its value on Q^P is

$$\sum [m_x(Q^{P'}) \mid l(P') = n,\ P' \leq P \text{ in lexicographical order}].$$

So, $x \longrightarrow \Phi(x) = (F \circ a(x),\ F_x \circ q(x))$ is measurable, and it is injective on the set $\mathfrak{X} \setminus (a^{-1} N_F \cup \bigcup_{x\in} (q^{-1} N_{F_x}) \cap C_x)$.

The last set is measurable since from the expression for N_{F_x} in (2.17, proof) it follows that

$$\bigcup_{x\in\mathfrak{X}} (q^{-1} N_{F_x}) \cap C_x =$$

$$= \liminf_{i\in\mathbb{N}} Q_i \cup \liminf_{n\in\mathbb{N}} \bigcup_{l(P)=n} ((Q^P \cup Q^{\tilde{P}}) \cap \{x\in\mathfrak{X} \mid m_x(Q^P)=0\}$$

and it has measure 0 since clearly $m_x(q^{-1}N_{F_x} \cap C_x) = 0$.

It is easy to check that the transported measure Φm is the Lebesgue measure λ^2 on $[0,1]^2$. The invertibility of Φ follows from lemma (2.18). □

Ergodic Decomposition of the Entropy

(13.2) Definition: If $(\mathfrak{X},\Sigma,m,\Psi)$ is an m.t. dynamical system, $(m_x)_{x\in\mathfrak{X}}$ its ergodic decomposition and α a finite partition, then

$$h_x(\alpha,\Psi) = h_{m_x}(\alpha,\Psi),$$

$$h_x(\Psi) = h_{m_x}(\Psi),$$

$$h_{\sup}(\alpha,\Psi) = \operatorname*{ess\,sup}_{x\in\mathfrak{X}} h_x(\alpha,\Psi),$$

$$h_{\inf}(\alpha,\Psi) = \operatorname*{ess\,inf}_{x\in\mathfrak{X}} h_x(\alpha,\Psi)\,,$$

$$h_{\sup}(\Psi) = \operatorname*{ess\,sup}_{x\in\mathfrak{X}} h_x(\Psi)\,,$$

$$h_{\inf}(\Psi) = \operatorname*{ess\,inf}_{x\in\mathfrak{X}} h_x(\Psi)\,.$$

(13.3) Theorem: If $(\mathfrak{X},\Sigma,m,\Psi)$ is an m.t. dynamical system and α a finite partition, then

a) $h_m(\alpha,\Psi) = \int h_x(\alpha,\Psi)dm(x)$

b) $h_m(\Psi) = \int h_x(\Psi)\, dm(x)$ (finite or not) .

If α is a generator, then $h_x(\alpha,\Psi) = h_x(\Psi)$ a.e..

Proof: For a measure ν on a top. dynamical system (X,T) we can interpret the equations $\int(\int g\, d\mu_x)\, d\nu(x) = \int g\, d\nu$ $(g \in C(X))$ as follows: $\tau : x \longrightarrow \mu_x$ is an a.e. defined, Borel measurable, T-invariant mapping $X \longrightarrow \mathfrak{M}_T(X)$, and

$$\nu = \int_X \mu_x\, d\nu(x) = \int_{\mathfrak{M}_T(X)} \rho\, d(\tau\nu)\,(\rho),$$

i.e. ν is the centre of mass of the measure $\tau\nu$ on the compact convex set $\mathfrak{M}_T(X)$. $\tau\nu$ is concentrated on the extreme points of $\mathfrak{M}_T(X)$, but we do not need this.

If $(X,T) = (S^{\mathbb{Z}},\sigma)$ is a shift space with a finite state space, we can use the fact (see (16.7) and (10.13) that $\rho \longrightarrow h_\rho(\sigma)$ is an upper semi-continuous, affine mapping, and derive from Choquet's theorem:

$$h_\nu(\sigma) = \int_{\mathfrak{M}_\sigma(S^{\mathbb{Z}})} h_\rho(\sigma)\, d(\tau\nu)(\rho) = \int_{S^{\mathbb{Z}}} h_x(\sigma)\, d\nu(x).$$

Now we return to the setting of the theorem. Φ_α is a mapping into a finite shift space, and for almost all $x \in \mathfrak{X}$ the measure $\mu_{\alpha,x} = \Phi_\alpha\, m_x$ exists; because it is a factor of an ergodic measure it is itself ergodic, and

$h_x(\alpha,\Psi) = h_{\mu_{\alpha,x}}(\sigma)$. So we obtain

$$h_m(\alpha,\Psi) = h_{\Phi_\alpha m}(\sigma) = \int_{S^Z} h_{\mu_y}(\sigma)\, d(\Phi_\alpha m)(y) = \int h_{\mu_{\Phi_\alpha x}}(\sigma) dm(x) =$$

$$= \int h_x(\alpha,\Psi)\, dm(x), \text{ thus a) is proved.}$$

Now let $\alpha_1 \subset \alpha_2 \subset \ldots$ be a generating sequence of finite partitions. Then a.e. $(\alpha_i | C_x)_{i\in\mathbb{N}}$ is a generating sequence for (C_x, Σ_x, m_x), so

$$h_x(\alpha,\Psi) \nearrow h_x(\Psi) \quad \text{a.e.}$$

$$h_m(\alpha,\Psi) \nearrow h_m(\Psi).$$

This together with a) proves b). The rest is obvious. □

The Theorem of Shannon-McMillan-Breiman

We include this theorem in this section because we shall use the limit in the theorem as an entropy on ergodic fibres. In the proof we only show that this interpretation is correct, and refer to the standard literature for the proof of the main part.

Recall that for a partition α, $\bar{I}(\alpha) = -\sum 1_{A_i} \log m(A_i)$, and if $\alpha \subset \alpha'$, then $\bar{I}(\alpha') \geq \bar{I}(\alpha)$. We write $I_n(\alpha) = n^{-1}\bar{I}((\alpha)_o^{n-1})$. If D is Ψ-invariant and $m(D) > 0$, then

$$I_n(\alpha_D) = n^{-1}\bar{I}((\alpha_D)_o^{n-1}) =$$
$$= n^{-1}[\bar{I}((\alpha)_o^{n-1} \vee (\mathfrak{X}\setminus D, D)) + \log m(D)] \qquad \text{on } D$$

(13.4) Theorem: Let $(\mathfrak{X}, m, \Psi)$ be an m.t. dynamical system, and α a finite partition. Then

a) there is a non-negative, Ψ-invariant, m-integrable function f_α such that $I_n(\alpha) \longrightarrow f_\alpha$ a.e. and in L_1, and $\int f_\alpha\, dm = h_m(\alpha,\Psi)$.

b) for a.e. $x \in \mathfrak{X}$, $f_\alpha(x) = h_x(\alpha,\Psi)$.

<u>Proof:</u> A proof of a) is in [149].

b) Identification of the limit: Let D be Ψ-invariant and $m(D) > 0$. Using (13.3),

$$\int_D (f_\alpha(x) - h_x(\alpha,\Psi))dm(x) = (\int_D f_\alpha dm) - m(D)\cdot h_{m_D}(\alpha_D,\Psi_D) =$$

$$= \int_D (f_\alpha - I_n(\alpha))dm + \int_D (I_n(\alpha) - I_n(\alpha_D))dm +$$

$$+ m(D)\cdot(\int I_n(\alpha_D)dm_D - h_{m_D}(\alpha_D,\Psi_D)).$$

By part a) of the theorem, the first and the last summand tend to 0 when $n \longrightarrow \infty$, and for the middle term we have

$$nI_n(\alpha) - nI_n(\alpha_D) =$$

$$= \bar{I}((\alpha)_o^{n-1}) - \bar{I}((\alpha)_o^{n-1} \vee (\mathfrak{X}\backslash D,D)) - \log m(D) \quad \text{on } D,$$

hence

$$-\log m(D) \geq n(I_n(\alpha) - I_n(\alpha_D)) \geq$$

$$\geq \bar{I}((\alpha)_o^{n-1}) - \bar{I}((\alpha)_o^{n-1} \vee (\mathfrak{X}\setminus D,D)) \quad \text{on } D$$

and

$$\frac{1}{n} \geq -\frac{1}{n} m(D) \log m(D) \geq \int_D (I_n(\alpha) - I_n(\alpha_D))\, dm \geq$$

$$\geq \frac{1}{n}\int_D [\bar{I}((\alpha)_o^{n-1}) - \bar{I}((\alpha)_o^{n-1} \vee (\mathfrak{X}\setminus D,D))]dm \geq$$

$$\geq \frac{1}{n}\int_{\mathfrak{X}} [\ldots]dm =$$

$$= \frac{1}{n}[H_m((\alpha)_o^{n-1}) - H_m((\alpha)_o^{n-1} \vee (\mathfrak{X}\setminus D,D))] =$$

$$= -\frac{1}{n} H_m((\mathfrak{X}\setminus D,D)\,|\,(\alpha)_o^{n-1}) \geq -\frac{1}{n}.$$

This means that $\int_D (f_\alpha(x) - h_x(\alpha,\Psi))\, dm = 0$ for all Ψ-invariant sets D, i.e. $f_\alpha(x) = h_x(\alpha,\Psi)$ a.e. □

What we shall use later is a consequence of this theorem:

<u>(13.5) Corollary:</u> Let α be a finite partition and $\epsilon > 0$. Then for sufficiently large n

$$m(\bigcup [\text{atoms } A \text{ of } (\alpha)_0^{n-1} | \exp\{-n(h_{sup}(\alpha,\Psi) + \epsilon)\} <$$

$$< m(A) < \exp\{-n(h_{inf}(\alpha,\Psi) - \epsilon)\}]) > 1 - \epsilon .$$

(In the ergodic case, of course, h_{sup} and h_{inf} are replaced by h_m).

<u>Proof:</u> Since a.e. $h_{inf}(\alpha,\Psi) \leq h_x(\alpha,\Psi) \leq h_{sup}(\alpha,\Psi)$ and $\bar{I}((\alpha)_0^{n-1})$ has the constant value $-\log m(A)$ on the atom A, the result follows from the stochastic convergence in the theorem. □

14. Topological entropy

In 1965, Adler, Konheim and McAndrew [2] introduced the topological equivalent of entropy, with open covers playing the role of partitions. A special case was defined earlier by Parry [146] as absolute entropy.

Let (X,T) be a top. dynamical system.

(14.1) Definition: Let $\mathfrak{U} = (U_\alpha)$ and $\mathfrak{U}' = (U'_\beta)$ be two open covers of X. $\mathfrak{U}'$ is said to be finer than $\mathfrak{U}$ ($\mathfrak{U} \leq \mathfrak{U}'$) if every U'_β is contained in some U_α. $\mathfrak{U}'$ is a subcover of $\mathfrak{U}$ if $\mathfrak{U}'$ is a cover $\subset \mathfrak{U}$. For any two open covers $\mathfrak{U}$ and $\mathfrak{U}'$ of X one defines $\mathfrak{U} \vee \mathfrak{U}'$ as the open cover whose elements are $U_\alpha \cap U'_\beta$. One denotes by $T^{-1}\mathfrak{U}$ the open cover whose elements are the $T^{-1}U_\alpha$, and by $(\mathfrak{U})_o^N$ the cover $\mathfrak{U} \vee T^{-1}\mathfrak{U} \vee \ldots \vee T^{-N}\mathfrak{U}$.

(14.2) Definition: If $\mathfrak{U} = (U_\alpha)$ is an open cover of X, one writes

$$H(\mathfrak{U}) = \log N(\mathfrak{U})$$

where $N(\mathfrak{U})$ denotes the smallest cardinality of a subcover of $\mathfrak{U}$.

$N(\mathfrak{U})$ is finite. One has obviously:

(14.3) Proposition:

(a) $H(T^{-1}\mathfrak{U}) = H(\mathfrak{U})$

(b) $H(\mathfrak{U} \vee \mathfrak{U}') \leq H(\mathfrak{U}) + H(\mathfrak{U}')$

Remark that $N((\mathfrak{U})_o^{N-1})$ is just the smallest cardinality of a family of N-tuples of elements of $\mathfrak{U}$ such that for any $x \in X$ there exists an N-tuple $(U_{\alpha_o},\ldots,U_{\alpha_{N-1}})$ of this family with $T^k x \in U_{\alpha_k}$ for $0 \leq k < N$.

(14.4) Proposition: For any open cover $\mathfrak{U}$

$$\lim \frac{1}{N} H((\mathfrak{U})_o^{N-1})$$

exists.

Proof: This follows from (14.3b) by just the same argument as in (10.6). $\square$

<u>(14.5) Definition:</u> For any open cover $\mathfrak{U}$, the expression

$$H(\mathfrak{U},T) = \lim \frac{1}{N} H((\mathfrak{U})_0^{N-1})$$

is called the <u>topological entropy of $\mathfrak{U}$ with respect to T.</u>

Clearly one has

<u>(14.6) Proposition:</u>

(a) $0 \leq H(\mathfrak{U},T) \leq H(\mathfrak{U})$

(b) If $\mathfrak{U} \leq \mathfrak{U}'$ then $H(\mathfrak{U},T) \leq H(\mathfrak{U}',T)$

<u>(14.7) Definition:</u> The expression

$$h_{top}(T) = \sup \{H(\mathfrak{U},T) | \mathfrak{U} \text{ open cover of } X\}$$

is called the <u>topological entropy of T.</u>

Clearly it is enough to consider finite covers. Also

<u>(14.8) Proposition:</u> If $\mathfrak{U}_n$ is a sequence of open covers of X such that $\mathfrak{U}_n \leq \mathfrak{U}_{n+1}$ and such that for any finite open cover $\mathfrak{U}$ one has $\mathfrak{U} \leq \mathfrak{U}_n$ for some n, then

$$h_{top}(T) = \lim_{n\to\infty} H(\mathfrak{U}_n,T).$$

Another definition of topological entropy has been given by Bowen [22] and Dinaburg [53]. It was inspired by the definition of ε-entropy and ε-capacity of Kolmogorov and Tihomirov [111]. This definition makes sense for noncompact metric spaces, a fact which will be useful later (see section 24).

Thus let d be a metric on the (not necessarily compact) space X and let T be a uniformly continuous map from X onto itself (we write $T \in UC(X,d)$).

<u>(14.9) Definition:</u> A subset $E \subset X$ is said to be <u>(n,ε)-separated</u> if for any two distinct points $x,y \in E$, there is a k with $0 \leq k < n$ such that $d(T^k x, T^k y) > \varepsilon$.

For compact $K \subset X$ one denotes by $s_n(\varepsilon,K)$ the largest cardinality of an (n,ε)-separated subset of K. If X is compact one writes $s_n(\varepsilon)$ instead of $s_n(\varepsilon,X)$. One sometimes writes $s_n(\varepsilon,K,T)$ or $s_n(\varepsilon,T)$.

(14.10) Definition: $s(\epsilon,K) = \limsup \frac{1}{n} \log s_n(\epsilon,K)$

and $$s(\epsilon) = \limsup \frac{1}{n} \log s_n(\epsilon)$$

if X is compact.
(One sometimes writes $s(\epsilon,K,T)$ or $s(\epsilon,T)$.)

(14.11) Proposition: $\lim_{\epsilon\to 0} s(\epsilon,K)$ exists.

Proof: If $\epsilon_1 < \epsilon_2$ then $s_n(\epsilon_1,K) \geq s_n(\epsilon_2,K)$. □

(14.12) Definition: One writes

$$h(T,K) = \lim_{\epsilon\to 0} s(\epsilon,K)$$

and

$$h_d(T) = \sup\{h(T,K) \mid K\subset X \text{ compact}\}.$$

Clearly if X is compact then $h_d(T) = h(T,X)$.

(14.13) Proposition: If the two metrics d and d' on X are uniformly equivalent, then $h_d(T) = h_{d'}(T)$.

This is easy to check. In particular, if X is compact, and d and d' are two metrics on X, then $h_d(T) = h_{d'}(T)$.

There is an alternative definition of $h_d(T)$ which is useful for some proofs.

(14.14) Definition: A set $F \subset X$ is said to be (n,ε)-spanning for $K \subset X$ if for each $x \in K$ there is a $y \in F$ with $d(T^k x, T^k y) \leq \epsilon$ for $0 \leq k < n$. One denotes by $r_n(\epsilon,K)$ the minimal cardinality of a set which (n,ϵ)-spans K and writes

$$r(\epsilon,K) = \limsup \frac{1}{n} \log r_n(\epsilon,K).$$

(One sometimes writes $r_n(\epsilon,K,T)$ or $r(\epsilon,K,T)$.)

(14.15) Proposition: If $K \subset X$ is compact,

$$h(T,K) = \lim_{\epsilon\to 0} r(\epsilon,K).$$

Proof: This is an immediate consequence of

$$r_n(\epsilon,K) \leq s_n(\epsilon,K) \leq r_n(\tfrac{\epsilon}{2},K).$$

The left inequality is obvious: any maximal (n,ϵ)-separated subset of K is (n,ϵ)-spanning for K. Let $J \subset X$ $(n,\frac{\epsilon}{2})$-span K. For each $x \in K$ there is a $j(x) \in J$ such that $d(T^k x, T^k j(x)) \leq \frac{\varepsilon}{2}$ for $0 \leq k < n$. If x_1, x_2 are two distinct points of an (n,ε)-separated subset E of K, then $j(x_1) \neq j(x_2)$ since otherwise $d(T^k x_1, T^k x_2) \leq \varepsilon$ for $0 \leq k < n$.

Hence card $E \leq$ card J, and therefore $s_n(\varepsilon, k) \leq r_n(\frac{\varepsilon}{2}, K)$. □

<u>(14.16) Theorem:</u> [22] If X is compact, then

$$h_d(T) = h_{top}(T).$$

<u>Proof:</u> (a) Let $\varepsilon > 0$ be given and let E be an (n,ε)-separated subset of X. Let $\mathfrak{U}$ be an open cover of X by sets U_α of diameter $< \varepsilon$. Two distinct points x_1, x_2 of E cannot lie in the same n-tuple $(U_{\alpha_o}, \ldots, U_{\alpha_{n-1}})$ of elements of $\mathfrak{U}$. Hence $s_n(\varepsilon) \leq N((\mathfrak{U})_o^{n-1})$, and therefore $s(\varepsilon) \leq H(\mathfrak{U}, T)$ and $h_d(T) \leq h_{top}(T)$.

(b) Let $\mathfrak{U} = (U_\alpha)$ be an open cover for X and let $\varepsilon > 0$ be a Lebesgue number for $\mathfrak{U}$. Thus for any $x \in X$, the closed ε-ball $B_\varepsilon(x)$ lies inside some U_α. Let $Q_n \subset X$ be an (n,ε)-spanning set with minimal cardinality $r_n(\varepsilon)$. For each $z \in Q_n$ and each k, $0 \leq k < n$, let $U_{\alpha_k}(z)$ be some element of $\mathfrak{U}$ containing $B_\varepsilon(T^k z)$. For any $x \in X$ there is a $z \in Q_n$ with $T^k x \in B_\varepsilon(T^k z)$ for $0 \leq k < n$. Thus, $T^k x \in U_{\alpha_k}(z)$, and the family

$$(U_{\alpha_o}(z) \cap \ldots \cap T^{-n+1} U_{\alpha_{n-1}}(z) \mid z \in Q_n)$$

is a subcover of $(\mathfrak{U})_o^{n-1}$. Hence

$$N((\mathfrak{U})_o^{n-1}) \leq \text{card } Q_n = r_n(\varepsilon)$$

and so $H(\mathfrak{U}, T) \leq r(\varepsilon)$, and $h_{top}(T) \leq h_d(T)$. □

<u>(14.17) Proposition:</u> If $T \in UC(X,d)$ then

$$h_d(T^k) = k\, h_d(T)$$

for all $k \geq 1$.

Proof: Any set $E \subset K$ which is (n,ϵ)-separated for T^k is obviously (nk,ϵ)-separated for T. Hence $s_n(\epsilon,T^k) \leq s_{nk}(\epsilon,T)$ and thus $h_d(T^k) \leq kh_d(T)$.

Since T is uniformly continuous, for each $\epsilon > 0$ there is a $\delta > 0$ such that $d(x,y) < \delta$ implies $d(T^j x, T^j y) < \epsilon$ for $0 \leq j < k$. So an (n,δ)-spanning set for K with respect to T^k is (nk,ϵ)-spanning for K with respect to T. Hence

$$r_n(\delta,K,T^k) \geq r_{kn}(\epsilon,K,T),$$

$$k \cdot r(\epsilon,K,T) \leq r(\epsilon,K,T^k)$$

and $$k \cdot h_d(T,K) \leq h_d(T). \qquad \square$$

If T^{-1} exists, it need not be uniformly continuous. But

(14.18) Proposition: If (X,T) is a top. dynamical system, then

$$h_{top}(T^k) = |k|\, h_{top}(T)$$

for all $k \in \mathbb{Z}$.

Proof: For $k \geq 1$ this follows from (14.17). For $k = 0$ it is obvious. $h_{top}(T) = h_{top}(T^{-1})$ since

$$\begin{aligned} H(T,\mathfrak{U}) &= \lim \frac{1}{N} H((\mathfrak{U})_o^{N-1}) \\ &= \lim \frac{1}{N} H(T^{N-1}(\mathfrak{U})_o^{N-1}) \\ &= \lim \frac{1}{N} H(\mathfrak{U} \vee T\mathfrak{U} \vee \dots \vee T^{N-1}\mathfrak{U}) \\ &= H(T^{-1},\mathfrak{U}). \end{aligned}$$

Hence $h_{top}(T^k) = |k|\, h_{top}(T)$ for all $k \leq -1$. $\square$

(14.19) Proposition: If $T \in UC(X,d)$ is periodic, then $h_d(T) = 0$. If there exists a metric d such that T is an isometry, or a contraction (i.e. if there is a $C < 1$ with $d(Tx,Ty) < Cd(x,y)$ for all $x,y \in X$) then $h_d(T) = 0$.

This is easy to check.

(14.20) Proposition: Let d be a metric on $\mathbb{R}^p$, $X \subset \mathbb{R}^p$ and $T \in UC(X,d)$ which is Lipschitz, i.e. there is a $C > 0$ with $d(Tx,Ty) < C \cdot d(x,y)$ for $x,y \in X$. Then

$$h_d(T) \leq \max (0, p \cdot \log C).$$

Proof: We may assume that d is the maximum metric in $\mathbb{R}^p$. Let $K \subset X$ be compact. Note that any (n,ϵ)-spanning set is $(n+1,C\epsilon)$-spanning.

Hence

$$r_n(\epsilon,X) \leq r_1(\epsilon \cdot C^{-n+1},X) \leq r_1(C^{-n+1},X)$$
$$\leq \left(\frac{\text{diam } X}{C^{-n+1}}\right)^p \leq (\text{diam } X)^p \, C^{p(n-1)} .$$

Therefore $r(\epsilon,X) \leq p \log C$ if $C \geq 1$. □

(14.21) Proposition: Let $T \in UC(X,d)$.

(a) If $K_1,K_2 \subset X$ are compact, then $h_d(T,K_1 \cup K_2) \leq \max\{h_d(T,K_1),\ h_d(T,K_2)\}$.

(b) If $X_1,X_2 \subset X$ are such that $TX_1 = X_1$, $TX_2 = X_2$ and $X_1 \cup X_2 = X$, then

$$h_d(T) = \max \{h_d(T|X_1),\ h_d(T|X_2)\}.$$

Proof: If $a_n,b_n > 0$ such that $\limsup n^{-1} \log a_n = a$ and $\limsup n^{-1} \log b_n = b$, then for any $c > \max (a,b)$ there is an n_o such that for all $n > n_o$ one has $a_n < e^{nc}$ and $b_n < e^{nc}$. Thus $a_n + b_n < 2e^{nc}$ and so $n^{-1} \log(a_n + b_n) < c + 2n^{-1}$. Hence

$$\limsup n^{-1} \log(a_n + b_n) \leq \max(a,b).$$

(a) follows from this inequality and the fact that

$$s_n(\epsilon,K_1 \cup K_2) \leq s_n(\epsilon,K_1) + s_n(\epsilon,K_2).$$

As for (b), let $K \subset X$ be compact. Since

$$s_n(\epsilon,K,T) \leq s_n(\epsilon,K \cap X_1,T|X_1) + s_n(\epsilon,K \cap X_2,T|X_2),$$

one obtains

$$h(T,K) \leq \max \{h(T|X_1, K \cap X_1), h(T|X_2, K \cap X_2)\}.$$

Thus $h_d(T) \leq \max \{h_d(T|X_1), h_d(T|X_2)\}$.

The converse is obvious. □

(14.22) Proposition: Let (X,T) and (X',T') be top.dynamical systems.

(a) $h_{top}(T \times T') = h_{top}(T) + h_{top}(T')$

(b) if (X',T') is a factor of (X,T), then $h_{top}(T') \leq h_{top}(T)$.

Proof: (a) Let d be a metric on X, d' on X' and define the maximum metric $\bar{d}$ on $X \times X'$ by $\bar{d}((x,x'),(y,y')) = \max(d(x,y), d'(x',y'))$. Let E (resp. E') be a minimal (n,ϵ)-spanning set for X (resp. X'). Then $E \times E'$ (n,ϵ)-spans $X \times X'$. Hence

$$r_n(\epsilon, T \times T') \leq r_n(\epsilon,T) \cdot r_n(\epsilon,T')$$

and so $\quad h_{\bar{d}}(T \times T') \leq h_d(T) + h_{d'}(T')$,

i.e. $\quad h_{top}(T \times T') \leq h_{top}(T) + h_{top}(T')$

Now let E (resp. E') be maximal (n,ϵ)-separated sets for X (resp. X'). $E \times E'$ is (n,ϵ)-separated, hence

$$s_n(\epsilon,T) \cdot s_n(\epsilon,T') \leq s_n(\epsilon, T \times T').$$

By the first inequality in the proof of (14.15)

$$r_n(\epsilon,T) \cdot r_n(\epsilon,T') \leq s_n(\epsilon, T \times T').$$

Let $\mathfrak{U}$ be an open cover of X and ϵ a Lebesgue number. As in the proof of (14.16) one sees that $N((\mathfrak{U})_o^{n-1}) \leq r_n(\epsilon, T)$. Similarly if $\mathfrak{U}'$ is an open cover of X' with Lebesgue number ϵ, $N((\mathfrak{U}')_o^{n-1}) \leq r_n(\epsilon,T')$. Thus

$$H(\mathfrak{U},T) + H(\mathfrak{U}',T') \leq s_n(\epsilon, T \times T')$$

and hence

$$h_{top}(T) + h_{top}(T') \leq h_{\bar{d}}(T \times T') = h_{top}(T \times T').$$

(b) is easy to check. □

The first few lines in the proof of (14.22.a) show that the following holds

<u>(14.23) Proposition:</u> If $T \in UC(X,d)$ and $T' \in UC(X',d')$ then

$$h_{\bar{d}}(T \times T') \leq h_d(T) + h_{d'}(T')$$

where $\bar{d}$ is the maximum metric on $X \times X'$.

<u>(14.24) Proposition:</u> Let $\mathbb{T}^1 = \mathbb{R}^1/\mathbb{Z}^1$ be the one dimensional torus and $T: \mathbb{T}^1 \longrightarrow \mathbb{T}^1$ a homeomorphism. Then $h_{top}(T) = 0$.

<u>Proof:</u> We follow the proof by Walters [185]. Choose $0 < \varepsilon < \frac{1}{4}$ such that $d(x,y) < \varepsilon$ implies $d(T^{-1}x, T^{-1}y) \leq \frac{1}{4}$. We claim that

$$r_n(\varepsilon, \mathbb{T}^1) \leq n([\tfrac{1}{\varepsilon}] + 1).$$

The proof goes by induction. $r_1(\varepsilon, \mathbb{T}^1) < [\frac{1}{\varepsilon}] + 1$ is obvious. Let F be minimal $(n - 1, \varepsilon)$-spanning. The points of $T^{n-1}F$ determine intervals. By adding points to $T^{n-1}F$ one subdivides these intervals. There exists a set $E \subset \mathbb{T}^1$ of cardinality $\leq [\frac{1}{\varepsilon}] + 1$ such that the intervals defined by $F \cup E$ have all length $\leq \varepsilon$. Write

$$F' = F \cup T^{-(n-1)}E .$$

We claim that F' is (n,ε)-spanning. Indeed, if $x \in \mathbb{T}^1$, there exists a $y \in F$ such that

$$d(T^j x, T^j y) \leq \varepsilon \quad \text{for } 0 \leq j < n - 1.$$

Remark that T^{-1} maps intervals to intervals, since intervals are just the connected subsets of $\mathbb{T}^1$. Thus there is one interval $I = [T^{n-1}x, T^{n-1}y]$ which is mapped by T^{-1} to that interval I' defined by $T^{n-2}x$ and $T^{n-2}y$ which has length $\leq \varepsilon$. Choose $z \in F'$ such that $T^{n-1}z \in I$. Clearly $T^{n-2}z \in I'$, hence $d(T^{n-2}z, T^{n-2}x) \leq \varepsilon$. I' is mapped by T^{-1} to an interval of length $\leq \frac{1}{4}$, and hence to an interval of length $\leq \varepsilon$. Since $T^{n-3}z$ lies in this interval, $d(T^{n-3}z, T^{n-3}x) \leq \varepsilon$. Proceeding by induction, one gets

$$d(T^j z, T^j x) \leq \varepsilon \quad \text{for } 0 \leq j \leq n - 1.$$

Hence F' is (n,ε)-spanning. But card $F = r_{n-1}(\varepsilon, \mathbb{T}^1)$ and card $F' \leq$ card $F + ([\frac{1}{\varepsilon}] + 1$, so

$$r(\varepsilon,\mathbb{T}^1) \leq n([\tfrac{1}{\varepsilon}] + 1),$$

as was claimed. This implies

$$r(\varepsilon,\mathbb{T}^1) = \limsup \frac{1}{n} \log r_n(\varepsilon,\mathbb{T}^1) = 0$$

and so $h_{top}(T) = 0$. □

The present section on topological entropy only presents the basic facts and simplest examples in this theory. In the following sections much more is said about topological entropy and some further examples are given. However, in these notes we could not present all known facts about topological entropy, therefore we give some hints for further studies.

In the literature we tried to give a complete survey about the papers concerning this subject. In [138] Misiurewicz showed a generalization of 14.22: $h_{top}(\prod_{i\in\mathbb{N}} T_i) = \sum_{i\in\mathbb{N}} h_{top}(T_i)$ where (X_i,T_i) $(i\in\mathbb{N})$ are topological dynamical systems. Another theorem is due to Bowen in [22] (see [107] for a special version). If (Y,S) is a factor of (X,T) then

$$h_{top}(T) \leq h_{top}(S) + \sup_{y\in Y} h_{top}(T/\pi^{-1}\{y\}),$$

where $\pi : X \longrightarrow Y$ is the given projection.

Especially for S = id, $h_{top}(T) = \sup_{y\in Y} h_{top}(T/\pi^{-1}\{y\})$. This special case will be proved as a corollary of the Dinaburg-Goodman theorem in section 18. We note that the above theorem has applications to group extensions.

We noted in the beginning of this section that the notion of topological entropy is similar to that one of measure theoretic entropy. Since it is important to investigate upper semicontinuity and continuity of the measure theoretic entropy (as a function on $\mathfrak{M}_T(X)$) one might ask a similar question for topological entropy: Let X be a compact metric space and $\mathfrak{T}$ be a set of homeomorphisms on X. Is the topological entropy upper semicontinuous or continuous as a function on $\mathfrak{T}$? For this we have to use a topology on $\mathfrak{T}$, for instance a C^r topology if X is a manifold. The answer is no, in general, as Misiurewicz has shown in [136]. Also it is known for topological dynamical systems (X_n, T_n) $(n \geq 0)$ that (in general)

$$\lim_{n\to\infty} h_{top}(T_n) \neq h_{top}(T_o)$$

if X_n converges to X_o in the Hausdorff-metric (see the example after (19.5) and (16.12)).

Finally, the reader should pay attention to Bowen's paper [26], where he pointed out the connection between topological entropy and Hausdorff dimension. This shows the tight connections between dimension theory, topological entropy, measure theoretic entropy (see [17]), ε-capacity and ε-entropy (see [111]) and some problems in number theory (see [42]).

15 Topological Generators

In section 9 we gave a general outline of the theory of measurable partitions and generators for a measure $m \in \mathfrak{M}_T(X)$. It was proved there that a countable generator always exists (9.5). This generating property is equivalent to the following statement (9.6): There exists a subset $X_o \subset X$ of measure one such that every two points $x,y \in X_o$ $(x \neq y)$ are "separated" by the generator. The aim of this section is to **translate this concept into a topological setting.** By this we mean that the generator should have some nice topological properties. We shall start with the notion of a topological generator ([44], see also [109]).

(15.1) Definition: Let $X_o \subset X$. An (at most countable) partition α of X is called a (topological) generator for X_o if the following conditions are satisfied:

i) $\bigcup_{A\in\alpha} \text{int } A$ is dense in X and $\text{int } A \neq \emptyset$ for every $A \in \alpha$.

ii) for every $x \in X_o$ and every sequence $A_{i_k} \in \alpha$ $(k\in\mathbb{Z})$ such that

$$x \in \bigcap_{n\in\mathbb{Z}} \overline{\bigcap_{k=-n}^{n} T^k \text{ int } A_{i_k}}$$

it follows that

$$\{x\} = \bigcap_{n\in\mathbb{Z}} \overline{\bigcap_{k=-n}^{n} T^k \text{ int } A_{i_k}}$$

(Note that for $A \subset X$ $\text{int } A$ denotes the topological interior and $\overline{A}$ the topological closure of A. Also recall the definition of a partition from (9.1) and the remark afterwards.)

We remark that the notion of a topological generator is much stronger than that of a generator for measures. Thus - by the way - we get a new kind of measure-theoretic generator: Let $m \in \mathfrak{M}_T(X)$. We say that a measurable set $A \in \mathfrak{B}_m$ is an m-continuity set if the boundary of A has measure zero, and we denote by $\mathfrak{B}_m^o$ the ring of m-continuity sets. Then we call a measurable partition $\alpha = (A_n | n \in \mathbb{N})$ $(A_n \in \mathfrak{B}_m^o)$ an m-continuity partition if

$$m(\bigcup_{n\in\mathbb{N}} \text{int } A_n) = 1.$$

Finally an m-continuity partition α of X is called a topological generator for m if it is a topological generator for some subset $X_o \subset X$ of measure one.

Since T is a homeomorphism we always may assume that a topological generator separates points for a T-invariant subset $X_o \subset X$. Another trivial observation is this one: If α is a topological generator for X_o then $\bigcup_{A\in\alpha} \text{int } A$ is an open, dense subset of X, hence by Baire's theorem

$$X_\alpha := \bigcap_{k\in\mathbb{Z}} T^k[\bigcup_{A\in\alpha} \text{int } A]$$

is residual.

Finnally, let S denote the one point compactification of $\mathbb{N}$ by "0" and $(S^{\mathbb{Z}},\sigma)$ the shift defined by S.

At a first glance it is not apparent why in the definition of a topological generator α the closures of the sets int A ($A \in (\alpha)_{-n}^n$) are taken. This becomes clear in the following theorem where we have to use the fact that for $\{x\} = \bigcap_{k\in\mathbb{Z}} T^k A_{i_k} \cap X_\alpha$ the system $\{\bigcap_{k=-n}^{n} T^k \text{ int } A_{i_k} \mid n\in\mathbb{N} \}$ is a basis for the neighbourhood system of x.

<u>(15.2) Theorem:</u> Let $\alpha = (A_n \mid n \in \mathbb{N})$ be a topological generator for $X_o \subset X$. If $X_o \cap X_\alpha \neq \emptyset$ then there are a σ-invariant subset $\emptyset \neq Y_\alpha \subset \mathbb{N}^{\mathbb{Z}}$ and a continuous map

$$\pi : Y_\alpha \longrightarrow X$$

with the following properties:

1) $T \circ \pi = \pi \circ \sigma$

2) π is invertible on $\pi^{-1}(X_o \cap X_\alpha)$ and

$\pi^{-1} | X_o \cap X_\alpha$ is continuous (hence is a homeomorphism)

3) For $y = (y_k)_{k\in\mathbb{Z}} \in Y_\alpha$ we have

$$\pi(y) \in \bigcap_{n\in\mathbb{Z}} \overline{\bigcap_{k=-n}^{n} T^k \operatorname{int} A_{y_k}}$$

4) $$Y_\alpha = \{(y_k)_{k\in\mathbb{Z}} \in \mathbb{N}^{\mathbb{Z}} \mid \operatorname{card} \bigcap_{n\in\mathbb{Z}} \overline{\bigcap_{k=-n}^{n} T^k \operatorname{int} A_{y_k}} = 1\} \cap \overline{\pi^{-1}(X_o \cap X_\alpha)}$$

<u>Proof:</u> Define $f : X_\alpha \cap X_o \longrightarrow \mathbb{N}^{\mathbb{Z}}$ as follows:
Let $\alpha = (A_n \mid n \in \mathbb{N})$ (where possibly some $A_n = \emptyset$ if α is finite). Since by the hypothesis for $x \in X_\alpha \cap X_o$ there exists a unique sequence $(A_{n_k})_{k\in\mathbb{Z}}$ such that

$$\{x\} = \bigcap_{k\in\mathbb{Z}} T^k \operatorname{int} A_{n_k} = \bigcap_{m\in\mathbb{Z}} \overline{\bigcap_{k\in -m}^{m} T^k \operatorname{int} A_{n_k}},$$

set $f(x) = (n_k)_{k\in\mathbb{Z}}$. Clearly f is injective and continuous. f is a homeomorphism onto its image iff the system

$$\{\bigcap_{k=-m}^{m} T^k \operatorname{int} A_{(f(x))_k} \mid m \in \mathbb{N}\}$$ is a basis for the system of neighbourhoods of x for every $x \in X_o \cap X_\alpha$.

Clearly $f(X_o \cap X_\alpha) \subset Y_\alpha$. Therefore define π on $f(X_o \cap X_\alpha)$ by $\pi = f^{-1}$ and then extend π to Y_α by a general method. Let $y \in Y_\alpha$ and choose any sequence $y^n \in f(X_o \cap X_\alpha)$ converging to y. From the convergence in $\mathbb{N}^{\mathbb{Z}}$ it is easy to see that $\lim_{n\to\infty} \pi(y^n)$ exists and does not depend on the chosen sequence converging to y. Hence $\pi(y) = \lim_{n\to\infty} \pi(y^n)$ is well defined and obviously continuous.

Now the theorem follows easily. □

<u>Corollary:</u> If α is a finite topological generator for X, then Y_α is closed and π is onto X, hence (X,T) is a factor of (Y_α, σ).

In the preceding theorem let $X_\alpha \cap X_o$ be a residual set. Then the map π of the theorem gives an "almost" homeomorphism, almost in the topological sense. Obviously this is the best one can hope for (see the remark after (3.10)).

Before giving some applications of (15.2) we shall deal with the question of existence of generators. First we have to look for an analogon to Rohlin's lemma. Though we shall need weaker properties than those stated in the following propositions we think they are interesting showing the similarities of the topological and measure theoretical concepts.

<u>(15.3) Definition:</u> By $P^n(T)$ we denote the set of <u>periodic</u>

points of T with period n (i.e. the set of fixed points of T^n) and we define $\tilde{P}^n(T) := \bigcup_{j=1}^{n-1} P^j(T)$ and

$Per_n(T) :=$ card $P^n(T)$.

(15.4) Proposition [44]: Let $n \in \mathbb{N}$ and $A \subset X \setminus \tilde{P}^n(T)$. Then there exists a set $V \subset A$ such that

i) $V \subset \overline{\text{int } V}$ (with respect to the induced topology on A)

ii) $V \cap T^jV = \emptyset$ for every $1 \leq j \leq n-1$

iii) $A \subset \bigcup_{k=-2n+2}^{2n-2} T^kV$

Proof: We are considering the induced topology on A and we shall prove the proposition using Zorn's lemma.

Let $\Pi := \{V \subset A \mid V \cup T^jV = \emptyset\ (\forall\ 1 \leq j \leq n-1);\ V \subset \overline{\text{int } V}\}$. If $A = \emptyset$ then the proposition is trivial; if $A \neq \emptyset$ then clearly $\Pi \neq \emptyset$. For proving the hypothesis of Zorn's lemma, suppose $\Pi_o \subset \Pi$ is totally ordered by the usual set inclusion. Then clearly $V := \bigcup_{V' \in \Pi_o} V'$ satisfies $V \subset A$, $V \cap T^jV = \emptyset$ $(1 \leq j \leq n-1)$ and $V \subset \bigcup_{V' \in \Pi_o} \overline{\text{int } V'} \subset \overline{\text{int} \bigcup_{V' \in \Pi_o} V'}$ and therefore belongs to Π.

By Zorn's lemma there exists a maximal $V \in \Pi$, and it is left to show that V satisfies iii) of the proposition.

Assume that there is $x \in A$ such that

$$\{x, \dots, T^{n-1}(x)\} \cap \bigcup_{k=o}^{n-1} T^k\overline{V} = \emptyset.$$

Since $x \notin \tilde{P}^n(T)$, there exists an open set $U \subset A$ with $x \in U$,

$U \cap T^jU = \emptyset$ $(\forall\ 1 \leq j \leq n-1)$ and

$$T^jU \cap \bigcup_{k=o}^{n-1} T^k\overline{V} = \emptyset \quad (1 \leq j \leq n-1).$$

Define $V_o := V \cup U$. Since $V_o \in \Pi$, we get a contradiction. It follows that for every $x \in A$ there is some $j \in \{0,\dots,n-1\}$ such that

$$x \in T^{-j} \bigcup_{k=o}^{n-1} T^k\overline{V} \quad \text{i.e.} \quad A \subset \bigcup_{k=-n+1}^{n-1} T^k\,\overline{V}.$$

Assume now that $\overline{V} \setminus \bigcup_{k=-n+1}^{n-1} T^kV$ contains a point x. Since this implies for every $0 \leq i \leq n-1$

$$T^i(x) \in T^i\overline{V} \setminus \bigcup_{j=-n+1+i}^{n+1+i} T^jV,$$

it follows that $T^i(x) \notin \bigcup_{j=o}^{n-1} T^jV$ for every $0 \leq i \leq n-1$.

Let $V_o = V \cup \{x\}$. Since $x \in A$ it follows immediately that $V_o \subset A$, $\overline{\operatorname{int} V_o} \supset V_o$ and that for every $1 \leq j \leq n-1$

$$V_o \cap T^jV_o = \emptyset.$$

This fact contradicts the maximality of V, hence

$$\overline{V} \subset \bigcup_{k=-n+1}^{n-1} T^kV$$

and so iii) is proved:

$$A \subset \bigcup_{j=-n+1}^{n-1} T^j\overline{V} \subset \bigcup_{k=-2n+2}^{2n-2} T^kV. \qquad \square$$

For the next proposition we assume that the set of

aperiodic points of T is dense in X.

(15.5) Lemma: Let $p_n (n \in \mathbb{N})$ be a strictly increasing sequence of positive integers ≥ 2 satisfying

$$\sum_{n=1}^{\infty} (p_n + 1)^{-1} < \frac{1}{2}.$$

Let $c_n := 3 \sum_{k=1}^{n} p_k$. Then there exists a partition $\alpha = (A_n | n \geq 0)$ of X such that

i) $A_n \subset \overline{\text{int } A_n}$ $(n \in \mathbb{N})$ and $\overline{\bigcup_{n=0}^{\infty} \text{int } A_n} = X$

ii) $A_n \cap T^i A_n = \emptyset$ $(1 \leq i \leq p_n)$ $(n \in \mathbb{N})$

iii) $X_\alpha \subset \bigcup_{k=-c_n}^{c_n} T^k A_n$ $(n \in \mathbb{N})$

Proof: The sets A_n $(n \in \mathbb{N})$ will be defined inductively using proposition (15.4).

For $n = 1$ there is $A_1 \subset X \setminus \tilde{P}^{p_1+1}(T)$ such that $A_1 \subset \overline{\text{int } A_1}$, $A_1 \cap T^i A_1 = \emptyset$ $(1 \leq i \leq p_1)$ and $X \setminus \tilde{P}^{p_1+1}(T) \subset \bigcup_{-c_1}^{c_1} T^i A_1$. Clearly $X \setminus \tilde{P}^{p_1+1}(T) \subset$ $\subset \bigcup_{k=-c_1}^{c_1} T^i(X \setminus A_1)$ holds also.

Assume now that $A_1, \ldots, A_n$ are constructed such that for each $1 \leq k \leq n$: $A_k \subset \overline{\text{int } A_k}$, $A_k \cap T^i A_k = \emptyset$ $(1 \leq k \leq p_k)$ and $X_k \setminus \tilde{P}^{p_k+1}(T) \subset \bigcup_{i=-c_k}^{c_k} T^i A_k$, where

$$X_k := \{x \in X \mid T^i(x) \notin \text{bd } A_j \text{ for every } 1 \leq j \leq k \text{ and } -c_k \leq i \leq c_k\}$$

Suppose furthermore that there exists a $C_n \subset X \setminus [\bigcup_{k=1}^{n} A_k \cup \tilde{P}^{p_{n+1}+1}(T)]$ satisfying $C_n \subset \overline{\text{int } C_n}$ and $X_n \setminus \tilde{P}^{p_{n+1}+1}(T) \subset \bigcup_{i=-c_n}^{c_n} T^i C_n$.

Now, by (15.4), there exists a set $A_{n+1} \subset C_n$ satisfying $A_{n+1} \subset \overline{\text{int } A_{n+1}}$, $A_{n+1} \cap T^i A_{n+1} = \emptyset$ $(1 \leq i \leq p_{n+1})$ and

$$C_n \subset \bigcup_{i=-2p_{n+1}}^{2p_{n+1}} T^i A_{n+1}.$$

Note that it is possible to use the topology on X in (15.4). Clearly it follows that

$$X_{n+1} \setminus \tilde{P}^{p_{n+1}+1}(T) \subset \bigcup_{i=-c_n-2p_{n+1}}^{c_n+2p_{n+1}} T^i A_{n+1}.$$

Define $C_{n+1} := \overline{\text{int } (X \setminus \bigcup_{k=1}^{n+1} A_k)} \cap \left(X \setminus (\bigcup_{k=1}^{n+1} A_k \cup \tilde{P}^{p_{n+2}+1}(T))\right)$.

It is left to show that

$$X_{n+1} \setminus \tilde{P}^{p_{n+2}+1}(T) \subset \bigcup_{i=-c_{n+1}}^{c_{n+1}} T^i C_{n+1}.$$

Let $x \in X_{n+1} \setminus \tilde{P}^{p_{n+2}+1}(T)$. There exists a $-c_n-2p_{n+1} \leq j \leq c_n+2p_{n+1}$ with $T^j(x) \in A_{n+1}$. Therefore $T^{j+i}(x) \notin A_{n+1}$ for every $1 \leq i \leq p_{n+1}$. Since for every $1 \leq k \leq n$ there are at most $\frac{p_{n+1}}{p_k+1} + 1$ indices $1 \leq i \leq p_{n+1}$ satisfying $T^{j+i}(x) \in A_k$ and since $p_{n+1} \sum_{k=1}^{n} (p_k + 1)^{-1} < \frac{1}{2} p_{n+1}$, it follows that there exists an $1 \leq i \leq p_{n+1}$ satisfying

$$T^{j+i}(x) \notin \bigcup_{k=1}^{n+1} A_k.$$

Now it is easy to see that $T^{j+i}(x) \in C_{n+1}$, which implies $X_{n+1} \setminus \tilde{P}^{p_{n+2}+1}(T) \subset \bigcup_{i=-c_n-3p_{n+1}}^{c_n+2p_{n+1}} T^i C_{n+1}$.

The statement of the lemma follows easily from the construction. □

<u>Corollary:</u> Let α be a partition satisfying the conditions of the last lemma. Let $n_j \geq 0$ $(j \in \mathbb{Z})$ be integers satisfying $\bigcap_{j=-m}^{m} T^j \operatorname{int} A_{n_j} \neq \emptyset$ for every $m \in \mathbb{N}$. Then for every $r \in \mathbb{N}$ there exists a k such that $-c_r \leq k \leq c_r$ and $n_k = r$

or - equivalently -

$$T^{-k} \overline{\bigcap_{j=-m}^{m} T^j \operatorname{int} A_{n_j}} \subset \overline{A_r}$$

for every $m \geq c_r$.

<u>Proof:</u> Choose any $x \in X_\alpha \cap \bigcap_{j=-c_r}^{c_r} T^j \operatorname{int} A_{n_j}$. From condition iii) of the lemma it follows that there is $k \in \{-c_r,\dots,c_r\}$ such that $T^{-k}(x) \in \operatorname{int} A_r$, hence $n_k = r$. □

<u>(15.6.) Theorem</u> [44]: Let X_o be the set of aperiodic points under T. Then there exists a topological generator for $\overline{X_o}$. Especially if X_o is dense in X, then there exists a topological generator for X.

<u>Proof:</u> Let $\alpha = (A_n | n \geq 0)$ be a partition constructed in (15.5) assigned to the sequences $(p_n)_{n\in\mathbb{N}}$ and $(c_n)_{n\in\mathbb{N}}$ given there. Let $\epsilon_n > 0$ $(n \in \mathbb{N})$, $\lim_{n\to\infty} \epsilon_n = 0$. Partitioning every $A_n (n \in \mathbb{N})$ into sets A_n^j $(1 \leq j \leq s_n)$ satisfying $A_n^j \subset \overline{\text{int } A_n^j}$ and $\text{diam } T^k A_n^j < \epsilon_n$ for $-c_n \leq k \leq c_n$, a countable partition β is obtained such that $X = \overline{\bigcup_{B\in\beta} \text{int } B}$ and $\text{int } B \neq \emptyset$ $(B\in\beta)$. Now let $x \in \overline{X_o}$ and $B_{i_k} \in \beta$ satisfying $x \in \bigcap_{m\in\mathbb{Z}} \overline{\bigcap_{k=-m}^{m} T^k \text{ int } B_{i_k}}$. From the corollary of (15.5) it follows that for every $n \in \mathbb{N}$ there is an $l\in\{-c_n,\ldots,c_n\}$ such that $T^{-l} \bigcap_{m\in\mathbb{Z}} \overline{\bigcap_{k=-m}^{m} T^k \text{ int } B_{i_k}} \subset A_n$. Clearly it follows already that

$$T^{-l} \bigcap_{m\in\mathbb{Z}} \overline{\bigcap_{k=-m}^{m} T^k \text{ int } B_{i_k}} \subset \overline{A_n^j} \text{ for some } 1 \leq j \leq s_n ;$$

therefore the construction of β yields

$$\text{diam} \bigcap_{m\in\mathbb{Z}} \overline{\bigcap_{k=-m}^{m} T^k \text{ int } B_{i_k}} < \epsilon_n,$$

hence

$$\{x\} = \bigcap_{m\in\mathbb{Z}} \overline{\bigcap_{k=-m}^{m} T^k \text{ int } B_{i_k}} .$$

□

Returning to theorem (15.2) - the study of the relations between topological generators and subshifts - let α denote a topological generator for $X_o \subset X$.

We assume that $X_\alpha \cap X_o$ is dense, so that from (15.2) we get a σ-invariant subset $Y_\alpha \subset \mathbb{N}^{\mathbb{Z}}$ and a map $\pi: Y_\alpha \longrightarrow X$. We denote by Λ_α the closure of Y_α in $S^{\mathbb{Z}}$.

<u>(15.7) Proposition:</u> In the situation just described the following holds:

1) T and $\sigma|\Lambda_\alpha$ are either both topologically transitive or not.

2) T is topologically mixing iff $\sigma|\Lambda_\alpha$ has got the same property.

3) For every measure $m \in \mathfrak{M}_T(X)$ for which α is an m-continuity partition, $\pi^{-1}|X_o \cap X_\alpha$ is a measure theoretical isomorphism iff X_o is measurable with $m(X_o) = 1$. m is positive on open sets iff $\pi^{-1}m$ is.

4) Every measure $m \in \mathfrak{M}_{\sigma|\Lambda_\alpha}(\Lambda_\alpha)$ satisfying $m(Y_\alpha) = 1$ (which also postulates that Y_α is m-measurable) can be transported by π into a measure in $\mathfrak{M}_T(X)$. If in addition m is positive on open sets then πm is positive on open sets also.

The proof of this is straight forward using the neighbourhood system property for X_o of α and the density of $X_o \cap X_\alpha$ and Y_α.

A special situation arises in (15.2) when the map π can be defined on all of Λ_α. Clearly if α is a finite topological generator for X then Y_α is a subshift over a finite alphabet (by construction), and π is defined on all of Λ_α. Another example of this situation is given in [44]. If (X,T) is minimal then it is possible to modify theorem (15.6) such that the A_n's defined in the proof converge to a single point $x_o \in X$ (that is for arbitrary, $x_n \in A_n$ $(n \in \mathbb{N})$ $\lim_{n\to\infty} x_n = x_o$). It is not difficult to see that π can be defined on Λ_α by setting $\pi((\ldots i_{-1}, 0, i_1, \ldots)) = x_o$.

For the existence of finite topological generators α for some $X_o \subset X$ we refer to section 28 (see [51]). In some cases X_o will be residual, so that $X_o \cap X_\alpha$ is dense and (15.2) and (15.7) apply. For expansive homeomorphisms T (see the following section) a finite topological generator for X always exists.

16. Expansive Homeomorphisms

In view of theorem (15.2) the interesting topological dynamical systems are those for which there exists a finite topological generator for X, because in this case the map π of (15.2) is defined on a (compact) finite subshift. We shall see that expansive homeomorphisms admit a finite topological generator for X.

However, this kind of homeomorphism has got stronger properties. We know already that the definitions of measure theoretic and topological entropy are quite similar. Later we shall see (section 28) that for many systems $(\mathfrak{X},m,\Psi)$ there exists a finite (measure theoretic) generator α, hence proposition (11.12) tells us that $h_m(\Psi,\alpha) = h_m(\Psi)$. The question we want to deal with in this section is that of finding all topological dynamical systems (X,T) which admit a "finite generator" $\mathfrak{U}$. (In this context it will be a finite open cover.) Furthermore the generator $\mathfrak{U}$ should have the property $H(T,\mathfrak{U}) = h_{top}(T)$.

As in the last sections (X,T) always stands for a topological dynamical system.

(16.1) Definition [109]: A finite open cover $\mathfrak{U} = (U_i \mid 1 \leq i \leq s)$ of X is said to be a generator for T if for every sequence $(k_i)_{i \in \mathbb{Z}}$ $(1 \leq k_i \leq s)$ the intersection

$$\bigcap_{i \in \mathbb{Z}} T^i \, \overline{U_{k_i}}$$

contains at most one point. We call s the power of the generator $\mathfrak{U}$.

It is easy to see that, if a topological dynamical system (X,T) admits a generator for T of the power s then there exists a topological generator for X as well, which contains exactly s sets. We shall return to this question later.

In order to characterize the topological dynamical systems which admit a generator for T we give the definition of an expansive homeomorphism, which is due to Utz [183].

(16.2) Definition: A homeomorphism T is called expansive if there exists an $\epsilon > 0$ such that for every pair of distinct points $x,y \in X$ there is an $n \in \mathbb{Z}$ with

$$d(T^n(x), T^n(y)) \geq \varepsilon.$$

Every positive number having this last property is called an <u>expansive constant</u> for T.

Before giving examples of expansive homeomorphisms we shall study the connection between the properties of being expansive, of admitting a generator for T and of admitting a topological generator for X.

<u>(16.3) Theorem</u> [109]: Let T be expansive and let $\varepsilon > 0$ be an expansive constant for T. Then every open cover $\mathfrak{U} = (U_i \mid 1 \leq i \leq s)$ of X satisfying

$$\operatorname{diam} U_i < \varepsilon \qquad (1 \leq i \leq s)$$

is a generator for T.

Conversely, if $\mathfrak{U} = (U_i \mid 1 \leq i \leq s)$ is a generator for T then T is expansive and every Lebesgue number of $\mathfrak{U}$ is an expansive constant for T.

<u>Remark:</u> This theorem shows that the property of being expansive does not depend on the metric used.

<u>Proof:</u> The first part of the theorem is trivial.

For the second one let $\mathfrak{U} = (U_i \mid 1 \leq i \leq s)$ be a generator for T and $\varepsilon > 0$ a Lebesgue number of $\mathfrak{U}$. Assume there are $x \neq y \in X$ such that $d(T^n(x), T^n(y)) < \varepsilon$ for every $n \in \mathbb{Z}$. Then for every $n \in \mathbb{Z}$ there is $U_{i_n} \in \mathfrak{U}$ such that $B_\varepsilon(T^{-n}(x)) \subset U_{i_n}$. Clearly this implies

$$x, y \in \bigcap_{n \in \mathbb{Z}} T^n U_{i_n},$$

a contradiction. □

(16.4) Proposition: Let T be expansive and let $\epsilon > 0$ be an expansive constant for T. Then every partition $\alpha = (A_1, \dots, A_s)$ of X satisfying

$$X = \overline{\bigcup_{i=1}^{s} \operatorname{int} A_i} \quad \text{and} \quad \operatorname{diam} A_i < \epsilon \qquad (1 \le i \le s)$$

is a topological generator for X.

Proof: There exists a $\delta > 0$ such that

$\max_i \operatorname{diam} A_i < \epsilon - \delta$.

Define the open cover $\mathfrak{U} := (B_\delta(\operatorname{int} A_i) \mid 1 \le i \le s)$ of X. By theorem (16.3) $\mathfrak{U}$ is a generator for T, hence α is a topological generator for X since

$$\bigcap_{n\in\mathbb{Z}} \overline{\bigcap_{k=-n}^{n} T^{-k} \operatorname{int} A_{i_k}} \subset \bigcap_{m\in\mathbb{Z}} T^{-m} \overline{B_\delta(\operatorname{int} A_{i_k})}$$

for every choice of $(i_k)_{k\in\mathbb{Z}} \in S^{\mathbb{Z}}$.

Remark: The converse of the last proposition does not hold, so that the existence of a finite topological generator for X is a weaker condition than expansiveness. This can easily be seen by the following example. Let X be the unit circle and $T(x) := ax$ where $a \in X$ is no root of unity. Since T is isometric it is not expansive. Every nontrivial partition α of X with $\bigcup \operatorname{int} A$ being dense in X is a tpological generator for X, however.

In view of theorem (15.2) and (16.4) we can prove:

(16.5) Proposition [155] : Every topological dynamical system (X,T), where T is expansive, is a factor of some subshift over a finite symbol set. Moreover, if X is zero-dimensional, (X,T) is isomorphic to some subshift over a finite symbol set. The minimal number of symbols used is equal to the minimal power of a generator for T.

Proof: This follows directly from (15.2) and (16.4) if we can construct a finite topological generator for X. But this follows from (16.6). □

(16.6) Lemma: Let X be a compact metric space and $\mathfrak{U} = (U_i \mid 1 \le i \le s)$ an open cover of X. Then there exists a Borel partition $\alpha = (A_i \mid 1 \le i \le s)$ such that $A_i \subset U_i$ $(1 \le i \le s)$.

Also it is possible to find a partition $\alpha = (A_i \mid 1 \le i \le s)$ such that for every $1 \le i \le s$

$$A_i \subset U_i \text{ and } A_i \subset \overline{\operatorname{int} A_i}.$$

Moreover, if $m \in \mathfrak{M}_T(X)$ is a given measure we can choose the Borel partition α so that in addition α is an m-continuity partition.

Proof: First note that there is an open cover $\mathfrak{U}' = (U_i' \mid 1 \le i \le s)$ and $\delta > 0$ such that $\overline{B_\delta(U_i')} \subset U_i$ for every $1 \le i \le s$. If $m \in \mathfrak{M}_T(X)$ is given there is $0 < \delta_o < \delta$ such that

$$m(\overline{B_{\delta_o}(U_i')}) = m(B_{\delta_o}(U_i')) \qquad (1 \le i \le s).$$

Let $\mathfrak{U} = (B_{\delta_o}(U_i') \mid 1 \le i \le s)$ and define inductively

$$A_1 = \overline{B_{\delta_o}(U_1')}$$

$$A_i = \overline{B_{\delta_o}(U_i')} \setminus \bigcup_{k=1}^{i-1} A_k \qquad (2 \le i \le s).$$

Clearly $\overline{A_i} \subset U_i$ and A_i is Borel measurable and an m-continuity-set. To satisfy $A_i \subset \overline{\operatorname{int} A_i}$ we only have to note that for $1 \le i \le s$ $\bigcup_{k=1}^{i} A_k$ is closed. □

Remark: We shall say that a partition α is finer than an open cover $\mathfrak{U}$ if for every $A \in \alpha$ there exists an open set $U \in \mathfrak{U}$ such that $\overline{A} \subset U$.

Before continuing with the theory we give some examples and counterexamples:

1) Let S be a finite set and let σ be the shift on $S^{\mathbb{Z}}$. Every subshift $\Lambda \subset S^{\mathbb{Z}}$ is expansive. This is easy to see looking

at the metric

$$d(x,y) := \sum_{n\in\mathbb{Z}} 2^{-|n|}\,|x_n - y_n|\,;\left(\left.\begin{array}{l} x = (x_n)_{n\in\mathbb{Z}} \\ y = (y_n)_{n\in\mathbb{Z}} \end{array}\right\} \in S^{\mathbb{Z}}\right).$$

Clearly with this metric $\frac{1}{2}$ is an expansive constant for $\sigma|\Lambda$. A generator $\mathfrak{U}$ for $\sigma|\Lambda$ is given by

$$\mathfrak{U} = ({}_o[i] \cap \Lambda \mid i \in S).$$

This open cover $\mathfrak{U}$ is called the natural generator. (Note that it is a partition at the same time.)

2) Every axiom-A-diffeomorphism (see [180] for the definition) and every hyperbolic torus automorphism is expansive (see sections 23, 24 and the literature cited in this context).

3) Let $X = [0,1]$. There can not exist any expansive homeomorphism T on X. (This is not hard to see.) Also there is no expansive homeomorphism on S^1 and a closed 2-cell (see [99], [154]). However, there exist expansive homeomorphisms on every k-dimensional torus and on every S^k for $k \geq 2$.

(16.7) Proposition: If T is expansive, then the function $h_{.}(T) : \mathfrak{M}_T(X) \longrightarrow \mathbb{R}$ is upper-semicontinuous with respect to the weak topology on $\mathfrak{M}_T(X)$.

Proof: Let m_n, $m \in \mathfrak{M}_T(X)$ and m be the weak limit of $(m_n)_{n\in\mathbb{N}}$. Using lemma (16.6) we know that there exists a Borel partition $\alpha = (A_1,\ldots,A_s)$ which is a generator for every m and m_n, such that every A_i is an m-continuity set. Therefore (see (2.7))

$$\lim_{n\to\infty} m_n(A) = m(A)$$

for every $A \in (\alpha)_o^{k-1}$ and every $k \in \mathbb{N}$. Let $\varepsilon > 0$ and choose $k \in \mathbb{N}$ such that

$$h_m(T) = h_m(T,\alpha) \geq \frac{1}{k} H_m((\alpha)_o^{k-1}) - \varepsilon.$$

Since $(\alpha)_o^{k-1}$ is finite there is $N \in \mathbb{N}$ such that for every $n \geq N$

$$\frac{1}{k} \mid H_{m_n}((\alpha)_o^{k-1}) - H_m((\alpha)_o^{k-1}) \mid < \varepsilon.$$

Therefore it follows that for $n \geq N$

$$h_m(T) \geq \frac{1}{k} H_m((\alpha)_o^{k-1}) - \varepsilon \geq \frac{1}{k} H_{m_n}((\alpha)_o^{k-1}) - 2\varepsilon$$

$$\geq h_{m_n}(T,\alpha) - 2\varepsilon = h_{m_n}(T) - 2\varepsilon. \qquad \square$$

<u>(16.8) Proposition:</u>

a) Expansiveness is an invariant for topological conjugacy.

b) If T is an expansive homeomorphism then so is T^n for every $n \in \mathbb{Z}$ $(n \neq 0)$.

c) If T is an expansive homeomorphism on X and $X_o \subset X$ T-invariant and closed, then $T|X_o$ is expansive.

d) If T_i are expansive homeomorphisms on X_i $(1 \leq i \leq s)$, then their product $T_1 \times \ldots \times T_s$ is expansive on $X_1 \times \ldots \times X_s$.

The proof of this proposition is very easy and left to the reader.

We note that factors of expansive homeomorphisms need not be expansive and that infinite products of nontrivial expansive systems never are expansive.

In the remaining part of this section we consider the topological entropy of expansive homeomorphisms.

<u>(16.9) Proposition</u> [109]: If $\mathfrak{U}$ is a generator for T then $h_{top}(T) = H(\mathfrak{U},T)$. In particular, if T is expansive, then $h_{top}(T)$ is bounded by the minimal power of a generator for T.

<u>Proof:</u> By proposition (14.8) we have to show that for every open cover $\mathfrak{B} = (V_i \mid 1 \leq i \leq r)$ there is $n_o \in \mathbb{N}$ such that

$$\mathfrak{B} \leq \bigvee_{k=-n_o}^{n_o} T^k \mathfrak{U}$$

where $\mathfrak{U}$ is a generator for T.

Let $\delta > 0$ be a Lebesgue number for the arbitrarily chosen open cover $\mathfrak{B}$. Since for every bisequence $U_{i_n} \in \mathfrak{U}$ $(n \in \mathbb{Z})$

$\operatorname{card} \bigcap_{n\in\mathbb{Z}} T^n \overline{U_{i_n}} \leq 1$ we can choose $m = m((i_n)_{n\in\mathbb{Z}})$ minimal such that

$$\operatorname{diam} \bigcap_{n=-m}^{m} T^n U_{i_n} < \delta.$$

Since the function m is continuous on the compact space $\{1,\ldots,\operatorname{card} \mathfrak{U}\}^{\mathbb{Z}}$ it attains its maximum (call it m again), which implies

$$\operatorname{diam} \bigcap_{n=-m}^{m} T^n \overline{U_{i_n}} < \delta$$

for every choice of $i_n \in \{1,\ldots,\operatorname{card} \mathfrak{U}\}$. Therefore

$$\mathfrak{B} \leq (\mathfrak{U})_{-m}^{m}. \qquad \square$$

A similar proof shows:

<u>Corollary 1:</u> If T is expansive with expansive constant $\epsilon > 0$, then for every $0 < \delta < \epsilon$ there is an $n(\delta) \in \mathbb{N}$ such that $d(x,y) \geq \delta$ implies the existence of an $n \in \{-n(\delta),\ldots,n(\delta)\}$ satisfying

$$d(T^n(x), T^n(y)) \geq \epsilon .$$

<u>Corollary 2:</u> For every expansive homeomorphism T with expansive constant $\epsilon > 0$ let $\delta < \epsilon$. Then

$$h_{top}(T) = \lim_{n\to\infty} \frac{1}{n} \log r_n(\delta,T)$$

and

$$h_{top}(T) = \lim_{n\to\infty} \frac{1}{n} \log s_n(\delta,T).$$

<u>Proof:</u> For $\delta < \epsilon$ choose $x_1,\ldots,x_s \in X$ such that

$$X = \bigcup_{i=1}^{s} B_{\epsilon-\delta}(x_i).$$

Now $\mathfrak{U} = (B_\epsilon(x_i) \mid 1 \leq i \leq s)$ is a finite open cover with

Lebesgue number δ. From (14.16), part (b) of the proof, one obtains for $n \in \mathbb{N}$

$$N((\mathfrak{U})_0^{n-1}) \leq r_n(\delta,T),$$

hence this implies together with (16.9)

$$h_{top}(T) \leq \liminf \frac{1}{n} \log r_n(\delta,T).$$

(14.15), first line of the proof, shows on the other hand

$$\limsup \frac{1}{n} \log r_n(\delta,T) \leq h_{top}(T).$$

The other equality is easily seen to follow from the first line in the proof of (14.15):

$$\lim_{n\to\infty} \frac{1}{n} \log r_n(\delta) \leq \liminf_{n\to\infty} \frac{1}{n} \log s_n(\delta)$$

$$\leq \limsup_{n\to\infty} \frac{1}{n} \log s_n(\delta) \leq \limsup \frac{1}{n} \log r_n(\frac{1}{2}\delta). \quad \square$$

Recall that $Per_n(T)$ denotes the cardinality of the set of periodic points with period n.

See [39] also for the next proposition.

<u>(16.10) Proposition:</u> (cf. [21], [18]) If T is expansive then

$$h_{top}(T) \geq \limsup_{n\to\infty} \frac{1}{n} \log Per_n(T)$$

<u>Proof:</u> $P^n(T)$ - the set of fixed points of T^n - is an (n,δ)-separating set for every $n \in \mathbb{N}$, where 2δ is an expansive constant for T. Hence $Per_n(T) \leq r_n(\delta,T)$ and from the last corollary we conclude that

$$h_{top}(T) = r(\delta,T) \geq \limsup_{n\to\infty} \frac{1}{n} \log Per_n(T). \quad \square$$

Every subshift $\Lambda \subset S^{\mathbb{Z}}$ (where S is a finite set) is expansive and the open cover

$$\mathfrak{U} = ({}_o[i] \cap \Lambda \neq \emptyset \mid i \in S)$$

is the natural generator. If $\theta_n(\Lambda)$ denotes the number of distinct n-blocks occuring in Λ we get

<u>(16.11) Proposition</u> [147]: $h_{top}(\sigma|\Lambda) = \lim_{n\to\infty} \frac{1}{n} \log \theta_n(\Lambda)$.

<u>Proof:</u>

$$\bigvee_{i=o}^{n-1} (\sigma|\Lambda)^{-i}\, \mathfrak{U} = ({}_o[k_o,\dots,k_{n-1}] \cap \Lambda \neq \emptyset \mid k_o,\dots,k_{n-1} \in S).$$

□

<u>Corollary:</u> $h_{top}(\sigma) = \log \operatorname{card} S$.

<u>Corollary</u> [18]: If there exists an $m \in \mathbb{N}$ such that for every $n \in \mathbb{N}$ every cylinder of length n occuring in Λ contains a periodic point of period $n + m$, then

$$h_{top}(\sigma|\Lambda) = \lim_{n\to\infty} \frac{1}{n} \log \operatorname{Per}_n(\sigma|\Lambda)$$

<u>Proof:</u> Since $\theta_n(\Lambda) \leq \operatorname{Per}_{n+m}(\sigma|\Lambda)$,

$$\limsup_{n\to\infty} \frac{1}{n} \log \operatorname{Per}_n(T) \leq h_{top}(\sigma|\Lambda) = \lim_{n\to\infty} \frac{1}{n} \log \theta_n(\Lambda)$$

$$\leq \liminf_{n\to\infty} \frac{1}{n} \log \operatorname{Per}_{n+m}(\sigma|\Lambda)$$

$$= \liminf_{n\to\infty} \frac{1}{n} \log \operatorname{Per}_n(\sigma|\Lambda) .$$

□

For a given finite set S, a subshift $\Lambda \subset S^{\mathbb{Z}}$ "in general" has zero topological entropy:

For this denote by $\mathcal{L}(S^{\mathbb{Z}})$ the set of all subshifts of $S^{\mathbb{Z}}$ endowed with the usual <u>Hausdorff metric</u> $\bar{d}$ for closed subsets:

$$\bar{d}(\Lambda,\Lambda') = \max\{\max_{x\in\Lambda} \min_{y\in\Lambda'} d(x,y);\ \max_{x\in\Lambda'} \min_{y\in\Lambda} d(x,y)\}$$

where d denotes the metric on $S^{\mathbb{Z}}$ and $\Lambda, \Lambda' \in \mathcal{L}(S^{\mathbb{Z}})$. $\mathcal{L}(S^{\mathbb{Z}})$ is a compact metric space.

Now we can make our statement precise:

<u>(16.12) Proposition</u> [169], [191]: The set

$$\{\Lambda \in \mathcal{L}(S^{\mathbb{Z}}) \mid h_{top}(\sigma|\Lambda) = 0\}$$

is residual in $\mathcal{L}(S^{\mathbb{Z}})$.

<u>Proof:</u> It suffices to show that

(i) if $\Lambda_n, \Lambda \in \mathcal{L}(S^{\mathbb{Z}})$, $\lim_{n\to\infty} \bar{d}(\Lambda_n, \Lambda) = 0$

then

$$h_{top}(\sigma|\Lambda) \geq \lim_{n\to\infty} h_{top}(\sigma|\Lambda_n).$$

(ii) $\{\Lambda \mid h_{top}(\sigma|\Lambda) = 0\}$ is dense in $\mathcal{L}(S^{\mathbb{Z}})$.

Firstly we prove i).

Let $\Lambda, \Lambda_n \in \mathcal{L}(S^{\mathbb{Z}})$ $(n \in \mathbb{N})$ satisfying $\lim \bar{d}(\Lambda, \Lambda_n) = 0$.

Denote by $\mathfrak{U}$ the natural open cover of $S^{\mathbb{Z}}$; clearly $\mathfrak{U} \cap \Lambda' = (U \cap \Lambda' \mid U \in \mathfrak{U})$ is the natural open cover for every $\Lambda' \in \mathcal{L}(S^{\mathbb{Z}})$. Let $\varepsilon > 0$ and pick $m \in \mathbb{N}$ satisfying

$$h_{top}(\sigma|\Lambda) \geq \frac{1}{m} \log N\Big(\bigvee_{k=o}^{m-1} (\sigma|\Lambda)^{-k}(\mathfrak{U} \cap \Lambda)\Big) - \varepsilon.$$

Define $W = \bigcup \{U \in (\mathfrak{U})_o^{m-1} \mid U \cap \Lambda \neq \emptyset\}$. Then W is an open neighbourhood of Λ with respect to the metric d and by the definition of the Hausdorff metric there is $N \in \mathbb{N}$ such that for every $n \geq N$ $\Lambda_n \subset W$. This clearly implies

$$N\Big(\bigvee_{k=o}^{m-1} (\sigma|\Lambda_n)^{-k} (\mathfrak{U} \cap \Lambda_n)\Big) \leq N\Big(\bigvee_{k=o}^{m-1} (\sigma|\Lambda)^{-k}(\mathfrak{U} \cap \Lambda)\Big),$$

hence for $n \geq N$

$$h_{top}(\sigma|\Lambda) \geq \frac{1}{m} \log N(\bigvee_{k=o}^{m-1} (\sigma|\Lambda_n)^{-k}(\mathfrak{U} \cap \Lambda_n)) - \epsilon$$

$$\geq H(\mathfrak{U} \cap \Lambda_n, \sigma|\Lambda_n) - \epsilon$$

$$= h_{top}(\sigma|\Lambda_n) - \epsilon .$$

Now we prove ii).

Let $\Lambda \subset \mathcal{L}(S^{\mathbb{Z}})$ and $\epsilon > 0$. There exists an $N \in \mathbb{N}$ such that $d(x,y) < \epsilon$ if $x,y \in S^{\mathbb{Z}}$ and $x_k = y_k$ for every $-N < k < N$.

Let $\beta_1,\ldots,\beta_r$ denote the different $2N + 1$ blocks occuring in Λ. Choose any $x^i \in {}_o[\beta_i]$ $(1 \leq i \leq r)$. Then for every $1 \leq i \leq r$ there are blocks $\gamma_k(i) \in \{\beta_j | 1 \leq j \leq r\}$ $(k = 1,2)$ such that $\gamma_1(i)$ occurs at two places $n_i(1) > n_i(2) \geq 0$ in x^i and $\gamma_2(i)$ occurs at two places $m_i(1) < m_i(2) \leq 0$ in x^i. Now define $z^i \in S^{\mathbb{Z}}$ by

$$z^i_k = \begin{cases} x^i_k & m_i(1) \leq k \leq n_i(1) \\ x^i_j & k = l(m_i(1) - m_i(2)) + j \leq m_i(2);\ l \geq 0,\ j \leq 0 \\ x^i_j & k = l(n_i(1) - n_i(2)) + j \geq n_i(2);\ l \geq 0,\ j \geq 0. \end{cases}$$

Let Λ' be the orbit closure of all z^i $(1 \leq i \leq r)$. Since all blocks of length $2N+1$ occuring in Λ occur in Λ' and vice versa it follows that $d(\Lambda,\Lambda') < \epsilon$. Moreover it is easy to see that for $t \in \mathbb{N}$ $\theta_t(\Lambda') \leq r(t + n - m)$ where

$n = \max_{1 \leq i \leq r} n_i(1)$ and $m = \min_{1 \leq i \leq r} m_i(1)$. Therefore $h_{top}(\sigma|\Lambda') = 0$ by (16.11). □

<u>Corollary:</u> If $\Lambda, \Lambda_n \in \mathcal{L}(S^{\mathbb{Z}})$, $\Lambda_{n+1} \subset \Lambda_n$ $(\forall\ n \in \mathbb{N})$ and $\Lambda = \bigcap_{n=1}^{\infty} \Lambda_n$ then

$$h_{top}(\sigma|\Lambda) = \lim_{n\to\infty} h_{top}(\sigma|\Lambda_n)$$

<u>Proof:</u> If $\Lambda = \bigcap_{n=1}^{\infty} \Lambda_n$ then $\lim_{n\to\infty} \bar{d}(\Lambda,\Lambda_n) = 0$.
Therefore by part i) of the last proof

$$h_{top}(\sigma|\Lambda) \geq \limsup_{n\to\infty} h_{top}(\sigma|\Lambda_n).$$

Conversely since $\Lambda_n \supset \Lambda$ $h_{top}(\sigma|\Lambda_n) \geq h_{top}(\sigma|\Lambda)$. □

<u>(16.13) Proposition</u> [66]: The following conditions are equivalent:

1) (X,T) is topologically conjugate to $(S^{\mathbb{Z}},\sigma)$ where $S = \{1,\dots,p\}$
2) T is expansive and has a generator $\mathfrak{U}$ with power p and $h_{top}(T) = \log p$.

<u>Proof:</u> "1) ⟹ 2)" is trivial
"2) ⟹ 1)" Let Λ be a subshift of $(S^{\mathbb{Z}},\sigma)$ with $h_{top}(\sigma|\Lambda) = \log p$, where $S = \{1,\dots,p\}$. At first we show that $\Lambda = S^{\mathbb{Z}}$.

Assume $\Lambda \neq S^{\mathbb{Z}}$. There exist **an $n\in\mathbb{N}$** and $i_k \in S$ $(0 \leq k \leq n-1)$ such that the block $(i_k)_{-n\leq k\leq n}$ does not occur in Λ.

Let $\mathfrak{B}=({}_o[j_o,\dots,j_{n-1}] \cap \Lambda \neq \emptyset \mid j_l \in S\ (0 \leq l \leq n-1)$.

Clearly $\mathfrak{B}$ is a generator for $(\sigma|\Lambda)^n$ and $\operatorname{card}\mathfrak{B} \leq (\operatorname{card} S)^n - 1$. Therefore $\theta_n(\Lambda) \leq p^n - 1$ gives a contradiction.

Secondly let $f : \Lambda \longrightarrow X$ denote the projection defined by (15.2) using a partition $\alpha = (A_1,\dots,A_p)$ with $\overline{A_i} \subset U_i$ $(1 \leq i \leq p)$ (see (16.6)), where $\Lambda \subset \{1,\dots,p\}^{\mathbb{Z}}$ is a subshift and where $\mathfrak{U} = (U_1,\dots,U_p)$ denotes the given generator for T of the power p.

$h_{top}(T) = \log p = H(\mathfrak{U},T)$ implies

$$\log p^{2n+1} = H((\mathfrak{U})_{-n}^{n},\ T^{2n+1}) \quad \text{for every } n \in \mathbb{N}.$$

Hence for every $n \in \mathbb{N}$ and every choice of $i_k \in \{1,\dots,p\}$

for $-n \leq k \leq n$:

$$\bigcap_{k=-n}^{n} T^k U_{i_k} \neq \emptyset.$$

Therefore it is possible to define a map

$$g : S^{\mathbb{Z}} \longrightarrow X$$

by

$$\{g((\omega_k)_{k\in\mathbb{Z}})\} = \bigcap_{k\in\mathbb{Z}} T^k \overline{U_{\omega_k}}.$$

Since for every $1 \leq i \leq p$ $\overline{A_i} \subset U_i$ it follows that

$$\{f((\omega_k)_{k\in\mathbb{Z}})\} = \bigcap_{k\in\mathbb{Z}} T^k \overline{A_{\omega_k}} \subset \bigcap_{k\in\mathbb{Z}} T^k U_{\omega_k}.$$

Clearly this and the definition of g implies $f(\omega) = g(\omega)$ for all $\omega \in \Lambda$ and

$$g((\omega_k)_{k\in\mathbb{Z}}) = \bigcap_{k\in\mathbb{Z}} T^k U_{\omega_k} \quad ((\omega_k)_{k\in\mathbb{Z}} \in \Lambda).$$

The first part of the proof and (15.2) show that $\Lambda = S^{\mathbb{Z}}$, hence they imply

$$f(\{\omega \in S^{\mathbb{Z}} \mid \omega_o = i\}) = \overline{A_i}$$

and

$$g(\{\omega \in S^{\mathbb{Z}} \mid \omega_o = i\}) = U_i \quad \text{for every } 1 \leq i \leq p.$$

Since $f = g$, every A_i is open and closed. Therefore with the notation of (15.2) $X_o = X_\alpha = X$ and so (15.2) implies that f is a homeomorphism. □

The literature concerning expansive hemeomorphisms is enormeous. We tried to be as complete as possible in the bibliography, so the reader is invited to study further this interesting theory. Especially in our context the further study of periodic points of expansive maps (see [37] for example) is tightly connected with the aim of our investigations (or more general recurrence properties under expansive maps). Also we cited some papers concerning expansive flows and expansive diffeomorphisms, for which many of the results presented here hold as well.

17 Subshifts of Finite Type

In section 7 shifts and subshifts over a finite alphabet were introduced. Now we continue, studying the subshifts of finite type first defined by W. Parry in 1964 ([146]). He called these subshifts intrinsic Markov chains, which stands for the fact that they are characterized as dynamical systems by a topological Markov property. This property allows us to define a certain Markov measure on this subshift in a natural way, which is 'unique' for the most important class of such subshifts (namely the topologically transitive ones). Subshifts of finite type turned out to be the basic tool for solving many problems: In section 18 we need them to relate topological and measure theoretic entropies, and we will see later that there exist diffeomorphisms for which the non-wandering set is essentially a subshift of finite type (for this the Markov partitions are used (see section 25)). Moreover, for constructing examples of subshifts, for constructing finite generators for measure-preserving transformations and for imbedding a measure theoretic dynamical system $(\mathfrak{X}, m, \psi)$ into some strictly ergodic subshift the subshifts of finite type play the fundamental role (see sections 27-31).

In this section the finite set $S = \{1, \ldots, s\}$ is fixed, $\sigma : S^{\mathbb{Z}} \to S^{\mathbb{Z}}$ denotes the shift transformation and $\mathcal{L}(S^{\mathbb{Z}})$ the set of all subshifts of $S^{\mathbb{Z}}$. (We remark that the theory can be developped for countable S as well.)

<u>(17.1) Definition</u>: Let B be a set of blocks occuring in $S^{\mathbb{Z}}$. The '<u>subshift defined by excluding B</u>' is the set

$$\Lambda_B := \{ x \in S^{\mathbb{Z}} \mid \text{no block } \beta \in B \text{ occurs in } x \}.$$

(Clearly Λ_B is shift-invariant and closed for its complement is open.)

<u>(17.2) Proposition:</u> For every $\Lambda \in \mathcal{L}(S^{\mathbb{Z}})$ there is a system B of blocks such that

$$\Lambda = \Lambda_B.$$

We shall call such a B an <u>excluded block system</u> for Λ (of course there may be many such B's).

<u>Proof</u>: Define B to be the set of all blocks occuring in $S^{\mathbb{Z}}$ but not in Λ. Then clearly $\Lambda \subset \Lambda_B$. If equality does not hold, $\Lambda_B \setminus \Lambda$ is open and we can find a cylinder A such that $A \cap \Lambda_B \neq \emptyset$ is contained in $\Lambda_B \setminus \Lambda$. This cylinder corresponds to a block which occurs in Λ_B but not in Λ. □

<u>Corollary</u>: For every $\Lambda \in \mathcal{L}(S^{\mathbb{Z}})$ and for every excluded block system B for Λ we have

$$\Lambda = \bigcap_{\beta \in B} \Lambda_{\{\beta\}}.$$

Moreover for every exhausting sequence $(B_n)_{n\in\mathbb{N}}$ of B (i.e., $B_n \subset B_{n+1}$ $(n \in \mathbb{N})$ and $B = \bigcup_{n\in\mathbb{N}} B_n$) it follows that

$$\Lambda = \bigcap_{n\in\mathbb{N}} \Lambda_{B_n}.$$

<u>Proof</u>: Obviously in both equalities the right hand side is a closed, invariant set and B is a set of excluded blocks. □

<u>(17.3) Proposition</u>: For every system B of blocks occuring in $S^{\mathbb{Z}}$ there exists a sequence $(B_n)_{n\in\mathbb{N}}$ of finite block systems satisfying $B_n \subset B_{n+1} \subset B$ $(n \in \mathbb{N})$ and $\bigcup_{n=1}^{\infty} B_n = B$. Therefore

$$\Lambda_B = \bigcap_{n\in\mathbb{N}} \Lambda_{B_n}$$

<u>Proof</u>: We only have to remark that the set of all blocks is countable. □

(17.3) is one of the important observations we have to make, since it turns out that the Λ_{B_n} $(n \in \mathbb{N})$ are easy to handle. They are called subshifts of finite type.

<u>(17.4) Definition</u>: A subshift $\Lambda \in \mathcal{L}(S^{\mathbb{Z}})$ is called a <u>subshift of finite type</u> (shortly f.t. subshift) if there exists a finite excluded block system for Λ. We say that $N \in \mathbb{N}$ is an <u>order</u> of the f.t. subshift $\Lambda \in \mathcal{L}(S^{\mathbb{Z}})$ if there exists an excluded block system B for Λ, which only con-

tains blocks of length $\leq$ N.

Remark: It follows immediately from the definition of an f.t. subshift Λ of order N that Λ is an f.t. subshift of order N + k for every $k \geq 0$ as well.

It is easy to see that there are only finitely many f.t. subshifts of a given order N. It is not difficult to see that for subshifts Λ and Λ' where Λ is f.t. of order N and $\Lambda' \supsetneqq \Lambda$ $\bar{d}(\Lambda,\Lambda') \geq 2^{-N}$ holds.

Studying the properties of f.t. subshifts we may restrict our attention to the f.t. subshifts of order 2:

(17.5) Proposition [146]: Any subshift of finite type is topologically conjugate to an f.t. subshift of order 2.

Proof: Let $\Lambda \in \mathcal{L}(S^{\mathbb{Z}})$ be an f.t. subshift, and let B be an excluded block system for Λ possessing only blocks of the fixed length $N \geq 2$.

Define $S' = \{\beta | \beta$ is an N-block occuring in $\Lambda\}$ and denote by B' the set of all blocks (β, β') (where $\beta = (b_0,\dots,b_{N-1})$ and $\beta' = (b'_0,\dots,b'_{N-1})$ belong to S') such that there exists $1 \leq j \leq N - 1$ with $b_j \neq b'_{j-1}$.

Let $\Lambda' = \Lambda_{B'} \in \mathcal{L}(S'^{\mathbb{Z}})$ and σ' be the shift on $S'^{\mathbb{Z}}$. Define the map

$$\Pi : \Lambda \to S'^{\mathbb{Z}}$$

by $\Pi((x_n)_{n\in\mathbb{Z}}) = (\beta_k)_{k\in\mathbb{Z}}$

where $\beta_k = (x_k, x_{k+1},\dots,x_{k+N-1})$ $(k \in \mathbb{Z})$.

Clearly for every $x \in \Lambda$ every 2-block (β_k, β_{k+1}) occuring in $\Pi(x)$ does not belong to B', hence

$$\Pi : \Lambda \to \Lambda'.$$

Similar to the remark before (7.13) the proposition follows. □

There are different ways of saying what an f.t. subshift is:

(17.6) Proposition: Let $\Lambda \in \mathcal{L}(S^{\mathbb{Z}})$ be a subshift. The following conditions are equivalent:

1) Λ is an f.t. subshift of order N.
2) There exists a block system B' of blocks of length N such that

$\Lambda = \{(x_n)_{n\in\mathbb{Z}} \in S^{\mathbb{Z}} \mid \forall n \in \mathbb{Z} : (x_n, x_{n+1}, \ldots, x_{n+N-1}) \in B'\}$.

A set B' of blocks defining a subshift $\Lambda \in \mathcal{L}(S^{\mathbb{Z}})$ in this way will be called a <u>defining system of blocks</u> for Λ.

3) For every pair $x, y \in \Lambda$ and every $n \in \mathbb{Z}$ such that $x_k = y_k$ $(k = n, n+1, \ldots, n+N-2)$ the point $z = (z_k)_{k \in \mathbb{Z}}$ defined by

$$z_k = \begin{cases} x_k & k \leq n + N - 2 \\ y_k & k \geq n \end{cases} \qquad (k \in \mathbb{Z})$$

belongs to Λ.

4) For every $n \in \mathbb{N}$, $n \geq N - 1$ and every choice of $a_k \in S$ $(k = 0, 1, \ldots, n)$ such that $U := {}_0[a_0, \ldots, a_{n-1}] \cap \Lambda \neq \emptyset$ and $V = {}_{n-N+1}[a_{n-N+1}, \ldots, a_n] \cap \Lambda \neq \emptyset$ it follows that $U \cap V \neq \emptyset$

<u>Proof</u>: "1) $\Rightarrow$ 2)" Let B be an excluded system of N-blocks for Λ and define B' to be the set of all N-blocks which do not belong to B.

"2) $\Rightarrow$ 3)" and "3) $\Rightarrow$ 4)" are obvious.

"4) $\Rightarrow$ 1)" Let B be the set of all N-blocks not occuring in Λ. By proposition (17.2) $\Lambda \subset \Lambda_B$.

Assume $\Lambda \neq \Lambda_B$. Then there is a block $\mathcal{B} = (a_0, \ldots, a_n)$ where $n > N$, occuring in Λ_B but not in Λ. We may assume that $(a_0, \ldots, a_{n-1})$ occurs in Λ. Define

$$U = {}_0[a_0, \ldots, a_{n-1}] \cap \Lambda \quad \text{and}$$

$$V = {}_{n-N+1}[a_{n-N+1}, \ldots, a_n] \cap \Lambda.$$

Clearly $U \neq 0$. Since $(a_{n-N+1}, \ldots, a_n)$ occurs in Λ_B it occurs in Λ as well. So we have $V \neq 0$. From 4) we get

$$\emptyset \neq U \cap V = {}_0[a_0, \ldots, a_n] \cap \Lambda,$$

a contradiction. □

<u>(17.7) Definition</u>: Let Λ be an f.t. subshift of order 2. The $s \times s$ matrix $L = (l_{ij})_{1 \leq i, j \leq s}$ defined by

$$l_{ij} = \begin{cases} 1 & \text{if } (i,j) < \Lambda \\ 0 & \text{if } (i,j) \nless \Lambda \end{cases}$$

is called the <u>transition matrix</u> of Λ.

In view of (17.6) we remark that any square matrix with entries 0 and 1 defines a (possibly empty) f.t. subshift of order 2 in an obvious way, but it is not necessarily the transition matrix. However, postulating that $\sum_{j=1}^{s} l_{ij} \geq 1$ and $\sum_{i=1}^{s} l_{ij} \geq 1$ it is easy to see that L defines an f.t. subshift Λ in which every $\beta \in B' = \{(i,j) | l_{ij} = 1;\ i,j \in S\}$ occurs.

For $n \in \mathbb{N}$ recall that $L^n = (l_{ij}^{(n)})_{i,j \in S}$ denotes the n-th power of L.

<u>(17.8) Lemma</u>: For every $n \in \mathbb{N}$ and $i,j \in S$:

$$l_{ij}^{(n)} > 0 \Leftrightarrow \{x \in \Lambda \mid x_0 = i,\ x_n = j\} \neq \emptyset.$$

<u>Proof</u>: The proof is straight forward using proposition (17.6). □

<u>(17.9) Proposition</u>: Let Λ be an f.t. subshift of order 2 with transition matrix L. Then the following conditions are equivalent:

1) $(\Lambda, \sigma|\Lambda)$ is topologically transitive
2) L is irreducible (see (8.6))
3) For every $i,j \in S$ occurring in Λ there exists an $n \in N$ such that $\{x \in \Lambda | x_0 = i,\ x_n = j\} \neq \emptyset$

<u>Proof</u>: We only have to show "3) ⇒ 1)".

It suffices to show that for every pair

$$U = {}_0[a_0, \ldots, a_n] \cap \Lambda \neq \emptyset$$

and

$$V = {}_0[b_0, \ldots, b_m] \cap \Lambda \neq \emptyset$$

there exists an $M \in \mathbb{N}$ such that

$$U \cap \sigma^M V \neq \emptyset.$$

From 3) we get $M \in \mathbb{N}$ and $x \in \Lambda$ such that $x_n = a_n$ and $x_{n+M} = b_0$. Now proposition (17.6) (3) tells that there is $y \in U \cap \sigma^M V$. □

<u>(17.10) Proposition</u>: Let Λ be an f.t. subshift of order 2 with transition matrix L. Then the following is equivalent:

1) $(\Lambda, \sigma|\Lambda)$ is topologically mixing

2) L is aperiodic (see (8.14))

3) $(\Lambda, \sigma|\Lambda)$ is topologically transitive and there exist $i_0, j_0 \in S$ and $M \in \mathbb{N}$ such that
$$l^{(M)}_{i_0 j_0} > 0 \text{ and } l^{(M+1)}_{i_0 j_0} > 0.$$

<u>Proof</u>: Proposition (8.17) shows "1) ⇒ 2)"

"2) ⇒ 3)" follows from (8.14), (8.17) and (17.9).

"3) ⇒ 1)" First we know from the preceding proof that it suffices to show that for $i, j \in S$ there is $n_0 \in \mathbb{N}$ such that $l^{(n)}_{ij} > 0$ for every $n \geq n_0$. Secondly because of transitivity and (17.6) it is enough to show this for $i = i_0$ and $j = i_0$.

There is $n_1 \in \mathbb{N}$ such that $l^{(n_1)}_{j_0 i_0} > 0$. Choose $n_0 = (n_1+M)^2$. Then for arbitrary $n \geq n_0$ there exist an $a \in N$ and a $0 \leq b < n_1+M$ such that $n = a(n_1 + M) + b$. Clearly $a \geq n_1 + M$ yields

$n = a'(n_1 + M) + b(n_1 + M + 1)$, where $a' = a-b > 0$.

Using (17.6) (3) again there is $x \in \Lambda$ satisfying

$$x_{k(n_1+M)} = i_0 \qquad \text{for } k = 0,\dots,a'$$

and

$$x_{a'(n_1+M)+k(n_1+M+1)} = i_0 \qquad \text{for } k = 0,\dots,b.$$ □

<u>Remark</u>: Conditions 1) and 3) of the last proposition can be formulated for arbitrary f.t. subshifts as well. It reads as follows:

Let B be a finite defining block system for the f.t. subshift Λ. For two blocks $\beta, \beta' \in B$ call n a <u>transition length</u>

for β to β' if there exists $x \in \Lambda$ such that β occurs in x at the place 0 and β' occurs in x at the place $l(\beta) + n$. Then Λ is topologically mixing iff there exist two blocks β and β' in B and transition lengths n and n+1 for β to β' and Λ is topologically transitive.

Using (17.9) and (17.10) we are now in the position to prove the structure theorem for f.t. subshifts.

Let Λ be an f.t. subshift of order 2 with transition matrix $L = (l_{ij})_{i,j\in S}$. The set of wandering states is defined by

$$S_W := \{i \in S \mid \forall\, n > 0 \quad l_{ii}^{(n)} = 0\}.$$

Let $\Lambda_W := \{(x_k)_{k\in\mathbb{Z}} \in \Lambda \mid \exists\, n \in \mathbb{Z} : x_n \in S_W\}$, $\Lambda_o := \Lambda \setminus \Lambda_W$ and $S_o = S \setminus S_W$. It is not hard to see that $(\Lambda_o, \sigma|\Lambda_o)$ also is an f.t. subshift of order 2. It is well known that for every $d \geq 1$ the relation $\overset{d}{\longleftrightarrow}$ on $S_o \times S_o$ given by

$$i \overset{d}{\longleftrightarrow} j \iff \exists\, n,m > 0 \text{ such that } l_{ij}^{(nd)} > 0 \text{ and } l_{ji}^{(md)} > 0$$

defines an equivalence relation on S_o. Denote the equivalence classes of "$\overset{d}{\longleftrightarrow}$" by $S_1^d, \dots, S_{r_d}^d$ and define

$$\Lambda_i^d := \{x = (x_k)_{k\in\mathbb{Z}} \in \Lambda_o \mid \forall\, n \in \mathbb{Z} : x_{nd} \in S_i^d\} \quad (1 \leq i \leq r_d).$$

Finally let $\tilde{\Lambda}$ be the nonwandering set of Λ. (In general an f.t. subshift has got many wandering points.) It will be seen that $\tilde{\Lambda}$ is an f.t. subshift of order 2 and also is the nonwandering set of Λ_o.

In order to see what $\Lambda \setminus \tilde{\Lambda}$ looks like it is sufficient to consider $x \in \Lambda_o \setminus \tilde{\Lambda}$. There is a cylinder $A = {}_{-n}[a_{-n}, \dots, a_n]$ containing x such that $A \cap \sigma^m A \cap \Lambda_o = \emptyset$ for every $m \in \mathbb{Z}$, $m \neq 0$. Therefore a_{-n} and a_n must belong to different equivalence classes. On the other hand it is possible to define an ordering of the set of equivalence classes such that a transition from i to j is possible only if i belongs to a greater class than j. Therefore every $x \in \Lambda_o \setminus \tilde{\Lambda}$ has components belonging to at least two different equivalence classes and the natural ordering of the indices of the components is compatible with the ordering of the equivalence classes. Now in order to investigate the structure of Λ it suffices to consider $\tilde{\Lambda}$.

(17.11) Proposition:

1) Every Λ_i^d $(1 \leq i \leq r_d,\ d \geq 1)$ is a topologically transitive f.t. subshift of order 2 with respect to σ^d. Moreover if d' divides d then for $1 \leq i \leq r_d$ and $1 \leq j \leq r_{d'}$ either $\Lambda_j^{d'} \supset \Lambda_i^d$ or $\Lambda_j^{d'} \cap \Lambda_i^d = \emptyset$.

2) For every $d \geq 1$ the family $(\Lambda_i^d \mid 1 \leq i \leq r_d)$ consists of pairwise disjoint sets being open in the topology on $\tilde{\Lambda}$ and satisfying

$$\tilde{\Lambda} = \sum_{i=1}^{r_d} \Lambda_i^d .$$

Especially $\tilde{\Lambda}$ decomposes into the sets $\Lambda_i^1 = \Lambda_i$ $(1 \leq i \leq r_1)$, which are called the basic sets for Λ.

3) For every basic set Λ_i $(1 \leq i \leq r_1)$ and $d \geq 1$ let

$$K_i^d := \{1 \leq j \leq r_d \mid \Lambda_j^d \subset \Lambda_i\}.$$

Then the following properties hold:

(i) For every $j \in K_i^d$ there is $k \in K_i^d$ such that $\sigma(\Lambda_j^d) = \Lambda_k^d$.

Hence we may write $K_i^d = \{t_1,\ldots,t_f\}$ such that

$$\sigma\, \Lambda_{t_j}^d = \Lambda_{t_{j+1 \bmod f}}^d \qquad (1 \leq j \leq f).$$

(ii) $$\Lambda_i = \sum_{j \in K_i^d} \Lambda_j^d$$

(iii) There exists a $d>1$ such that $\sigma^d | \Lambda_j^d$ is topologically mixing for every $j \in K_i^d$.

Note that the f defined here depends on i and d. Their maximum over d is called the period of $\sigma|\Lambda_i$.

Proof:

1) Let $d \geq 1$. Clearly by definition Λ_i^d is a subshift with respect to σ^d (that is: σ^d is considered to be the shift on $S'^{\mathbf{Z}}$, where

$$S' = \{\beta = (b_o,\ldots,b_{d-1}) \mid \beta \text{ occurs in } \Lambda_o\}).$$

In order to show topological transitivity let ${}_{-md}[a_{-md},\ldots,a_{md}]$ and ${}_{-m'd}[b_{-m'd},\ldots,b_{m'd}]$ be two cylinders occuring in Λ_i^d. Since

$a_{md}, b_{-m'd} \in S_i^d$ there is $k > 0$ such that $1^{(kd)}_{a_{md}, b_{-m'd}} > 0$.
Thus using the finite type property of $\sigma|\Lambda_o$ it follows that $\Lambda_i^d \cap {}_{-md}[a_{-md},\dots,a_{md}] \cap \sigma^{kd}_{-m'd}[b_{-m'd},\dots,b_{m'd}] \neq \emptyset$.
The finite type property of σ^d on Λ_i^d is trivial, since this means that for any two blocks $\alpha = (a_o,\dots,a_{2d-1})$ and $\beta = (b_o,\dots,b_{2d-1})$ occuring in Λ_i^d such that $b_t = a_{d+t}$ $(0 \leq t \leq d-1)$ the composed block $(a_o,\dots,a_{2d-1}, b_d,\dots,b_{2d-1})$ occurs in Λ_i^d as well.

For the last property let d' divide d, $1 \leq i \leq r_d$ and $1 \leq j \leq r_{d'}$. Since for $k,l \in S_o$ $k \overset{d}{\longleftrightarrow} l$ implies $k \overset{d'}{\longleftrightarrow} l$, $S_i^d \subset S_j^{d'}$ or $S_i^d \cap S_j^{d'} = \emptyset$. In the latter case it follows at once that $\Lambda_i^d \cap \Lambda_j^{d'} = \emptyset$. In the first case we show that $\Lambda_i^d \subset \Lambda_j^{d'}$. For this let $x = (x_k)_{k\in\mathbb{Z}} \in \Lambda_i^d$. Now let $n \in \mathbb{Z}$ be arbitrary and choose $m \in \mathbb{Z}$ such that nd' lies between 0 and md. Since x_o and x_{md} belong to $S_i^d \subset S_j^{d'}$, it is easy to see that $x_{nd'} \overset{d'}{\longleftrightarrow} x_o$ hence $x_{nd'} \in S_j^{d'}$.

2) Let $d \geq 1$. $\Lambda_i^d \cap \Lambda_j^d = \emptyset$ for $1 \leq i \neq j \leq r_d$ follows immediately from the definition. Also it is easy to see that $\Lambda_i^d \subset \tilde{\Lambda}$ $(1 \leq i \leq r_d)$.

In view of part 1) we are done by proving
$\tilde{\Lambda} \subset \sum_i \Lambda_i^d$.

Let $x = (x_k)_{x\in\mathbb{Z}}$ be a nonwandering point. Let $n > 0$; then $1^{(nd)}_{x_o x_{nd}} > 0$. Since x is nonwandering there exist arbitrarily large $m \in \mathbb{N}$ such that for $U = \{(y_k)_{k\in\mathbb{Z}} \in \Lambda \mid y_o = x_o;\ y_{nd} = x_{nd}\}$. $U \cap \sigma^m U \neq \emptyset$. Hence there is $m \in \mathbb{N}$ such that $1^{(m)}_{x_{nd}x_o} > 0$, therefore $1^{(nd+m)}_{x_o x_o} > 0$. Thus for

$$t = m + (d-1)\,(nd+m)$$

it follows (using the finite type property again) that $1^{(t)}_{x_{nd}x_o} > 0$ and that d divides t; hence $x_o \overset{d}{\longleftrightarrow} x_{nd}$. It follows now that $x \in \Lambda_i^d$ where S_i^d is the equivalence class of x_o.

3) i) Let $d \geq 1$. It suffices to show that for every $1 \leq j \leq r_d$ there is $1 \leq k \leq r_d$ such that $\sigma(\Lambda_j^d) = \Lambda_k^d$. Let $x \in \Lambda_j^d$ and pick $1 \leq k \leq r_d$ such that $\sigma(x) \in \Lambda_k^d$ (by 2). Since σ^d is topologically

transitive on Λ_l^d for every $1 \leq l \leq r_d$ it has got the same property on every $\sigma(\Lambda_l^d)$, hence $\sigma\Lambda_j^d \subset \Lambda_k^d$. Replacing σ by σ^{-1} and interchanging j and k, $\sigma(\Lambda_j^d) \supset \Lambda_k^d$ follows.

ii) This is obvious from 1) and 2).

iii) Choose any state $a \in S_i^1$ and $d \geq 1$ such that

$$l_{aa}^{(d)} > 0.$$

Clearly by the finite type property of σ we know $l_{aa}^{(2d)} > 0$ as well.

Now - using (17.6) together with part 1) - $\sigma^d|\Lambda_j^d$ is topologically mixing where Λ_j^d belongs to the equivalence class S_j^d of $a \in S_i^1$. We only have to remark that $\sigma^d|\sigma^l\Lambda_j^d$ is topologically mixing for every $l \geq 0$ in order to conclude from part 3) i) and ii) that σ^d is topologically mixing on every Λ_k^d for $k \in K_i^d$. □

<u>Corollary:</u> Let Λ be an f.t. subshift. Then there exist topologically transitive disjoint f.t. subshifts $\Lambda_1,\ldots,\Lambda_r$ of the same order such that $\tilde{\Lambda} = \sum \Lambda_i$, where $\tilde{\Lambda}$ denotes the non-wandering set of Λ. The sets Λ_i are called the basic sets for Λ. For each basic set Λ_i $(1 \leq i \leq r)$ there exist $d \geq 1$ and a decomposition $\Lambda_i = \sum_{k=1}^{q-} \bar{\Lambda}_k$ into closed, disjoint and σ^d-invariant subsets such that $\sigma^d|\bar{\Lambda}_k$ is topologically mixing. The sets $\bar{\Lambda}_k$ are called the <u>basic parts</u> for Λ_i.

<u>Remark:</u> Later on - in section 23 - we shall quote a more general theorem than the corollary of (17.11) for axiom-A-diffeomorphisms, which first was shown in [21] and [180].

Now we shall study the entropy properties of f.t. subshifts.

<u>(17.12) Proposition</u> [146]: Let Λ be an f.t. subshift of order 2 with transition matrix L. Then

$$h_{top}(\sigma|\Lambda) = \log \lambda$$

where λ is the spectral radius of L. λ is the eigenvalue of maximal modulus of L. If $\Lambda_1,\ldots,\Lambda_r$ denote the basic sets of Λ defined in (17.11) then

$$h_{top}(\sigma|\Lambda) = \max_{1\le k\le r} h_{top}(\sigma|\Lambda_k)$$

Proof: Let us first show that λ is an eigenvalue of the transition matrix L_k of Λ_k for some $1 \le k \le r$.

Re-arranging the states of S we may assume that L has got the form

$$L = \begin{pmatrix} E_1 & & & 0 \\ & E_2 & & \\ & & \ddots & \\ l_{ij} & & & E_r \end{pmatrix}$$

where each E_i is either the transition matrix of some Λ_k $(1 \le k \le r)$ or zero (corresponding to the wandering states). (This can be done as follows: Consider the set

$$S = \{\Lambda_i, \{k\} \mid 1 \le i \le r,\ k \in S_W\}.$$

Define an order $>$ on S by $a > b$ if there exist $i \in a$, $j \in b$ and $n \in \mathbb{N}$ such that $l^n_{ij} > 0$.

Clearly ">" is a partial ordering and every re-arranging of S such that the states of $a \in S$ are smaller than the states of $b \in S$ iff $a > b$ gives such an announced form of L.)

Evaluating det $(L - \lambda I)$ one sees that

$$\det (L - \lambda I) = (-\lambda)^{\operatorname{card} S_W} \prod_{i=1}^{r} \det (L_i - \lambda I)$$

(where L_i denotes the transition matrix of Λ_i). Therefore the eigenvalues of L are possibly 0 and the eigenvalues of L_i. But L_i is irreducible ((17.9), (17.11)) hence the Perron-Frobenius theorem applies (see (8.7)) and the spectral radius of L is the maximal eigenvalue λ, which is real.

Recall now that in (16.11) $\theta_n(\Lambda)$ denoted the number of distinct n-blocks in Λ and that

$$h_{top}(\sigma|\Lambda) = \lim_{n\to\infty} \frac{1}{n} \log \theta_n(\Lambda).$$

Since a block $(a_o,\dots,a_{n-1})$ occurs in Λ if and only if $\prod_{k=o}^{n-2} l_{a_k a_{k+1}} = 1$ we have

$$\theta_{n+1}(\Lambda) = \sum_{a_o,\ldots,a_n \in S} \prod_{k=o}^{n-1} l_{a_k a_{k+1}} = \sum_{i,j\in S} l_{ij}^{(n)} = \|L^n\|$$

where $\|\ \|$ denotes the 1-norm in $\mathbb{R}^{S^2}$.

Using the spectral radius theorem ([197]) one knows that $\|L^n\|^{1/n} \longrightarrow \lambda$, hence

$$h_{top}(\sigma|\Lambda) = \lim_{n\to\infty} \frac{1}{n} \log \|L^{n-1}\| = \log \lambda. \qquad \square$$

(17.13) Proposition [18]: Let Λ be a topologically mixing f.t. subshift of order 2. Then

$$h_{top}(\sigma|\Lambda) = \lim_{n\to\infty} \frac{1}{n} \log Per_n(\sigma|\Lambda).$$

Proof: By the corollary of (16.12) we have to show that there is $n_o \in \mathbb{N}$ such that for every n-block β occuring in Λ there is a periodic point in $_o[\beta]$ of period $n + n_o$. Take n_o so large that $l_{ij}^{(n_o)} > 0$ for every $i,j \in S$. Now if $\beta = (a_o,\ldots,a_{n-1})$ is an n-block occuring in Λ, there is an $n+n_o$ block $\beta' = (a_o,\ldots,a_{n-1},b_o,\ldots,b_{n_o-2},a_o)$ occuring in Λ. Using (17.6) 3) it is easy to define a periodic point in β' of period $n+n_o$. $\square$

Corollary: Let Λ be an f.t. subshift (of order 2). Then

$$h_{top}(\sigma|\Lambda) = \limsup_{n\to\infty} \frac{1}{n} \log Per_n(\sigma|\Lambda) .$$

Proof: Use (17.11), (14.20), (17.13). $\square$

Note that $h_{top}(\sigma|\Lambda) = \lim \frac{1}{nd} \log Per_{nd}(\sigma|\Lambda_i)$ where d is the period of Λ_i and $h_{top}(\sigma|\Lambda) = h_{top}(\sigma|\Lambda_i)$. For further information on topological entropy see [81] and [82].

(17.14) Proposition [146]: Let Λ be a topologically transitive f.t. subshift of order 2. Then there exists an ergodic Markov measure $\mu_\Lambda \in \mathfrak{M}_\sigma(\sigma|\Lambda)$ such that

$$h_{\mu_\Lambda}(\sigma|\Lambda) = h_{top}(\sigma|\Lambda).$$

Proof: Let $L = (l_{ij})_{1\leq i,j\leq s}$ denote the transition matrix of Λ, which by (17.9) is irreducible. Therefore (8.7) tells that for λ - the eigenvalue of maximal modulus - there are a right eigenvector $u = (u_1,\ldots,u_s)$ and a left eigenvector $v = (v_1,\ldots,v_s)$ such that $u_i > 0$, $v_i > 0$ for every $1 \leq i \leq s$. We may assume that $\sum_{i\in S} u_i v_i = 1$.

We define a stationary Markov measure μ_Λ as follows: The stochastic matrix $P = (p_{ij})_{i,j\in S}$ is defined by

$$p_{ij} = \frac{u_j l_{ij}}{\lambda\, u_i} \qquad (i,j \in S)$$

and the initial distribution $\Pi = (p_1,\dots,p_s)$ by

$$p_i = u_i\, v_i \qquad (i \in S).$$

(Note that for $i \in S$ $\sum_{j\in S} p_{ij} = \frac{1}{\lambda\, u_i} \sum_{j\in S} l_{ij}\, u_j = \frac{1}{\lambda u_i}(L\,u)_i = 1$, $\sum_{i\in S} p_i = \sum_{i\in S} u_i v_i = 1$ and

$$\Pi P = \Big(\sum_{i\in S} p_i p_{ij}\Big)_{j\in S} = \left(\sum_{i\in S} \frac{u_i v_i u_j l_{ij}}{\lambda\, u_i}\right)_{j\in S}$$

$$= (v_j u_j)_{j\in S} = \Pi.\ )$$

Since $l_{ij} = 0$ iff (i,j) does not occur in Λ and $p_{ij} = 0$ iff $l_{ij} = 0$ we see that

$$\mu_\Lambda(\Lambda) = 1,$$

hence μ_Λ is a Markov measure on Λ. Using (12.3) we conclude that

$$\begin{aligned} h\mu_\Lambda(\sigma|\Lambda) &= -\sum_{i,j\in S} p_i p_{ij} \log p_{ij} = \\ &= -\sum_{i,j\in S} u_i v_i \frac{u_j l_{ij}}{\lambda u_i} \log \frac{n_j l_{ij}}{\lambda u_i} = \\ &= (\log\lambda)\Big[\sum_{i\in S} \frac{v_i}{\lambda} \sum_{j\in S} u_j l_{ij}\Big] + \sum_{i\in S} \frac{v_i \log u_i}{\lambda} \sum_{j\in S} u_j l_{ij} \\ &\quad - \sum_{j\in S} \frac{u_j \log u_j}{\lambda} \sum_{i\in S} v_i l_{ij} - \sum_{i,j\in S} v_i u_j \frac{l_{ij}}{\lambda} \log l_{ij} \\ &= \log\lambda \end{aligned}$$

(since $l_{ij} \log l_{ij} = 0 \ \ \forall\, i,j \in S$ and $\sum_{i\in S} u_i v_i = 1$).

□

(17.15) Definition: The measure μ_Λ defined in proposition (17.14) is called the Parry measure assigned to the top. transitive f.t. subshift Λ.

Corollary 1: Let Λ be an f.t. subshift. Then there exists a measure μ_Λ such that

$$h_{\mu_\Lambda}(\sigma|\Lambda) = h_{top}(\sigma|\Lambda).$$

Proof: Use (17.5), (17.11), (17.12). □

Corollary 2: Let Λ be a subshift. Then there exists a measure μ_Λ such that

$$h_{\mu_\Lambda}(\sigma|\Lambda) = h_{top}(\sigma|\Lambda).$$

Proof: Use the last corollary, (17.2) corollary, (16.2) corollary, and (16.7). □

Corollary 3: The Parry measure is positive on open sets.

Proof: This follows immediately from the definition. □

Corollary 4: The Parry measure on a topologically mixing f.t. subshift is a Bernoulli measure.

Proof: In this case the Parry measure is a mixing Markov measure, hence the corollary follows from (12.14). □

In section 19 we shall study further properties of the Parry measure.

18. The Variational Principle for Topological Entropy

Let (X,T) denote a topological dynamical system. In 1965, Adler, Konheim and McAndrew conjectured that

$$(*) \qquad h_{top}(T) = \sup_{m \in \mathfrak{M}_T(X)} h_m(T)$$

and thus also raised the question in which cases there exists a measure $m \in \mathfrak{M}_T(X)$ satisfying $h_m(T) = h_{top}(T)$. In the last section we dealt with this problem for subshifts.

In 1969, Goodwyn ([64]) proved that $h_m(T) \leq h_{top}(T)$ for all $m \in \mathfrak{M}_T(X)$. Bowen gave a simpler proof for the finite dimensional case in [22], which was generalized to arbitrary compact spaces in [45]. Other proofs appeared in [65] and [138]. Here we shall follow the proof given in [45] and [138].

In 1970, Dinaburg ([53]) proved (*) for the finite dimensional case and Goodman ([60]) adapted the proof to the general metric case. Another proof is given in [26] by Bowen.

The equation (*) is called the variational principle for topological entropy. Ruelle ([161]) and Walters ([186]) generalized (*) by introducing the notion of pressure for $f \in C(X)$ and proved the variational principle in this case also. Their approach will be described in the appendix of this section. (Bowen gave another proof of this principle in [31].) Finally, Misiurewicz found a remarkable short and elegant proof of the variational principle ([208]), which we present in the appendix.

It is essential for solving the problem to consider for a partition α the number $p(\alpha)$ explained in:

<u>(18.1) Definition</u>: Let $\alpha = (A_1,...,A_s)$ be a finite partition or a finite open cover of X. Then for $x \in X$

$$p(x,\alpha) := \text{card}\,\{1 \le i \le s \mid x \in \overline{A_i}\}$$

and

$$p(\alpha) := \max_{x \in X} p(x,\alpha).$$

<u>(18.2) Lemma:</u> For every finite partition $\alpha = (A_1,\dots,A_s)$ of X we have

$$\lim_{N\to\infty} \frac{1}{N} \log\,[\text{card}\,(\alpha)_o^{N-1}] \le h_{top}(T) + \log p(\alpha).$$

<u>Proof:</u> For every $x \in X$ there exists an open neighbourhood $W(x)$ of x satisfying

$$\text{card}\,\{1 \le i \le s \mid A_i \cap W(x) \neq \emptyset\} \le p(x,\alpha).$$

Let $\mathfrak{W}$ be a fixed finite subcover of $(W(x) \mid x \in X)$, and $\varepsilon > 0$ a Lebesgue number of $\mathfrak{W}$.

There exists a function $F : X \longrightarrow \mathfrak{W}$ such that for every $x \in X$ the ε-ball $B_\varepsilon(x)$ around x is contained in $F(x)$.

Now we choose $N \in \mathbb{N}$ arbitrarily and an (N,ε)-spanning set E of minimal cardinality, i.e. card $E = r_N(\varepsilon,T) =: r_N$ (see (14.14)).

Define

$$S := \{(i_o,\dots,i_{N-1}) \mid \exists y \in E \text{ s.th. } A_{i_j} \cap F(T^j y) \neq \emptyset \ (\forall\ 0 \le j \le N-1)\}.$$

Since for fixed $y \in E$ by the construction of F

$$\text{card}\,\{(i_o,\dots,i_{N-1}) \mid A_{i_j} \cap F(T^j y) \neq \emptyset \quad \forall\ 0 \le j \le N-1\} \le [p(\alpha)]^N$$

it follows that

$$\text{card}\, S \le \text{card}\, E[p(\alpha)]^N = r_N \cdot [p(\alpha)]^N.$$

It remains to compare S with the following set

$$R := \{(k_o,\dots,k_{N-1}) \mid \bigcap_{j=o}^{N-1} T^{-j} A_{k_j} \neq \emptyset\}.$$

Obviously we must have $R \subset S$, since for $(k_o,\dots,k_{N-1}) \in R$ there is $x \in \bigcap_{j=o}^{N-1} T^{-j} A_{k_j}$ and clearly we find $y \in E$ with

$d(T^jx,T^jy) < \epsilon$, i.e. $T^j(x) \in B_\epsilon(T^jy) \subset F(T^jy)$.

Now we are done:

$$\frac{1}{N} \log \text{card } (\alpha)_0^{N-1} = \frac{1}{N} \log \text{card } R \leq \frac{1}{N} \log \text{card } S$$

$$\leq \frac{1}{N} \log r_N + \log p(\alpha).$$

Letting $N \longrightarrow \infty$, the lemma follows. □

<u>Corollary:</u> For every measure m and every finite measurable partition α of X

$$h_m(T,\alpha) \leq h_{top}(T) + \log p(\alpha).$$

<u>Proof:</u> For every probability vector $(q_1,\ldots,q_s)$ $-\sum_{i=1}^{s} q_i \log q_i \leq \log s$ yields

$$-\frac{1}{N} \sum_{A \in (\alpha)_0^{N-1}} m(A) \log m(A) \leq \log \text{card } (\alpha)_0^{N-1}.$$

Now apply the Lemma. □

<u>(18.3) Lemma:</u> For $m \in \mathfrak{M}_T(X)$ there exists a sequence $(\alpha_n)_{n\in\mathbb{N}}$ of finite m-measurable partitions of X with $p(\alpha_n) \leq 2$ ($\forall\, n \in \mathbb{N}$) and $\lim_{n\to\infty} h_m(T^k,\alpha_n) = h_m(T^k)$ for every $k \in \mathbb{N}$.

<u>Proof:</u> Choose any refining sequence $(\beta_n = (B_1^n,\ldots,B_{s_n}^n))_{n\in\mathbb{N}}$ of finite measurable partitions satisfying $\bigvee_{n\in\mathbb{N}} \Sigma(\beta_n) = \mathfrak{B}$. Let $\epsilon_n > 0$, $\lim_{n\to\infty} \epsilon_n = 0$. Now fix n. Because of regularity, for given $\eta > 0$ there exists a compact set $K_j^\eta \subset B_j^n$ for every $2 \leq j \leq s_n$ such that $m(B_j^n \setminus K_j^\eta) < \eta$. Therefore, if $\alpha_n^\eta := \{K_j^\eta, X \setminus \bigcup K_j^\eta \mid 2 \leq j \leq s_n\}$ then

$H(\beta_n|\alpha_n^\eta) + H(\alpha_n^\eta|\beta_n) \longrightarrow 0$ as $\eta \longrightarrow 0$ (see (11.10)). Hence there is $\eta > 0$ such that for $\alpha_n := \alpha_n^\eta$ $H(\beta_n|\alpha_n) + H(\alpha_n|\beta_n) < \epsilon_n$.

Clearly, α_n satisfies $p(\alpha_n) \leq 2$ for every $n \in \mathbb{N}$, and for every $k,n \in \mathbb{N}$ we have

$$|h_m(T^k,\alpha_n) - h_m(T^k,\beta_n)| < \varepsilon_n. \quad \text{(cf. (11.7))}.$$

Since the sequence $(\beta_n)_{n\in\mathbb{N}}$ generates $\mathfrak{B}$ we have for $k \in \mathbb{N}$ ((11.11))

$$h_m(T^k) = \lim_{n\to\infty} h_m(T^k,\beta_n) = \lim_{n\to\infty} h_m(T^k,\alpha_n). \qquad \square$$

(18.4) Theorem [64]: Every measure $m \in \mathfrak{M}_T(X)$ satisfies

$$h_m(T) \leq h_{top}(T).$$

Proof: Let $m \in \mathfrak{M}_T(X)$ and $(\alpha_n)_{n\in\mathbb{N}}$ be the sequence of (18.3).

Using lemma (18.2) for every $k \in \mathbb{N}$ one gets

$$h_m(T^k) = \lim_{n\to\infty} h_m(T^k,\alpha_n) \leq \lim_{n\to\infty} [h_{top}(T^k) + \log p(\alpha_n)]$$

$$= h_{top}(T^k) + \log 2.$$

Apply proposition (14.18) to get

$$h_m(T) \leq h_{top}(T) + \frac{1}{k} \log 2.$$

The theorem follows if $k \longrightarrow \infty$. $\square$

Remark: A topological space has covering dimension at most m if for every open cover $\mathfrak{U}$ of X there exists a refinement $\mathfrak{U}'$ of $\mathfrak{U}$ in the sense of (14.1) such that $p(\mathfrak{U}') \leq m + 1$. For these spaces it is not necessary to prove lemma (18.3) in order to show Goodwyn's theorem. (see Bowen [22]).

We now start with a preliminary lemma and proposition for the proof of the Dinaburg-Goodman theorem. Dinaburg [53] showed the theorem in the finite dimensional case and Goodman [60] proved it in general. Essentially we shall follow the proofs given in [60] and [186].

(18.5) Lemma: Let $m \in \mathfrak{M}_{T^n}(X)$ for some $n \in \mathbb{N}$. Then

$$\bar{m} = \frac{1}{n} \sum_{j=0}^{n-1} T^j m \in \mathfrak{M}_T(X) \text{ and}$$

$$h_{\bar{m}}(T^n) = h_m(T^n).$$

Proof: $m \in \mathfrak{M}_T(X)$ is obvious and since the entropy function $h_\mu(T)$ is affine the equality follows (see (10.13)). □

We remark here that if β is a partition then for every $0 \leq j \leq n-1, h_{T^j m}(T^n, (\beta)_0^{n-1}) = h_m(T^n, (\beta)_0^{n-1})$. We shall make use of this fact in section 20.

The following proposition actually is stronger than what is needed:

(18.6) Proposition: For every finite open cover $\mathfrak{U}$ of X and every $m \in \mathbb{N}$ there exists a finite open cover $\mathfrak{B}$ finer than $(\mathfrak{U})_0^{m-1}$ satisfying $p(\mathfrak{B}) \leq m$ card $\mathfrak{U} + 1$.

Proof: The proof is given by constructing a factor (see [64]). Write $\mathfrak{U} = \{U_1, \ldots, U_q\}$ and define

$$f_i : X \longrightarrow I := [0,1] \qquad (1 \leq i \leq q)$$

by $f_i(x) = [\text{diam } X]^{-1} d(x, X \setminus U_i)$ and

$$f := (f_1, \ldots, f_q) : X \longrightarrow I^q.$$

Clearly the maps $f_i (1 \leq i \leq q)$ and f are continuous. Let $S = I^q$ and σ be the shift on $S^{\mathbb{Z}}$. Define $\pi : X \longrightarrow S^{\mathbb{Z}}$ by $\pi(x) = (f(T^t x))_{t \in \mathbb{Z}}$ $(x \in X)$. Clearly π is a continuous surjection onto $Y := \pi(X)$ and commutes with T and σ $(\pi \circ T = \sigma \circ \pi)$. Hence $(Y, \sigma|Y)$ is a factor of (X,T).

Since $\pi((\mathfrak{U})_0^{m-1})$ is an open cover of $S^{\mathbb{Z}}$ which depends only on the coordinates from 0 up to m - 1, think of it as an open cover $\mathfrak{W}$ of I^{qm}. From a well-known fact of dimension theory it follows that there is an open cover $\mathfrak{W}'$ of I^{qm} finer than $\mathfrak{W}$ such that $p(\mathfrak{W}') \leq mq + 1$. Again consider $\mathfrak{W}'$ to be an open cover of $S^{\mathbb{Z}}$ depending on the coordinates from 0 up to m - 1 and define $\mathfrak{B} = \pi^{-1}\mathfrak{W}'$. Clearly $\mathfrak{B}$ is finer than $(\mathfrak{U})_0^{m-1}$ and satisfies $p(\mathfrak{B}) \leq mq + 1$. □

Corollary: Let $\mathfrak{U}$ be a finite open cover of X and $m \in \mathbb{N}$. Then there exists a Borel partition α such that $p(\alpha) \leq m$ card $\mathfrak{U} + 1$, $A \subset \overline{\text{int } A}$ for every $A \in \alpha$ and α is finer than $(\mathfrak{U})_0^{m-1}$.

(Recall that this means: the closure $\bar{A}$ of every $A \in \alpha$ is contained in some member of $(\mathfrak{U})_0^{m-1}$).

Proof: Use (16.6) and the proposition. □

We note that in the last corollary we could have replaced α by a closed cover. In this case we could use the corollary in the same way as we do it in (18.8).

(18.7) Proposition: Let $\mathfrak{U}$ be a finite open cover and $N \in \mathbb{N}$. For every finite Borel partition α finer than $(\mathfrak{U})_0^{N-1}$ there exists a measure $m \in \mathfrak{M}_{T^N}(X)$ such that for every finite m-measurable partition β of X the following inequality holds:

$$h_m(T^N,\beta) \geq H(T^N,(\mathfrak{U})_0^{N-1}) - \log p(\beta|\alpha)$$

where

$$p(\beta|\alpha) := \max_{B\in\beta} \text{card}\,\{A \in \alpha | \bar{A} \cap \bar{B} \neq \emptyset\}.$$

Proof: Write $\alpha = (A_1,\dots,A_s)$. Let $S = \{1,\dots,s\}$ and let $\Lambda \subset S^{\mathbb{Z}}$ be the subshift defined by

$$\Lambda := \{(y_n)_{n\in\mathbb{Z}} \in S^{\mathbb{Z}} \mid \bigcap_{n\in\mathbb{Z}} T^{nN}\,\overline{A_{y_n}} \neq \emptyset\}.$$

Define the topological dynamical system (Z,R) by

$$Z := \{(x,y) \in X \times \Lambda \mid x \in \bigcap_{n\in\mathbb{Z}} T^{nN}\,\overline{A_{y_n}}\}$$

and

$$R((x,y)) := (T^N(x), (\sigma|\Lambda)(y)).$$

Denote by π_i $(i = 1,2)$ the projections from Z onto the i-th coordinate. Clearly each π_i is continuous if Z is equipped with the induced product topology. Therefore $\hat{\gamma} := \pi_2^{-1}\{{}_0[i] | i \in S\}$ is a Borel partition on Z. Furthermore π_2 induces a surjective map (see (3.10))

$$\pi_2 : \mathfrak{M}_R(Z) \longrightarrow \mathfrak{M}_\sigma(\Lambda).$$

After having introduced these notions we prove the proposition.

Since Λ is a subshift the corollary 2 of (17.14) gives a measure $\mu_\Lambda \in \mathfrak{M}_\sigma(\Lambda)$ satisfying

$$h_{\mu_\Lambda}(\sigma|\Lambda) = h_{top}(\sigma|\Lambda) = \lim_{n\to\infty} \frac{1}{n} \log \theta_n(\Lambda)$$

$$\geq \lim_{n\to\infty} \frac{1}{n} \log \text{card} \bigvee_{k=o}^{n-1} T^{-kN} \alpha.$$

Because α is finer than $(\mathfrak{U})_o^{N-1}$ one obtains

$$h_{\mu_\Lambda}(\sigma|\Lambda) \geq \lim_{n\to\infty} \frac{1}{n} \log N(\bigvee_{k=o}^{n-1} T^{-Nk}(\mathfrak{U})_o^{N-1})$$

$$= H(T^N, (\mathfrak{U})_o^{N-1}).$$

There exists a $\nu \in \mathfrak{M}_R(Z)$ such that $\pi_2\nu = \mu_\Lambda$. Define $m := \pi_1\nu$.

Now let β be an m-measurable partition of X and let $B \in \beta$. If $c_1, \ldots, c_r \in S$ denote all elements satisfying $\overline{B} \cap \overline{A_{c_k}} \neq \emptyset$ $(1 \leq k \leq r)$, then

$$\pi_1^{-1} B = \{(x,y) \in Z | x \in B\} \subset \bigcup_{k=1}^{r} \{(x,y) \in Z | y_o = c_k\}$$

$$= \bigcup_{k=1}^{r} \pi_2^{-1}\ {}_o[c_k].$$

Therefore, it follows that - using $r \leq p(\beta|\alpha)$ and (11.2)-

$$H_\nu(\hat{\gamma}|\gamma) = - \sum_{C\in\gamma} \nu(C) \sum_{\hat{C}\in\hat{\gamma}} \frac{\nu(C \cap \hat{C})}{\nu(C)} \log \frac{\nu(C \cap \hat{C})}{\nu(C)}$$

$$\leq \log p(\beta|\alpha)$$

for every $\nu \in \mathfrak{M}_R(Z)$, where $\gamma = \pi_1^{-1}\beta$. Clearly this estimate shows

$$h_\nu(R,\gamma) \geq h_\nu(R,\hat{\gamma}) - \log p(\beta|\alpha),$$

and it follows easily that

$$h_m(T^N,\beta) = h_\nu(R,\gamma) \geq h_\nu(R,\hat{\gamma}) - \log p(\beta|\alpha)$$

$$= h_{\mu_\Lambda}(\sigma|\Lambda) - \log p(\beta|\alpha)$$

$$\geq H(T^N, (\mathfrak{U})_0^{N-1}) - \log p(\beta | \alpha). \qquad \square$$

(18.8) Theorem [53], [60]:

$$h_{top}(T) = \sup_{m \in \mathfrak{M}_T(X)} h_m(T).$$

Proof: Let $\epsilon > 0$ and $N \in \mathbb{N}$ be given. Choose a finite open cover $\mathfrak{U}$ such that

$$h_{top}(T) \leq H(\mathfrak{U}, T) + \epsilon.$$

Using (18.6) one finds a Borel partition $\alpha = (A_1, \ldots, A_s)$ finer than $(\mathfrak{U})_0^{N-1}$ satisfying $A_i \subset \overline{\text{int } A_i}$, and $p(\alpha) \leq N \text{ card } \mathfrak{U} + 1$. Note that N and $\mathfrak{U}$ are chosen independently.

Using (16.6) pick any Borel partition $\beta \supset \alpha$ such that $p(\beta | \alpha) \leq p(\alpha)$. This is possible since there is an open cover $\mathfrak{B}$ such that every $V \in \mathfrak{B}$ intersects at most $p(\alpha)$ elements of α.

Now proposition (18.7) gives a measure $m \in \mathfrak{M}_{T^N}(X)$ satisfying $h_m(T^N, \beta) \geq H(T^N, (\mathfrak{U})_0^{N-1}) - \log p(\alpha)$. By (18.5)

$$\bar{m} := \frac{1}{N} \sum_{j=0}^{N-1} T^j m \in \mathfrak{M}_T(X) \text{ and}$$

$$Nh_{\bar{m}}(T) = h_{\bar{m}}(T^N) = h_m(T^N) \geq H(T^N, (\mathfrak{U})_0^{N-1}) - \log p(\alpha)$$

$$= N(h_{top}(T) - \epsilon) - \log p(\alpha).$$

Dividing by N yields

$$h_{\bar{m}}(T) \geq h_{top}(T) - \epsilon - \frac{1}{N} \log(N \text{ card } \mathfrak{U} + 1).$$

Letting $N \longrightarrow \infty$ (N is independent of $\mathfrak{U}$!) gives

$$\sup_{m \in \mathfrak{M}_T(X)} h_m(T) \geq h_{top}(T) - \epsilon.$$

Now let $\epsilon \longrightarrow 0$ to get the theorem. $\square$

We finish this section by stating some corollaries.

Corollary 1: Let Ω denote the nonwandering set of T. Then

$$h_{top}(T) = h_{top}(T|\Omega).$$

Proof: Every $m \in \mathfrak{M}_T(X)$ satisfies $m(\Omega) = 1$ (see (6.20)). □

Corollary 2: $h_{top}(T) = \sup\{h_m(T) \mid m \in \mathfrak{M}_T(X) \text{ is ergodic}\}$

Proof: This follows immediately from (13.3). □

Corollary 3: Suppose $X = \bigcup_{i \in I} X_i$ (I arbitrary) where each X_i is closed and T-invariant. Then

$$h_{top}(T) = \sup_{i \in I} h_{top}(T|X_i).$$

Proof: By corollary 2 $h_{top}(T) = \sup\{h_m(T) \mid m \text{ is ergodic}\}$ and for every $m \in \mathfrak{M}_T(X)$ ergodic there exists an $i \in I$ such that $m(X_i) = 1$. □

Corollary 4: If T is expansive, then there is a measure $m \in \mathfrak{M}_T(X)$ satisfying

$$h_{top}(T) = h_m(T).$$

Proof: Use (16.7). □

Corollary 5: If the set of ergodic measures is finite then there exists a measure $m \in \mathfrak{M}_T(X)$ satisfying $h_m(T) = h_{top}(T)$.

In particular this holds for uniquely ergodic (X,T).

Proof: Corollary 2. □

Appendix: The pressure function of a homeomorphism

Recently Ruelle [161] and Walters [186] found a generalization of topological entropy.

It is well known in statistical mechanics that an equilibrium phase m of a gas - that is a measure on the space of all microscopic configurations - minimizes the free energy. There exists a continuous function f - defined by the given potential of the gas - such that m minimizes the free energy if and only if m maximizes the expression

$$h_m(T) + \int f \, dm,$$

where T denotes the translation group on the lattice and h the entropy.

Walters studied this question for continuous transformations. Here we shall give a short survey about his work without going into details and without giving all proofs. The theory turns out to be similar to that one of topological entropy though there are some more difficulties to overcome. Let (X,T) be a topological dynamical system. We shall present the theory only for these systems. (In [186] it is also done for continuous transformations.)

(18.9) Definition:

1) Let $f \in C(X)$, $n \in \mathbb{N}$ and $\varepsilon > 0$.

Define

$$Q_n(T,f,\varepsilon) := \inf \{\sum_{x \in E} \exp(\sum_{i=0}^{n-1} f \circ T^i(x) \mid E \text{ is } (n,\varepsilon)\text{-spanning}\}$$

and

$$P_n(T,f,\varepsilon) := \sup \{\sum_{x \in E} \exp(\sum_{i=0}^{n-1} f \circ T^i(x) \mid E \text{ is } (n,\varepsilon)\text{-separating}\}.$$

Now let

$$Q(T,f,\varepsilon) = \limsup_{n \to \infty} \frac{1}{n} \log Q_n(T,f,\varepsilon)$$

and

$$P(T,f,\varepsilon) = \limsup_{n \to \infty} \frac{1}{n} \log P_n(T,f,\varepsilon).$$

2) Let $f \in C(X)$, $n \in \mathbb{N}$ and $\mathfrak{U}$ be an open cover.

Define

$$\hat{Q}_n(T,f,\mathfrak{U}) := \inf \left\{ \sum_{A\in\mathfrak{B}} \inf_{x\in A} \exp\left(\sum_{i=o}^{n-1} f\circ T^i(x)\right) \middle| \begin{array}{l}\mathfrak{B} \text{ is a finite sub-}\\ \text{cover of } (\mathfrak{U})_o^{n-1}\end{array} \right\}$$

and

$$\hat{P}_n(T,f,\mathfrak{U}) := \inf \left\{ \sum_{A\in\mathfrak{B}} \sup_{x\in A} \exp\left(\sum_{i=o}^{n-1} f\circ T^i(x)\right) \middle| \begin{array}{l}\mathfrak{B} \text{ is a finite}\\ \text{subcover of } (\mathfrak{U})_o^{n-1}\end{array} \right\}.$$

Let $\hat{Q}(T,f,\mathfrak{U}) := \limsup\limits_{n\to\infty} \frac{1}{n} \log \hat{Q}_n(T,f,\mathfrak{U})$

and $\hat{P}(T,f,\mathfrak{U}) := \limsup\limits_{n\to\infty} \frac{1}{n} \log \hat{P}_n(T,f,\mathfrak{U})$.

(18.10) Proposition: For $f \in C(X)$ the following limits exist and are all equal:

$$P(T,f) := \lim_{\varepsilon\to o} P(T,f,\varepsilon)$$

$$Q(T,f) := \lim_{\varepsilon\to o} Q(T,f,\varepsilon)$$

$$\hat{Q}(T,f) := \sup\{\hat{Q}(T,f,\mathfrak{U}) \mid \mathfrak{U} \text{ is an open cover of } X\}$$

$$\hat{P}(T,f) := \lim_{\delta\to o} \sup\{\hat{P}(T,f,\mathfrak{U}) \mid \mathfrak{U} \text{ is an open cover of } X \text{ with diam} < \delta\}.$$

The function $f \longrightarrow P(T,f)$ defines a map on $C(X)$ and is called the pressure function of T. [Indeed, one can show in statistical mechanics that its dimension really is that one of the usual pressure.] Setting $f = 0$ in the definition of $P(T,f)$ one can see that

$$P(T,0) = h_{top}(T).$$

There are some nice properties of $P(T,\cdot)$. For instance this function is positive, continuous, convex and subadditive. The generators for T play the same role for $P(T,\cdot)$ as for the topological entropy, hence it is not surprising that it is possible to get similar theorems for f.t. subshifts Λ of order 2. ((17.12), (17.14)), however, it is necessary to replace the transition matrix L by another one. Starting with $f \in C(\Lambda)$ depending only on the 0-th coordinate one defines the new matrix $L' = L \cdot (a_{ik})_{1\leq i,k\leq s}$, where $a_{ik} = 0$ for $i \neq k$ and $a_{kk} = \exp(f|_o[k])$. Now the theory goes through for this f and one easily can extend the result to general f and to arbitrary subshifts. This gives (see [186] and also [182]):

(18.11) Proposition: For every subshift $\Lambda \subset S^{\mathbb{Z}}$ and every $f \in C(\Lambda)$ there exists a measure $m \in \mathfrak{M}_\sigma(\Lambda)$ such that

$$h_m(\sigma|\Lambda) + \int f dm = P(\sigma|\Lambda, f).$$

The variational principle was first proved by Ruelle [161] under some strong conditions, then by Walters [186] in full generality. For other proofs see Bowen [31] and Misiurewicz [208].

(18.12) Proposition: (Variational principle)
For $f \in C(X)$

$$P(T,f) = \sup\{h_m(T) + \int f dm | m \in \mathfrak{M}_T(X)\}.$$

We shall show this proposition following the elegant proof of Misiurewicz in [208].

Proof: A) $P(T,f) \geq \sup\{h_m(T) + \int f dm | m \in \mathfrak{M}_T(X)\}$. Let $m \in \mathfrak{M}_T(X)$. For given $\varepsilon > 0$ choose $M \in \mathbb{N}$ so large that $\log 2 \leq \varepsilon \cdot M$. Let α be a Borel partition and write $(\alpha)_o^{M-1} = (A_1, \ldots, A_s)$. Approximating every A_i by compact sets $B_i \subset A_i$ one can assume that

$$\beta = (B_i\ ,\ X \setminus \bigcup_{t=1}^{s} B_t \mid 1 < i < s) \quad \text{satisfies}$$

$$H_m((\alpha)_o^{M-1}|\beta) \leq \varepsilon \quad \text{(see (2.3))}.$$

Clearly for every $k \in \mathbb{Z}$ one has also

$$H_m(T^k(\alpha)_o^{M-1}|T^k\beta) \leq \varepsilon$$

and so for $n \in \mathbb{N}$

$$H_m((\alpha)_o^{nM-1}|\bigvee_{k=o}^{nM-1} T^{-k}\beta) \leq n\varepsilon \qquad \text{(cf. (11.7), proof)}.$$

Setting $\gamma = \bigvee_{k=o}^{nM-1} T^{-k}\beta$ it follows that

(1) $H_m((\alpha)_o^{nM-1}) \leq H_m(\gamma) + n\varepsilon$.

Pick any $0 < \eta < \inf\{d(B_i,B_j) \mid 1 \leq i \neq j \leq s\}$ and $\delta < 2^{-1}\eta$ such that $|f(x) - f(y)| \leq \varepsilon$ whenever $d(x,y) \leq \delta$.

Let n be fixed and let E be a maximal (nM,δ)-separated set.

Write $f_n = \sum_{j=o}^{nM-1} f\circ T^j$ and for $C \in \gamma$ define $\alpha(C) = \sup\{f_n(x) \mid x \in C\}$.

Since $\int 1_C f_n dm \leq m(C)\ \alpha(C)$ one obtains

$$(2) \quad H_m(\gamma) + \int f_n dm \leq \sum_{C\in\gamma} m(C)\ (\alpha(C) - \log m(C))$$

$$= - b \sum_{C\in\gamma} b^{-1}\ e^{\alpha(C)}\ h(m(C)e^{-\alpha(C)})$$

$$\leq bh(\sum_{C\in\gamma} b^{-1}e^{\alpha(C)}m(C)\ e^{-\alpha(C)}) = \log b\ ,$$

where h is the concave function $h(x) = -x \log x$ and $b = \sum_{C\in\gamma} e^{\alpha(C)}$.

For that $x \in \overline{C}$ with $f_n(x) = \alpha(C)$ choose $y(C) \in E$ satisfying $d(T^jx,T^jy(C)) \leq \delta$ $(0 \leq j \leq n\ M-1)$. Therefore $\alpha(C) \leq f_n(y(C)) + nM\varepsilon$.

Since $\delta < 2^{-1}\eta$, every δ-ball intersects at most two elements in β. Hence we have

$$\text{card}\{C \in \gamma \mid \exists x \in \overline{C} \text{ with } d(T^jx,T^jy)\leq\delta\ (0\leq j\leq nM-1)\} \leq 2^n$$

for every $y \in X$, and thus for $y \in E$

$$\text{card}\{C \in \gamma \mid y(C) = y\} \leq 2^n.$$

But now we easily can get the result:

$$2^n \sum_{y\in E} \exp f_n(y) \geq \sum_{C\in\gamma} \exp f_n(y(C)) \geq \sum_{C\in\gamma} \exp(\alpha(C) - nM\varepsilon)$$

i.e. $\quad n \log 2 + \log \sum_{y\in E} \exp f_n(y) \geq \log b - nM\varepsilon.$

Using (1), (2) and dividing by nM gives

$$\frac{1}{nM} H_m((\alpha)_o^{nM-1}) + \int f dm \leq \frac{1}{nM} H_m(\gamma) + \frac{1}{nM} \int f_n dm + \varepsilon$$

$$\leq \frac{1}{nM} \log b + \varepsilon \leq \frac{1}{nM} \log \sum_{y\in E} \exp f_n(y) + 3\varepsilon.$$

Now take sup over E, then letting $n\to\infty$ (not M!):

$$h_m(T,\alpha) + \int f dm \leq P(T,f,\delta) + 3\varepsilon < P(T,f) + 3\varepsilon.$$

Since $\varepsilon > o$ and α where chosen arbitrary it follows that $h_m(T) + \int f dm \leq P(T,f)$.

B. We note that for $f = 0$ the proof simplifies considerably. Each of the elements $U_i = B_i \cup X \setminus \bigcup B_t$ intersects at most two elements of β, hence for $\mathfrak{U} = (U_1,\ldots,U_s)$ one gets

$$H_m((\beta)_o^{n-1}) \leq \log N((\mathfrak{U})_o^{n-1}) + n \log 2.$$

Therefore (if the elements B_i approximate well enough):

$$h_m(T,\alpha) \leq h_m(T,\beta) + 1 \leq h(T,\mathfrak{U}) + \log 2 + 1$$

$$\leq h_{top}(T) + 1 + \log 2.$$

Replacing T by $T^N (N \in \mathbb{N})$ the result follows using (10.12) and (14.18).

C. $P(T,f) \leq \sup\{h_m(T) + \int f dm \mid m \in \mathfrak{M}_T(X)\}$.

Let $\varepsilon > 0$. For $n \in \mathbb{N}$ let E_n be an (n,ε)-separating set such that $\left(f_n = \sum_{j=o}^{n-1} f\circ T^j \right)$

$$\log \sum_{y \in E_n} \exp f_n(y) \geq \log P_n(T,f,\varepsilon) - 1.$$

Define $m_n := \frac{1}{n} \sum_{j=o}^{n-1} T^j \mu_n$ where

$$\mu_n = \sum_{y \in E_n} \exp(f_n(y) - \log \sum_{y \in E_n} \exp f_n(y))\, \delta(y).$$

Choose any subsequence n_i $(i \in \mathbb{N})$ satisfying

$$\lim_{i \to \infty} \frac{1}{n_i} \log P_{n_i}(T,f,\varepsilon) = P(T,f,\varepsilon),$$

and we may assume that m_{n_i} converges weakly to a measure $m \in \mathfrak{M}_T(X)$.

Let α be a finite m-continuity partition whose elements are of diameter less than ε (see (16.6)).

Since for $A \in (\alpha)_o^{n-1}$, $\mathrm{card}(E_n \cap A) \leq 1$ one obtains

$$(3) \quad H_{\mu_n}((\alpha)_o^{n-1}) + \int f_n d\mu_n = \sum_{y \in E_n} \mu_n(\{y\})\,(f_n(y) - \log \mu_n(\{y\}))$$
$$= \log \sum_{y \in E_n} \exp f_n(y).$$

Fix $n, M \in \mathbb{N}$ with $n \geq 2M$ and set $s(j) = [\frac{n-j}{M}]$ for $j = 0,\dots,M-1$. If $R_j = \{0,1,\dots,j-1, s(j)M+j,\dots,n-1\}$ then we can write

$$(\alpha)_o^{n-1} = \bigvee_{k=o}^{s(j)-1} T^{-kM-j}(\alpha)_o^{M-1} \vee \bigvee_{l \in R_j} T^{-l}\alpha$$

for $0 \leq j \leq M-1$. Therefore:

$$H_{\mu_n}((\alpha)_o^{n-1}) + \int f_n d\mu_n \leq \sum_{k=o}^{s(j)-1} H_{\mu_n}(T^{-kM-j}(\alpha)_o^{M-1})$$
$$+ \mathrm{card}\, R_j \log \mathrm{card}\, \alpha + \int f_n d\mu_n.$$

Using $H_\mu(T^{-1}\gamma) = H_{T^1\mu}(\gamma)$, $\mathrm{card}\, R_j \leq 2M$ and (3), it follows by summing over j that

$$\sum_{l=0}^{n-1} H_{T^l\mu_n}((\alpha)_0^{M-1}) + M\int f_n d\mu_n \geq \sum_{j=0}^{M-1}\sum_{k=0}^{s(j)-1} H_{\mu_n}(T^{-kM-j}(\alpha)_0^{M-1})$$

$$+ M\int f_n d\mu_n \geq M \log \sum_{y\in E_n} \exp f_n(y) - M^2 \log \text{card } \alpha.$$

Since the function $-x \log x$ is concave,

$$H_{m_n}((\alpha)_0^{M-1}) + M\int f dm_n \geq Mn^{-1} \log \sum_{y\in E_n} \exp f_n(y) - 2n^{-1}M^2 \log \text{card } \alpha$$

$$\geq M\, n^{-1} \log P_n(T,f,\varepsilon) - M\, n^{-1} - n^{-1}2M^2 \log \text{card } \alpha.$$

If n runs through the sequence (n_i) while M still is fixed it follows that

$$H_m((\alpha)_0^{M-1}) + M\int f dm \geq M\, P(T,f,\varepsilon).$$

Now let $M\to\infty$ and then $\varepsilon\to 0$. □

19. Measures with Maximal Entropy - Intrinsically Ergodic Systems

Continuing the study of the last section - the investigation of the variational principle - we want to know what transformations T admit a measure $m \in \mathfrak{M}_T(X)$ satisfying

$$h_m(T) = \sup_{m' \in \mathfrak{M}_T(X)} h_{m'}(T).$$

What is shown in section 18 enables us to give

(19.1) Definition: A measure $m \in \mathfrak{M}_T(X)$ is said to have maximal entropy (or to be an equilibrium state) if it maximizes entropy, i.e. if

$$h_m(T) = h_{top}(T).$$

The set of equilibrium states for T is denoted by $\mathfrak{M}_{max}(T)$.

We saw in sections 17 and 18 that for subshifts $\Lambda \subset S^{\mathbf{Z}}$ where S is finite $\mathfrak{M}_{max}(\sigma|\Lambda) \neq \emptyset$ and that for T expansive or for T uniquely ergodic $\mathfrak{M}_{max}(T) \neq \emptyset$. All these cases are contained in the following proposition which is a corollary of (18.8):

(19.2) Proposition: If the entropy function $h_m(T)$ is upper semicontinuous on $\mathfrak{M}_T(X)$ then

$$\mathfrak{M}_{max}(T) \neq \emptyset.$$

Two other observations about $\mathfrak{M}_{max}(T)$ are easy to make. One of them is

(19.3) Proposition: If $h_{top}(T) = \infty$ then

$$\mathfrak{M}_{max}(T) \neq \emptyset.$$

Proof: Using corollary 2 of (18.8) choose ergodic measures $m_n \in \mathfrak{M}_T(X)$ satisfying

$$\lim_{n\to\infty} h_{m_n}(T) = h_{top}(T).$$

We may assume as well that $2^{-n} h_{m_n}(T) \geq 1$ $(n \in \mathbb{N})$. Define a measure $m \in \mathfrak{M}_T(X)$ by

$$m = \sum_{n\in\mathbb{N}} 2^{-n} m_n .$$

Clearly the affinity of h implies

$$h_m(T) = \sum_{n\in\mathbb{N}} 2^{-n}\, h_{m_n}(T) = \infty,$$

hence $m \in \mathfrak{M}_{max}(T)$. □

The other easy observation is:

<u>(19.4) Proposition:</u> Suppose $h_{top}(T) < \infty$ and $\mathfrak{M}_{max}(T) \neq \emptyset$. Then there exists an ergodic measure $m \in \mathfrak{M}_{max}(T)$.

<u>Proof:</u> Choose any $m \in \mathfrak{M}_{max}(T)$ and decompose m into the ergodic fibres m_x (see (13.3)). Then $h_m(T) = \int h_{m_x}(T)\, dm(x) < \infty$, hence there is $x_o \in X$ such that

$$h_{m_{x_o}}(T) \geq h_{m_x}(T)\, m(dx) = h_{top}(T).$$ □

$\mathfrak{M}_{max}(T) \neq \emptyset$ does not always hold. This was shown first by an example of Gurevič [75]. His example is similar to the following one (see [46]):

As in (15.2) we take $S = \mathbf{Z}_+$ and the shiftspace $(S^{\mathbf{Z}},\sigma)$. Suppose $(\Lambda_n, \sigma|\Lambda_n)$ $(n \in \mathbb{N})$ are pairwise disjoint finite subshifts contained in $\mathbb{N}^{\mathbf{Z}}$ such that even the symbols of one subshift Λ_n do not occur in another one. In addition we can find Λ_n such that

$$h_{top}(\sigma|\Lambda_n) < h_{top}(\sigma|\Lambda_{n+1}) \quad (n \in \mathbb{N})$$

and

$$\lim_{n\to\infty} h_{top}(\sigma|\Lambda_n) < \infty.$$

Define $\Lambda := \overline{\bigcup_{n\in\mathbb{N}} \Lambda_n}$. Clearly $(\Lambda,\sigma|\Lambda)$ is a subshift and it is not hard to see that Λ is the one point compactification of $\bigcup_{n\in\mathbb{N}} \Lambda_n$ by $z^o = (z_k)_{k\in\mathbf{Z}}$ where $z_k = 0$ $(k \in \mathbf{Z})$.

From proposition (19.4) we know that $\mathfrak{M}_{max}(\sigma|\Lambda) = \emptyset$ iff there is no ergodic measure $m \in \mathfrak{M}_\sigma(\Lambda)$ satisfying $h_m(\sigma|\Lambda) = h_{top}(\sigma|\Lambda)$.

The latter condition is verified quite easily:

Suppose $m \in \mathfrak{M}_\sigma(\Lambda)$ is ergodic. Then there is $n \in \mathbb{N}$ such that $m(\Lambda_n) = 1$ unless m is the unit mass on z^o.

In both cases it follows from (18.4) that either

$$h_m(\sigma|\Lambda) \leq h_{top}(\sigma|\Lambda_n) < h_{top}(\sigma|\Lambda)$$

or

$$h_m(\sigma|\Lambda) = 0 < h_{top}(\sigma|\Lambda).$$

It is easy to modify this example in order to imbed $\bigcup_{n=1}^{\infty} \Lambda_n$ into a topologically transitive subshift. We only have to define a wandering point approximating every Λ_n. Clearly this does not change anything (see corollary 1 of (18.8)) concerning entropies.

Using the ideas of section 27 (see [198]) one can even prove more than the last example shows:

<u>(19.5) Proposition:</u> There exists a minimal topological dynamical system (X,T) with $\mathfrak{M}_{max}(T) = \emptyset$, and also a minimal system (X,T) with $h_{top}(T) = \infty$ and $h_\mu(T) < \infty$ for every ergodic $\mu \in \mathfrak{M}_T(X)$.

We give another example for $\mathfrak{M}_{max}(T) \neq \emptyset$.

<u>(19.6) Definition</u> [22]: A measure $m \in \mathfrak{M}_T(X)$ is called <u>T-homogeneous</u> iff for every $\epsilon > 0$ there exist a $\delta>0$ and a $c>0$ such that the following property holds: For every $n \in \mathbb{N}$ and $x,y \in X$

$$c \cdot m\left(\bigcap_{k=0}^{n-1} T^{-k}\overline{B_\epsilon(T^k x)}\right) \geq m\left(\bigcap_{k=0}^{n-1} T^{-k}\overline{B_\delta(T^k y)}\right)$$

where $B_\epsilon(z)$ denotes the open ball of radius ϵ around z.

Note that this definition is independent of the metric used.

<u>(19.7) Proposition</u> [22]: If $m \in \mathfrak{M}_T(X)$ is T-homogeneous, then $m \in \mathfrak{M}_{max}(T)$.

<u>Proof:</u> We shall write $B_\epsilon^n(x) := \bigcap_{k=0}^{n-1} T^{-k}\overline{B_\epsilon(T^k x)}$. Let $\epsilon > 0$ and pick $\delta > 0$ and $c > 0$ such that

$$cm(B^n_{\epsilon/2}(x)) \geq m(B^n_\delta(y)) \qquad (n \in \mathbb{N},\ x,y \in X).$$

Let E be a maximal (n,ϵ)-separating set. Then $B^n_{\epsilon/2}(x) \cap B^n_{\epsilon/2}(y) = \emptyset$ for every distinct pair $x,y \in E$. Since $1 \geq \sum_{x \in E} m(B^n_{\epsilon/2}(x)) \geq s_n(\epsilon,T)\ c^{-1} \sup_{z \in X} m(B^n_\delta(z))$, one obtains $s_n(\epsilon,T) \leq c[\sup_{z \in X} m(B^n_\delta(z))]^{-1}$.

Now let $\alpha = (A_1,\dots,A_N)$ be an m-measurable partition satisfying $\operatorname{diam} A_i < \delta$ $(1 \leq i \leq N)$. Since for every $A \in (\alpha)_o^{n-1}$ there exists a $z \in X$ such that $A \subset B^n_\delta(z)$ we get

$$m(A) \leq \sup_{z \in X} m(B^n_\delta(z)).$$

It follows that

$$-\sum_{A \in (\alpha)_o^{n-1}} m(A) \log m(A) \geq \log [\sup_{z \in X} m(B^n_\delta(z))]^{-1}$$

$$\geq \log [c^{-1}\ s_n(\epsilon,T)].$$

Dividing by n and letting $n \to \infty$ yields:

$$h_m(T) \geq h_m(T,\alpha) \geq s(\epsilon,T).$$

Now for $\epsilon \to 0$ one obtains the proposition. □

Example: ([22]) Let X be a compact, metric group. There exists a right invariant metric d (compatible with the topology on X) and a right invariant Haar measure m on X. Let $T : X \to X$ be defined by $T(x) = A(x)g$ where A is a group automorphism and $g \in X$.

It is well known that m is A-invariant and hence T-invariant as well. In order to show that m is T-homogeneous it suffices to prove

$$B^n_\epsilon(x) = \bigcap_{k=o}^{n-1} [A^{-k} B^1_\epsilon(e)]x$$

for every $n \in N$, $\epsilon > 0$ and $x \in X$, where e is the unit in X. Since $B^n_\epsilon(x) = \bigcap_{k=o}^{n-1} T^{-k} B^1_\epsilon(T^k(x))$ is suffices to prove by induction that $T^{-k} B^1_\epsilon(T^k(x)) = [A^{-k} B^1_\epsilon(e)]x$.

This is trivial for $k = 0$. Then

$$T^{-k-1} B^1_\epsilon(T^{k+1}(x)) = T^{-1} [[A^{-k} B^1_\epsilon(e)] T(x)] =$$

$$= A^{-1} \{[A^{-k}B^1_\epsilon(e)] A(x)g] g^{-1}\}$$

$$= [A^{-k-1} B^1_\epsilon(e)]x .$$

(One has to use $A^{-k}[B^1_\epsilon(e) A(x)] = [A^{-k} B^1_\epsilon(e)]x$.)

Hence we know from proposition (19.7) that

$$h_m(T) = h_{top}(T).$$

This result is due to Bowen [22], for group automorphisms we know it from Berg's paper [16]. We shall consider this example several times in the following.

We proceed investigating $\mathfrak{M}_{max}(T)$.

<u>(19.8) Proposition</u> [186]: For every homeomorphism $T, \mathfrak{M}_{max}(T)$ is a convex set. If $h_{top}(T) < \infty$ then the extremal points of $\mathfrak{M}_{max}(T)$ are precisely the ergodic measures in $\mathfrak{M}_{max}(T)$.

Furthermore, if the entropy function $h_m(T)$ is upper semi-continuous on $\overline{\mathfrak{M}_{max}(T)}$, then $\mathfrak{M}_{max}(T)$ is compact.

<u>Proof:</u> The affinity of the entropy function $h_m(T)$ shows convexity of $\mathfrak{M}_{max}(T)$. It also shows that a decomposition $m = \lambda m_1 + (1-\lambda)m_2$ for $m \in \mathfrak{M}_{max}(T)$ within $\mathfrak{M}_T(X)$ is already a decomposition within $\mathfrak{M}_{max}(T)$, so that an extreme point of $\mathfrak{M}_{max}(T)$ is extreme in $\mathfrak{M}_T(X)$ and hence ergodic. □

The preceding proof shows more:

Corollary: If $m \in \mathfrak{M}_{max}(T)$, $h_{top}(T) < \infty$ and if $\{m_x | x \in X\}$ denotes the ergodic decomposition of m, then for almost all $x \in X$ $m_x \in \mathfrak{M}_{max}(T)$.

Following Ruelle in [161] we can prove the following generalization. Define

$$I = \{ m \in \mathfrak{M}_T(X) \mid \exists\, m_n \in \mathfrak{M}_T(X)\ (n \in \mathbb{N}) \text{ such that } \lim_{n\to\infty} m_n = m \text{ and } \lim_{n\to\infty} h_{m_n}(T) = h_{top}(T) \}$$

and

$$J = \{m \in \mathfrak{M}_T(X) | \forall\ f \in C(X)\ \int f dm + h_{top}(T) \leq P(T,f)\}.$$

(19.9) Proposition [161]: I is a convex, closed set and contains $\mathfrak{M}_{max}(T)$. If the entropy function $h_m(T)$ is upper semicontinuous on $\mathfrak{M}_T(X)$ then $I = \mathfrak{M}_{max}(T)$. In any case, $I = J$.

Proof: It is obvious that both I and J are convex and closed. $\mathfrak{M}_{max}(T) \subset I$ is trivial and $\mathfrak{M}_{max}(T) = I$ in case of semicontinuity is obvious as well.

We show now $I = J$.

Let $m \in I$ and choose $m_n \in \mathfrak{M}_T(X)$ with $\lim_{n\to\infty} m_n = m$ and $\lim_{n\to\infty} h_{m_n}(T) = h_{top}(T)$. Recall the variational principle (18.12) in the appendix of the last section:

$$P(T,f) = \sup_{m' \in \mathfrak{M}_T(X)} [h_{m'}(T) + \int f dm'].$$

Now let $f \in C(X)$ and $\epsilon > 0$. Choose $n \in \mathbb{N}$ such that $| \int f dm - \int f dm_n| < \epsilon$ and $h_{top}(T) - h_{m_n}(T) < \epsilon$.

Then it follows that

$$\int f dm + h_{top}(T) \leq \int f dm_n + h_{m_n}(T) + 2\epsilon.$$

Letting $\epsilon \longrightarrow 0$

$$\int f dm + h_{top}(T) \leq P(T,f).$$

Thus we proved $I \subset J$.

Conversely we show $J \subset I$.

First of all we prove for $f \in C(X)$

$$\sup_{m\in J} \int f dm = \sup_{m\in I} \int f dm.$$

Let $m \in J$ and $\epsilon_n > 0$ $(n \in \mathbb{N})$ such that $\lim_{n\to\infty} \epsilon_n = 0$. For every $n \in \mathbb{N}$ there exists an $m_n' \in \mathfrak{M}_T(X)$ such that

$$\int \frac{1}{n} f\, dm + h_{top}(T) \leq \int \frac{1}{n} f dm_n' + h_{m_n'}(T) + \frac{1}{n} \epsilon_n.$$

We may assume that the sequence $(m_n')_{n\in\mathbb{N}}$ converges weakly to $m' \in \mathfrak{M}_T(X)$. Since

$$h_{top}(T) \leq h_{m_n'}(T) + \frac{1}{n} [\epsilon_n + \int f dm_n' - \int f dm],$$

$\lim_{n\to\infty} h_{m_n'}(T) = h_{top}(T)$ and thus $m' \in I$.

On the other hand

$$\int f dm = n \int \frac{1}{n} f dm \leq \int f dm_n' + \epsilon_n + n(h_{m_n'}(T) - h_{top}(T))$$
$$\leq \int f dm_n' + \epsilon_n.$$

For $n \longrightarrow \infty$ this inequality remains true and one gets $\int f dm \leq \int f dm'$.

Therefore $\sup_{m\in J} \int f dm \leq \sup_{m'\in I} \int f dm'$ and by the first part $(I \subset J)$ even equality holds.

Assume now that there exists a measure $m \in J \setminus I$. We shall show that the last equality does not hold for some $f \in C(X)$.

Since I is compact there exist continuous functions $f_1,\ldots,f_n \in C(X)$ and $\epsilon > 0$ such that

$$U := \{m' \in \mathfrak{M}_T(X) \mid |\int f_i dm' - \int f_i dm| < \epsilon \quad (1 \leq i \leq n)\}$$

does not intersect I. The convex function

$$F : \mathfrak{M}_T(X) \longrightarrow \mathbb{R}^n$$

defined by $F(m') = (\int f_i dm')_{1\leq i\leq n}$ maps I onto a convex set in the reflexive space $\mathbb{R}^n$ which does not contain $F(m)$. Hence we can separate $F(m)$ and $F(I)$ by a linear functional which is given by

$\alpha_i \in \mathbb{R}$ $(1 \le i \le n)$ such that

$$\max_{m' \in I} \sum_{i=1}^{n} \int \alpha_i f_i \, dm' < \sum_{i=1}^{n} \int \alpha_i f_i \, dm.$$

Therefore $f = \sum_{i=1}^{n} \alpha_i f_i$ gives the contradiction, and $J \subset I$ follows. □

Remark: In [161] Ruelle proved this proposition for special transformation groups and he did it not only for $P(T,0)$ but also for $P(T,g)$ where $g \in C(X)$. Our proof works in this case also.

(19.10) Proposition [161], [186]: Let the entropy function be upper semicontinuous and $h_{top}(T) < \infty$. Then

$$(*) \qquad h_m(T) = \inf_{f \in C(X)} [P(T,f) - \int f dm]$$

for every $m \in \mathfrak{M}_T(X)$.

Conversely if (*) holds for every $m \in \mathfrak{M}_T(X)$, then the entropy function $h_m(T)$ is upper semicontinuous on $\mathfrak{M}_T(X)$.

Proof: The converse is trivial since the right hand side of (*) is an upper semicontinuous function on $\mathfrak{M}_T(X)$.

Now let $m \in \mathfrak{M}_T(X)$. One inequality is trivial from the definition of $P(T,f)$:

$$h_m(T) \le \inf_{f \in C(X)} [P(T,f) - \int f dm].$$

For the converse inequality let $h > h_m(T)$ be arbitrary and define $C = \{(m',t) \in \mathfrak{M}_T(X) \times \mathbb{R} \mid 0 \le t \le h_{m'}(T)\}$. Since $h_{m'}(T)$ is upper-semicontinuous, C is a compact, convex set which does not contain (m,h). Therefore, there exist $f_1 \dots f_n \in C(X)$ and $\epsilon > 0$ such that

$$U := \{(m',t) \in \mathfrak{M}_T(X) \times \mathbb{R} \mid |t-h| < \epsilon; |\int f_i dm' - \int f_i dm| < \epsilon (1 \le i \le n)\}$$

does not intersect C. Again if we define a map F on $\mathfrak{M}_T(X) \times \mathbb{R}$ by $F((m',t)) = (t, \int f_1 dm', \dots, \int f_n dm')$, then F is convex and continuous, hence $F(C)$ is convex and does not contain (m,h). Therefore one finds $\alpha, \alpha_1, \dots, \alpha_n \in \mathbb{R}$ such that

$$\sup_{(m',t)\in C} [\alpha t + \sum_{i=1}^{n} \alpha_i \int f_i \, dm'] < \alpha h + \sum_{i=1}^{n} \alpha_i \int f_i \, dm.$$

Define $f = \frac{1}{\alpha} \sum_{i=1}^{n} \alpha_i f_i$ (note that $\alpha > 0$!). Then

$$\sup_{(m',t)\in C} [t + \int f dm'] < h + \int f dm.$$

For $t = h_{m'}(T)$ one gets

$$P(T,f) < h + \int f dm.$$

Letting $h \longrightarrow h_m(T)$ it follows that

$$h_m(T) \geq \inf_{f\in C(X)} [P(T,f) - \int f dm]. \qquad \square$$

In order to clarify proposition (19.10) we shall give several examples.

One example assumes $h_{top}(T) = \infty$. Then $P(T,f) = \infty$ for every $f \in C(X)$ and for every measure $m \in \mathfrak{M}_T(X)$ with finite entropy we have

$$h_m(T) < \inf_{f\in C(X)} [P(T,f) - \int f dm] = \infty.$$

The other extreme of (19.10) is given by a uniquely ergodic system (X,T). Clearly in this case for the unique measure $m \in \mathfrak{M}_T(X)$ and every $f \in C(X)$ we get

$$h_m(T) = P(T,f) - \int f dm.$$

In order to give an example where (*) does not hold and $h_{top}(T) < \infty$, consider the example before (19.5). For simplicity suppose the subshifts $(\Lambda_n, \sigma|\Lambda_n)$ considered there are all equal (but still have different symbols in use) and uniquely ergodic with

$$0 < h_{top}(\sigma|\Lambda_n) < \infty \qquad (n \in \mathbb{N}).$$

Denote by m_n the unique measure in $\mathfrak{M}_\sigma(\Lambda_n)$ $(n \in \mathbb{N})$. Now let $f \in C(\Lambda)$ be arbitrary. Since $h_{m_n}(\sigma|\Lambda) = h_{top}(\sigma|\Lambda)$ for every

$n \in \mathbb{N}$ (corollaries 2,3 of (18.8) and (19.4)) it follows that

$$P(\sigma|\Lambda,f) \geq h_{m_n}(\sigma|\Lambda_n) + \int f\, dm_n =$$

$$= h_{top}(\sigma|\Lambda) + \int f\, dm_n.$$

Suppose now for $m_o = \delta_{z^o}$ (the point mass on z^o) we would have

$$0 = h_{m_o}(\sigma) = \inf_{f \in C(\Lambda)} [P(\sigma,f) - f(z^o)].$$

Then for given $\varepsilon > 0$ there exists an $f \in C(\Lambda)$ such that for every $n \in \mathbb{N}$

$$0 \geq P(\sigma,f) - \int f\, dm_o - \varepsilon$$

$$\geq h_{top}(\sigma|\Lambda) + \int f\, dm_n - \int f dm_o - \varepsilon.$$

Letting $n \to \infty$ (that is $(m_n)_{n \in \mathbb{N}}$ converges weakly to m_o) we obtain $h_{top}(\sigma|\Lambda) \leq \varepsilon$. Since $\varepsilon > 0$ was arbitrary, $h_{top}(\sigma|\Lambda) = 0$, a contradiction.

Turning towards another question let us consider a uniquely ergodic topological dynamical system (X,T). In this case there is a unique measure m maximizing the entropy function and this equilibrium state is the "natural" measure in $\mathfrak{M}_T(X)$. It often happens that one has a **distinct feeling which measure is the** "most appropriate" one, for instance the Haar measure, the Lebesgue measure and the Parry measure. It is quite obvious that they are "best", but most of their properties, like being positive on all open sets, or ergodic, or non-atomic may be shared by many other invariant measures. In all these cases, however, they turn out to be the unique measure with maximal entropy.

<u>(19.11) Definition</u> [188]: The topological dynamical system (X,T) is said to be <u>intrinsically ergodic</u> if $\mathfrak{M}_{max}(T)$ consists of a single measure.

By (19.4) this measure must be ergodic (because $h_{top}(T) = \infty$ and intrinsic ergodicity imply unique ergodicity).

Intrinsic ergodicity can be viewed as a generalization of unique ergodicity. It should be noted, however, that neither products nor factors of intrinsically ergodic (X,T) have to be intrinsically ergodic.

There are two possibilities why a system (X,T) can fail to be intrinsically ergodic. Either $\mathfrak{M}_{max}(T)$ is empty (cf (19.5)) or $\mathfrak{M}_{max}(T)$ contains more than one element. One obvious example for this second possibility is given by the disjoint union of two copies of the same topological dynamical system (see (14.20)). There exist more interesting examples. In [167] Shtilman constructs for any $N > 0$ a topologically transitive subshift with exactly N ergodic measures of maximal entropy. In [121] Krieger constructs a transformation T such that $\mathfrak{M}_{max}(T)$ contains two Bernoulli measures positive on all open sets. Kornfeld [112] and Goodman [62] show by examples that:

(19.12) Proposition: There exist minimal transformations $T : X \longrightarrow X$ admitting several measures with maximal entropy.

The constructions in sections 27,31 give stronger results: There exist minimal systems (X,T) with $\mathfrak{M}_{max}(T) = \mathfrak{M}_T(X)$ and with an arbitrary prescribed number (finite, countable or uncountable) of ergodic measures.

The basic example of an intrinsically ergodic topological dynamical system is given by the shift $\sigma: S^{\mathbb{Z}} \longrightarrow S^{\mathbb{Z}}$ (S finite).

(19.13) Proposition: $(S^{\mathbb{Z}},\sigma)$ is intrinsically ergodic.

Proof: Let $\alpha = ({}_o[1],\dots,{}_o[s])$ denote the natural generator. $(\alpha)_o^{k-1}$ has s^k elements. By (10.4.b) one has $H_\mu((\alpha)_o^{k-1}) \leq k \log s$ for all $\mu \in \mathfrak{M}_\sigma(S^{\mathbb{Z}})$. Equality holds for all k if $\mu = \mu_\pi$, the Bernoulli measure with probability vector $\pi = (s^{-1},\dots,s^{-1})$. For any $\mu \neq \mu_\pi$ the inequality is strict for some k, and thus $h_\mu(\sigma) < \log s = h_{top}(\sigma)$. □

The notion of intrinsic ergodicity was first exploited by Parry in 1964.

(19.14) Theorem [146]: Let Λ be a subshift of finite type which is topologically transitive. Then $\sigma|\Lambda$ is intrinsically ergodic.

Proof: We shall follow the proof given by Adler and Weiss [6]. We have seen in (17.14) that the Parry measure μ_Λ has maximal entropy. We shall use the same notations as in the proof of (17.14).

Let k denote the minimum and K the maximum of

$\{\lambda\cdot u_i\cdot v_j \mid i,j = 1,\ldots,s\}$. It is easy to check that for any cylinder C of length n occuring in Λ, one has

$$\frac{k}{\lambda^n} \leq \mu_\Lambda(C) \leq \frac{K}{\lambda^n} \ .$$

Recall that $\Sigma((\alpha)_o^{n-1})$ denotes the algebra generated by the partition $(\alpha)_o^{n-1}$ in $S^{\mathbb{Z}}$. Any element E in $\Sigma((\alpha)_o^{n-1})|\Lambda$ is a disjoint union of cylinders of length n occuring in Λ. Let # E denote the number of those cylinders. Clearly

$$(*) \qquad \frac{k\cdot \# E}{\lambda^n} \leq \mu_\Lambda(E) \leq \frac{K\cdot \# E}{\lambda^n} \ .$$

Now suppose that $\mathfrak{M}_{max}(\sigma|\Lambda)$ does not only consist of μ_Λ. Then by (19.8) there exists an ergodic measure $\nu \neq \mu_\Lambda$ with maximal entropy $\log \lambda$. By (5.4) ν and μ_Λ are mutually singular. Since the sequence $\Sigma((\alpha)_o^{n-1})|\Lambda$ generates $\Sigma|\Lambda$, there exists a sequence of sets $E_n \in \Sigma((\alpha)_o^{n-1})|\Lambda$ with $\lim_{n\to\infty} \mu_\Lambda(E_n) = 0$ and $\lim_{n\to\infty} \nu(E_n) = 1$.

By assumption

$$\inf_{n\in\mathbb{N}} \frac{1}{n} H_\nu((\alpha)_o^{n-1}) = \log \lambda .$$

Hence one obtains

$$\begin{aligned}\log \lambda^n &\leq H_\nu((\alpha)_o^{n-1}) \leq H_\nu((E_n, X\setminus E_n)) + H_\nu((\alpha)_o^{n-1}|(E_n, X\setminus E_n)) \\ &= H_\nu((E_n, X\setminus E_n)) + \nu(E_n)\, H_{\nu|E_n}((\alpha)_o^{n-1}|E_n) \\ &\qquad + (1-\nu(E_n))\, H_{\nu|X\setminus E_n}((\alpha)_o^{n-1}|X\setminus E_n) \\ &\leq \nu(E_n) \log \frac{\# E_n}{\nu(E_n)} + (1-\nu(E_n)) \log \frac{\#(X\setminus E_n)}{1-\nu(E_n)} \ .\end{aligned}$$

(For the last inequality use (10.4.b).)

Applying the first half of (*) one obtains

$$0 \leq \nu(E_n) \log \frac{\mu_\Lambda(E_n)}{k\nu(E_n)} + (1-\nu(E_n)) \log \frac{1-\mu_\Lambda(E_n)}{k(1-\nu(E_n))} .$$

Since $\mu_\Lambda(E_n) \to 0$ and $\nu(E_n) \to 1$, the first summand of the right hand side converges to $-\infty$, a contradiction. □

Other classes of intrinsically ergodic subshifts can be found in the papers by B.Weiss ([188], [190]) and Krieger ([121]). Also section 25 gives examples for the last theorem.

As a corollary of the Dinaburg-Goodman theorem we saw in (19.2) that semicontinuity of the measure theoretic entropy always implies $\mathfrak{M}_{max}(T) \neq \emptyset$. Obviously this condition is far from being necessary.

So the question arises to find a necessary and sufficient condition for $\mathfrak{M}_{max}(T) \neq \emptyset$.

Let us first ask **another question:** what topological property is necessary and sufficient for the entropy function to be upper semicontinuous?

We are dealing with both problems in this section, starting with the second one, however only with the sufficient part. For doing this we need the notions of entropy-expansiveness and asymptotic entropy-expansiveness. The first one is due to Bowen [23], the second one has first been defined by **Misiurewicz** [135]. Here we follow [138]. The idea is to get a uniform estimate for conditional entropy, which can be made sufficiently small. Therefore, we start to define conditional topological entropy in a similar way as it is done in the measure theoretic case.

In this section (X,T) denotes a fixed top. dynamical system.

(20.1) Definition: For open covers $\mathfrak{U}$ and $\mathfrak{V}$ of X define

$$N(\mathfrak{U}|\mathfrak{V}) : \max_{V\in\mathfrak{V}} \min\{\text{card } \mathfrak{U}' \mid \mathfrak{U}' \text{ is a subcover of } \mathfrak{U}|V\}.$$

$\log N(\mathfrak{U}|\mathfrak{V})$ is called the conditional topological entropy of $\mathfrak{U}$ with respect to $\mathfrak{V}$.

The following lemma is easy to show:

(20.2) Lemma: Let $\mathfrak{U},\mathfrak{U}',\mathfrak{V}$ and $\mathfrak{V}'$ be open covers of X. Then

$$N(T^{-1}\mathfrak{U}|T^{-1}\mathfrak{V}) = N(\mathfrak{U}|\mathfrak{V})$$

and

$$N(\mathfrak{U} \vee \mathfrak{U}'|\mathfrak{V} \vee \mathfrak{V}') \leq N(\mathfrak{U}|\mathfrak{V})\, N(\mathfrak{U}'|\mathfrak{V}').$$

Corollary: Let $\mathfrak{U},\mathfrak{V}$ be open covers of X. Then

$$\lim_{n\to\infty} \frac{1}{n} \log N((\mathfrak{U})_0^{n-1} | (\mathfrak{B})_0^{n-1}) =: h(T,\mathfrak{U}|\mathfrak{B})$$

exists and is called the <u>conditional topological entropy of T on $\mathfrak{U}$ with respect to $\mathfrak{B}$</u>.

<u>Proof:</u> Lemma (20.2) shows that the sequence $N((\mathfrak{U})_0^{n-1} | (\mathfrak{B})_0^{n-1})$ is submultiplicative. Then apply (10.7). □

<u>(20.3) Lemma</u>: If $\mathfrak{U}, \mathfrak{U}'$ and $\mathfrak{B}$ are open covers of X and $\mathfrak{U} \geq \mathfrak{U}'$ then

$$h(T, \mathfrak{U}|\mathfrak{B}) \geq h(T, \mathfrak{U}'|\mathfrak{B}).$$

<u>Proof</u>: Obvious. □

<u>(20.4) Definition</u>: Let $\mathfrak{B}$ be an open cover of X.

$$h(T|\mathfrak{B}) := \sup \{h(T,\mathfrak{U}|\mathfrak{B}) \mid \mathfrak{U} \text{ open cover of } X\}$$

$$= \lim \{h(T,\mathfrak{U}|\mathfrak{B}) \mid \mathfrak{U} \text{ open cover of } X\}$$

is called the <u>conditional topological entropy of T with respect to $\mathfrak{B}$</u>.

<u>(20.5) Lemma</u>: If $\mathfrak{U}$ and $\mathfrak{B}$ are open covers of X and $\mathfrak{U} \geq \mathfrak{B}$, then

$$h(T|\mathfrak{U}) \leq h(T|\mathfrak{B}).$$

<u>Proof:</u> If $\mathfrak{U}'$ is an open cover and $n \in \mathbb{N}$, then

$$N((\mathfrak{U}')_0^{n-1} | (\mathfrak{U})_0^{n-1}) \leq N((\mathfrak{U}')_0^{n-1} | (\mathfrak{B})_0^{n-1}).$$ □

<u>(20.6) Definition:</u>

$$h^*(T) := \inf \{h(T|\mathfrak{B}) | \mathfrak{B} \text{ open cover of } X\}$$

$$= \lim \{h(T|\mathfrak{B}) | \mathfrak{B} \text{ open cover of } X\}$$

is called the <u>conditional topological entropy of T</u>.

(20.7) Definition: ([23], [138]) The homeomorphism T is called h-expansive (entropy-expansive) if there exists an open cover $\mathfrak{B}$ of X such that

$$h(T|\mathfrak{B}) = 0.$$

T is called **asymptotically** h-expansive (asymptotically entropy-expansive) if

$$h^*(T) = 0.$$

Since we are only giving a short survey about the properties of $h^*(T)$ and $h(T,\mathfrak{B})$, the reader should contact [138] for further details. In [23] Bowen originally gave the definition of h-expansive homeomorphisms. His definition reads as follows:

Recall that $d(\cdot,\cdot)$ denotes the metric on X. Define an equivalent metric d_n by

$$d_n(x,y) = \max_{0\leq i\leq n-1} d(T^i(x), T^i(y))$$

for every $n \in \mathbb{N}$. Let (cf. (19.6), (19.7))

$$B^n_\epsilon(x) := \{y \in X \mid d_n(x,y) \leq \epsilon\}$$

for $n \in \mathbb{N}$, $\epsilon > 0$ and $x \in X$.

Recall furthermore that $r_l(\bigcap_{n=0}^{\infty} B^n_\epsilon(x), \delta, T)$ denotes the minimal cardinality of an (l,δ)-spanning set for $\bigcap_{n=0}^{\infty} B^n_\epsilon(x)$ in the usual d-**metric** (cf. (14.14)) and define

$$r(\bigcap_{n=0}^{\infty} B^n_\epsilon(x), \delta, T) := \limsup_{l\to\infty} \frac{1}{l} \log r_l(\bigcap_{n=0}^{\infty} B^n_\epsilon(x), \delta, T),$$

$$h(T, \bigcap_{n=0}^{\infty} B^n_\epsilon(x)) := \lim_{\delta\to 0} r(\bigcap_{n=0}^{\infty} B^n_\epsilon(x), \delta, T)$$

and

$$h^*(\varepsilon) = \sup_{x \in X} h(T, \bigcap_{n=0}^{\infty} B_\varepsilon^n(x)).$$

Then we can prove:

(20.8) Proposition [138]: T is h-expansive iff there exists an $\varepsilon > 0$ such that

$$h^*(\varepsilon) = 0.$$

T is asymptotically h-expansive iff

$$\lim_{\varepsilon \to 0} h^*(\varepsilon) = 0.$$

Proof: Obviously we are finished by showing the following: Let $\mathfrak{U}$, $\mathfrak{B}$ be open covers of X and $\varepsilon > 0$ such that diam $U < \varepsilon$ ($\forall\ U \in \mathfrak{U}$) and such that ε is a Lebesgue number of $\mathfrak{B}$. Then

$$h(T|\mathfrak{U}) \leq h^*(\varepsilon) \leq h(T|\mathfrak{B}).$$

1) We shall show the second inequality first.

Let $x \in X$, $\delta > 0$ and $\mathfrak{U}'$ be an open cover of X with diameter less than δ. Let $n \in \mathbb{N}$. Since ε is a Lebesgue number of $\mathfrak{B}$, there exists a $V \in (\mathfrak{B})_0^{n-1}$ such that $x \in V$ and

$$\bigcap_{k=0}^{\infty} B_\varepsilon^k(x) \subset B_\varepsilon^n(x) \subset V.$$

Since for every $U \in (\mathfrak{U}')_0^{n-1}|V$ and for every $z, y \in U$

$$d(T^k(z), T^k(y)) < \delta \qquad \text{for all } 0 \leq k \leq n-1,$$

it follows for every subcover $\hat{\mathfrak{U}}$ of $(\mathfrak{U}')_0^{n-1}|V$ that

$$r_n(\bigcap_{k=0}^{\infty} B_\varepsilon^k(x), \delta, T) \leq \text{card}\ \hat{\mathfrak{U}},$$

hence

$$r_n\Big(\bigcap_{k=0}^{\infty} B_{\varepsilon}^{k}(x),\delta,T\Big) \le N((\mathfrak{U}')_0^{n-1} \mid (\mathfrak{B})_0^{n-1}).$$

It follows by definition that

$$r\Big(\bigcap_{k=0}^{\infty} B_{\varepsilon}^{k}(x),\delta,T\Big) \le h(T,\mathfrak{U}' \mid \mathfrak{B}) \le h(T \mid \mathfrak{B}).$$

Letting $\delta \longrightarrow 0$ yields

$$h\Big(T, \bigcap_{k=0}^{\infty} B_{\varepsilon}^{k}(x)\Big) \le h(T \mid \mathfrak{B}).$$

Since $x \in X$ was chosen arbitrarily

$$h^*(\varepsilon) \le h(T \mid \mathfrak{B}).$$

2) The other inequality is more difficult.

We show that for every $\delta > 0$ there exist open covers $\mathfrak{U}_t'$ such that $h(T, \mathfrak{U}_t' \mid \mathfrak{U}) < h^*(\varepsilon) + \delta$ and $\lim_{t\to\infty} \operatorname{diam} \mathfrak{U}_t = 0$.
Let $\gamma > 0$ be arbitrary. Define $\mathfrak{U}' = \{B_{2\gamma}(x) \mid x \in X\}$.

Since for every $n \in \mathbb{N}$, $U \in (\mathfrak{U})_0^{n-1}$, $x,y \in U$ and $0 \le i \le n-1$ we have $d(T^i x, T^i y) < \varepsilon$, it follows that $U \subset B_{\varepsilon}^{n}(x)$ for every $x \in U$.

Let $\hat{\mathfrak{U}}$ be a subcover of $(\mathfrak{U}')_0^{n-1} \mid U$ of minimal cardinality. The inequality

$$\operatorname{card} \hat{\mathfrak{U}} \le r_n(U,\gamma,T) \le r_n(B_{\varepsilon}^{n}(x), \gamma,T)$$

$$\le \sup_{x\in X} r_n(B_{\varepsilon}^{n}(x), \gamma,T)$$

(use (14.16) part b of the proof) implies

$$N((\mathfrak{U}')_0^{n-1} \mid (\mathfrak{U})_0^{n-1}) \le \sup_{x\in X} r_n(B_{\varepsilon}^{n}(x), \gamma,T).$$

Thus it is left to show that

$$\limsup_{n\to\infty} \frac{1}{n} \log \sup_{x\in X} r_n(B^n_\epsilon(x),\gamma,T) \leq h^*(\epsilon) + \delta$$

for every $\gamma > 0$. We do this following [23].

Let $y \in X$ be fixed. By the definition of $h^*(\epsilon)$, there exists an $m_y \in \mathbb{N}$ such that

$$\frac{1}{m_y} \log \operatorname{card} E_y \leq h^*(\epsilon) + \delta ,$$

where E_y is a minimal $(m_y, \frac{1}{4}\gamma)$-spanning set for $\bigcap_{k=0}^{\infty} B^k_\epsilon(y)$. Define

$$W_y := \{u \in X \mid \exists\, z \in E_y : d(T^k z, T^k u) < \tfrac{1}{2}\gamma (0 \leq k < m_y)\}.$$

Then there exists an $N_y > m_y$ such that $B_\epsilon^{N_y}(y) \subset W_y$, **and since W_y is open we can find an $\epsilon' > \epsilon$ with $B_{\epsilon'}^{N_y}(y) \subset W_y$.** Now we pick an open set $V_y \ni y$ such that for $u \in V_y$ and $0 \leq k \leq N_y - 1$ $d(T^k u, T^k y) < \epsilon' - \epsilon$. Clearly this implies for $u \in V_y$ that $B_\epsilon^{N_y}(u) \subset W_y$.

Since X is compact it is possible to find $y_1,\ldots,y_s$ such that $X = \bigcup_{i=1}^{s} V_{y_i}$.

In order to show the statement let $x \in X$ and $n \in \mathbb{N}$.

Let F_i denote an $(m_{y_i}, \frac{1}{2}\gamma)$-spanning set for W_{y_i} and let F denote an $(N, \frac{1}{2}\gamma)$-spanning set for X where $N = \max_{1\leq i\leq s} N_{y_i}$.

We shall split the set of integers $\{0,1,\ldots,n-1\}$ into disjoint subsets $\{t_0,\ldots,t_1\}$, $\{t_1 + 1,\ldots,t_2\}$, ..., $\{t_{r-1} + 1,\ldots,t_r\}$ where $t_0 = 0 < t_1 < \ldots < t_r = n$. Define $t_0 = 0$ and let y_{i_0} satisfy $x \in V_{y_{i_0}}$. Suppose now that $t_o < t_1 < \ldots < t_k$, $y_{i_o}, \ldots, y_{i_k}$ are defined such that $T^{t_j}(x) \in V_{y_{i_j}}$ $(0 \leq j \leq k)$. If $t_k > n - N$ then we set $t_{k+1} = n$; if $t_k \leq n - N$ then we set $t_{k+1} = t_k + m_{y_{i_k}}$.

In this way we get a finite sequence $0 = t_o < t_1 < \ldots < t_r = n$. By construction we know that $T^{t_j}(x) \in V_{y_{i_j}}$ $(0 \leq j \leq r - 2)$, hence

$$B_\epsilon^{N_{y_{i_j}}}(T^{t_j}x) \subset W_{y_{i_j}}.$$

Because of $T^{t_j}(B_\epsilon^n(x)) \subset \{z \in X \mid d(T^k(z), T^{k+t_j}(x)) < \epsilon \ (0 \leq k \leq N_{y_{i_j}} - 1)\}$

it follows that

$$T^{t_j}(B_\epsilon^n(x)) \subset W_{y_{i_j}} \qquad (0 \leq j \leq r - 2).$$

Therefore F_{i_j} is a $(t_{j+1} - t_j, \frac{1}{2}\gamma)$-spanning set for $T^{t_j}(B_\epsilon^n(x))$ $(0 \leq j \leq r - 2)$ and F is a $(t_r - t_{r-1}, \frac{1}{2}\gamma)$-spanning set for $T^{t_{r-1}}(B_\epsilon^n(x))$. Considering the open cover

$$\mathfrak{U} = \{T^{-t_{r-1}}(B_{\frac{1}{2}\gamma}(z_{r-1})) \cap \bigcap_{j=o}^{r-2} T^{-t_j}(B_{\frac{1}{2}\gamma}(z_j)) \mid z_j \in F_{i_j} \ (0 \leq j \leq r-2)$$
$$z_{r-1} \in F\}$$

of $B_\epsilon^n(x)$ we see that

$$r_n(B_\epsilon^n(x), \gamma, T) \leq \operatorname{card} \mathfrak{U} \leq \operatorname{card} F \prod_{j=o}^{r-2} \operatorname{card} F_{i_j}.$$

Choosing F minimal and setting $F_i = E_{y_i}$ $(1 \leq i \leq s)$ it follows that

$$r_n(B^n_\varepsilon(x),\gamma,T) \le r_n(X,\tfrac{1}{2}\gamma,T) \prod_{j=o}^{r-2} r_{m_{y_{i_j}}} \left(\bigcap_{m=o}^{\infty} B^m_\varepsilon(y_{i_j}),\tfrac{1}{4}\gamma,T\right)$$

$$\le r_N(X,\tfrac{1}{2}\gamma,T) \prod_{j=o}^{r-2} \exp\,(h^*(\varepsilon) + \delta)\, m_{y_{i_j}}$$

$$\le r_N(X,\tfrac{1}{2}\gamma,T)\ \exp\,(h^*(\varepsilon) + \delta)n.$$

Since $n \in \mathbb{N}$ and $x \in X$ were choosen arbitrarily it follows:

$$\frac{1}{n} \log \sup_{x\in X} r_n(B^n_\varepsilon(x),\gamma,T) \le [\tfrac{1}{n} \log r_N(X,\tfrac{1}{2}\gamma,T)] + h^*(\varepsilon) + \delta.$$

Letting $n \longrightarrow \infty$ we get the statement. □

(20.9) Theorem [138]: If T is asymptotically h-expansive, then the measure theoretic entropy is an upper semicontinuous function on $\mathfrak{M}_T(X)$.

Proof: Let $m \in \mathfrak{M}_T(X)$ and assume that the sequence $m_n \in \mathfrak{M}_T(X)$ converges weakly to m.

1) Let $\varepsilon > 0$ and pick any open cover $\mathfrak{B}$ such that

$$h(T|\mathfrak{B}) < \varepsilon.$$

Using lemma (16.6) there is an m-continuity Borel partition α which is finer than $\mathfrak{B}$ (see the remark after (16.6)).

By definition of $h(T,\mathfrak{B})$ for any open cover $\mathfrak{U}$ there exists an $2 \le n = n(\mathfrak{U})$ such that

$$(*) \qquad \frac{1}{2n+1} \log N((\mathfrak{U})^{n-1}_{-n+1} \mid (\mathfrak{B})^{n-1}_{-n+1}) < 2\varepsilon$$

and

$$H((\tfrac{1}{n},1-\tfrac{1}{n})) := -\frac{1}{n} \log \frac{1}{n} - (1 - \frac{1}{n}) \log (1-\frac{1}{n}) < \varepsilon.$$

2) We claim that

$$h_{m'}(T,\alpha) \geq h_{m'}(T) - 6\varepsilon$$

for every $m' \in \mathfrak{M}_T(X)$.

Let $m' \in \mathfrak{M}_T(X)$ and let $(\mathfrak{U}_k)_{k\in\mathbb{N}}$ be a sequence of open covers satisfying

$$\sup_{U\in\mathfrak{U}_k} \operatorname{diam} U < \varepsilon_k,$$

where $\varepsilon_k > 0$ $(k \in \mathbb{N})$ and $\lim_{k\to\infty} \varepsilon_k = 0$.

Let - for a moment - k be fixed and set $\mathbf{n_k := n(\mathfrak{U}_k)}$. Writing $(\alpha)_{-n_k+1}^{n_k-1} = (A_1,\ldots,A_s)$ choose sets $V_i \in (\mathfrak{B})_{-n_k+1}^{n_k-1}$ with $\overline{A_i} \subset V_i$ $(1 \leq i \leq s)$. Applying (*) to $\mathfrak{U}_k$ for every $1 \leq i \leq s$ there exists a subcover $\mathfrak{U}_k(V_i)$ of $(\mathfrak{U}_k)_{-n_k+1}^{n_k-1}|V_i$ satisfying

$$\operatorname{card} \mathfrak{U}_k(V_i) \leq \exp 2(2n_k+1)\,\varepsilon.$$

Using (16.6) again for every $1 \leq i \leq s$ one obtains a partition α^i of V_i finer than $\mathfrak{U}_k(V_i)$ such that

$$\operatorname{card} \alpha^i = \operatorname{card} \mathfrak{U}_k(V_i) \leq \exp(2\varepsilon(2n_k+1)).$$

Defining a partition $\tilde{\alpha}_k$ piecewise by setting

$$\tilde{\alpha}_k|A_i = \alpha^i|A_i \quad (1 \leq i \leq s)$$

one gets

$$\max_{A\in(\alpha)_{-n_k+1}^{n_k-1}} \operatorname{card}\{A' \in \tilde{\alpha}_k | A' \cap A \neq \emptyset\} \leq \exp(2\varepsilon(2n_k+1)).$$

Let W_k be a Rohlin set such that $W_k \cap T^i W_k = \emptyset$ for $1 \leq i \leq n_k - 1$ and $m' \left(\bigcup_{i=-n_k+1}^{n_k-1} T^i W_k \right) = 1$ (see (1.18)) and define $\alpha_k := \tilde{\alpha}_k \cap W_k$ (see (9.2)). Note that for $A \in (\alpha)_{-n_k+1}^{n_k-1} \cap W_k$

$$\text{card}\,\{A' \in \alpha_k \mid A' \cap A \neq \emptyset\} \leq \exp 2(2n_k + 1)\varepsilon,$$

hence it follows using $m'(W_k) \leq n_k^{-1}$ that

$$H_{m'}(\alpha_k \mid (\alpha)_{-n_k+1}^{n_k-1} \cap W_k) \leq 5\varepsilon.$$

Therefore by (11.7),

$$\begin{aligned} h_{m'}(T,\alpha_k) - 5\varepsilon &\leq h_{m'}(T,(\alpha)_{-n_k+1}^{n_k-1} \cap W_k) \\ &\leq h_{m'}(T,(\alpha)_{-n_k+1}^{n_k-1}) + H\left(\left(\frac{1}{n_k}, 1 - \frac{1}{n_k}\right)\right) \end{aligned}$$

yields

$$h_{m'}(T,\alpha) \geq h_{m'}(T,\alpha_k) - 6\varepsilon.$$

This last inequality holds for every $k \in \mathbb{N}$. Therefore $h_{m'}(T,\alpha) \geq h_{m'}(T) - 6\varepsilon$ is shown if the sequence $(\alpha_k)_{k\in\mathbb{N}}$ generates the σ-algebra under T (see (9.4)) or - what is equivalent by **a simple corollary of (9.6) - if the sequence** $(\alpha_k)_{k\in\mathbb{N}}$ separates under T all points m' - a.e.

Let $W := \bigcap_{k\in\mathbb{N}} \bigcup_{i=-n_k+1}^{n_k-1} T^i W_k$ which clearly satisfies $m'(W) = 1$. Let $x \neq y \in W$ and pick $k \in \mathbb{N}$ such that $d(x,y) > \varepsilon_k$. There is $-n_k < t < n_k$ such that $T^t x \in W_k$. Assume that $T^t x$ and $T^t y$ belong to the same $A \in \alpha_k$. Since

$A \subset \bigcap_{i=-n_k+1}^{n_k-1} T^{-i} U_{l_i}$, where each $U_{l_i} \in \mathfrak{U}_k \vee \mathfrak{B}$ is chosen suitably, $x, y \in U_{l_{-t}}$ implies $d(x,y) < \varepsilon_k$, a contradiction.

3) Returning to m, m_n $(n \in \mathbb{N})$, $\varepsilon > 0$ and α in 1), proposition (2.7) shows that for $A \in (\alpha)_o^{p-1}$ $(p \in \mathbb{N})$

$$\lim_{n\to\infty} m_n(A) = m(A).$$

Choose $p \in \mathbb{N}$ such that $h_m(T,\alpha) \geq \frac{1}{p} H_m((\alpha)_o^{p-1}) - \varepsilon$.
There exists an $n_o \in \mathbb{N}$ such that for $n > n_o$

$$\frac{1}{p} \mid H_m((\alpha)_o^{p-1}) - H_{m_n}((\alpha)_o^{p-1}))| < \varepsilon .$$

It follows that (using 2))

$$h_m(T) \geq h_m(T,\alpha) \geq \frac{1}{p} H_{m_n}((\alpha)_o^{p-1}) - 2\varepsilon$$

$$\geq h_{m_n}(T,\alpha) - 2\varepsilon \geq h_{m_n}(T) - 8\varepsilon$$

for every $n \geq n_o$.

Since $\varepsilon > 0$ was chosen arbitrarily it follows that

$$h_m(T) \geq \limsup_{n\to\infty} h_{m_n}(T). \qquad \Box$$

Example: The converse of the last proposition does not hold - necessarily because $h_{top}(T) = \infty$ clearly implies $h(T|\mathfrak{B}) = \infty$ for every $\mathfrak{B}$. Also if we assume $h_{top}(T) < \infty$ then upper semicontinuity of the entropy function does not imply $h^*(T) = 0$.

We give an example modifying the standard example for (19.5) explained in the beginning of section 19. **Misiurewicz's** example works in the same manner (see [138]).

Let the Λ_n's be equal (but using different symbols) and strictly ergodic subshifts with $0 < h_{top}(\sigma|\Lambda_n) < \infty$. Choose

$\omega \in \Lambda_1$. For $n \in \mathbb{N}$ define $\omega^n = (\omega_j^n)_{j\in\mathbb{Z}}$ setting $\omega_{nk+j}^n = \omega_j$ $(0 \leq j \leq n-1,\ k \in \mathbb{Z})$ where $\omega = (\omega_j)_{j\in\mathbb{Z}}$. Now for $n \in \mathbb{N}$ let $X_n = \bigcup_{j=0}^{n-1} \{\sigma^j(\omega^n)\} \times \Lambda_n$ **- the discrete topology on** the first factor - and let $X_o = \Lambda_1 \times \{z^o\}$, where Λ_1 has got the usual topology.

So we can define

$$X = \bigcup_{n=o}^{\infty} X_n \text{ and } T((\eta,x)) = (\sigma(\eta),\ \sigma(x)),$$

where the topology on X is defined as follows: If $\eta \in \Lambda_1$ and $(\eta^n,\ x_n) \in X_n$ then $\lim\limits_{n\to\infty} (\eta^n, x_n) = (\eta, 0)$ if $\lim\limits_{n\to\infty} \eta^n = \eta$. Clearly X is compact and T a homeomorphism.

Corollary 3 of theorem (18.8) implies

$$h_{top}(T) = h_{top}(\sigma|\Lambda_1).$$

It is easy to see that every measure $m \in \mathfrak{M}_{T|X_n}(X_n)$ satisfies $h_m(T) = h_{top}(\sigma|\Lambda_1)$: If π denotes the projection $\pi : X_n \longrightarrow \Lambda_n$ (resp. $\pi : X_o \longrightarrow \Lambda_1$) then (recall that $\sigma|\pi(X_n)$ is strictly ergodic)

$$h_{top}(\sigma|\pi(X_n)) = h_{\pi m}(\sigma|\pi(X_n)) \leq h_m(T|X_n) \leq h_{top}(T|X_n)$$

$$= h_{top}(\sigma|\pi(X_n)).$$

From the affinity of the entropy function (see (10.13)) it follows that $h_m(T) = h_{top}(\sigma|\Lambda_1)$ for every $m \in \mathfrak{M}_T(X)$, hence the entropy function is continuous.

In order to see that T is not asymptotically h-expansive, let $\mathfrak{B}$ be an open cover of X, which - by lemma (20.5) - can be assumed to contain sets of the form

$$\{\sigma^j(\omega^n)\} \times W_k^n \quad (0 \leq j \leq n-1,\ 1 \leq k \leq r_n,\ \ n \leq n_o)$$

where $(W_k^n | 1 \leq k \leq r_n)$ is an open cover of Λ_n, and sets of the form

$$\Big(\bigcup_{n=n_o+1}^{\infty} \Lambda_n \cup \{z^o\}\Big) \times W_k \quad (1 \leq k \leq r)$$

where $(W_k' | 1 \leq k \leq r)$ is an open cover of Λ_1 and where $W_k := W_k' \cup \{\sigma^j(\omega^n) \mid n > n_o,\ \sigma^j(\omega) \in W_k'\}$.

Now let $s \in \mathbb{N}$. There exists a $V \in (\mathfrak{B})_o^{s-1}$ and a $0 \leq j \leq n-1$ (where $n > n_o$) such that $\{\sigma^j(\omega^n)\} \times \Lambda_n \subset V$. Since for every open cover $\mathfrak{U}$ of X

$$\text{card } \mathfrak{U} | V \geq \text{card } \pi(\mathfrak{U}) | \Lambda_n$$

holds, it follows that

$$N((\pi(\mathfrak{U}) | \Lambda_n)_o^{s-1}) = N((\pi(\mathfrak{U}))_o^{s-1} | \Lambda_n)$$
$$\leq N((\mathfrak{U})_o^{s-1} | (\mathfrak{B})_o^{s-1}).$$

Therefore taking log and then limits over s and $\mathfrak{U}$ it follows that

$$0 < h(\sigma | \Lambda_n) \leq h(T, \mathfrak{B}).$$

So far we studied the question **what** is a sufficient (topological) condition for upper semicontinuity of the entropy function. It is still an open problem to find a necessary and sufficient condition.

We know from theorem (18.8) - the Dinaburg-Goodman-theorem - that asymptotical h-expansiveness of T always implies $\mathfrak{m}_{max}(T) \neq \emptyset$, however, the converse does not hold. In view of proposition (19.9) we only have to assure that for some sequence $m_n \in \mathfrak{m}_T(X)$ with $m_n \longrightarrow m$ and $h_{m_n}(T) \longrightarrow h_{top}(T)$ the entropy function is upper semicontinuous (that is $h_m(T) \geq \limsup_{n\to\infty} h_{m_n}(T)$). In the following we shall look for conditions in terms of open covers to assure this property.

<u>(20.10) Definition</u> [46]: A homeomorphism T is called <u>locally entropy-expansive</u> (locally h-expansive) if there exists a sequence $(\mathfrak{U}_n)_{n\in\mathbb{N}}$ of finite open covers of X satisfying

$$\sum_{n=1}^{\infty} H(T, \mathfrak{U}_n) < \infty$$

and

$$\lim_{n\to\infty} H(T, \bigvee_{k=1}^{n} \mathfrak{U}_k) = h_{top}(T).$$

<u>Remark:</u> Suppose there exists an open cover $\mathfrak{U}$ of X satisfying $H(T, \mathfrak{U}) = h_{top}(T) < \infty$. Then T obviously is locally h-expansive. Thus local h-expansiveness can be viewed as a generalization of expansiveness.

It is clear from the definition of $h_{top}(T)$, that topological entropy is finite for locally h-expansive T's. The expression "locally" is used to indicate that such a T ensures the local upper semi-continuity of entropy in the sense above.

In order to prepare this theorem recall what was done in the propositions (18.6) and (18.7):

Let $\mathfrak{U}$ be a finite open cover of X and $N \in \mathbb{N}$. Then there exist a finite open cover $\mathfrak{B} \geq (\mathfrak{U})_0^{N-1}$ and a Borel partition α' finer than $\mathfrak{B}$ such that

$$\text{card } \alpha' = \text{card } \mathfrak{B},$$

$$p(\mathfrak{B}) \leq N \text{ card } \mathfrak{U} + 1$$

and

$$p(\alpha') \leq N \text{ card } \mathfrak{U} + 1.$$

On the other hand for any Borel partition α finer than $(\mathfrak{U})_0^{N-1}$ there exists a measure $m \in \mathfrak{M}_{T^N}(X)$ such that for every m-measurable partition β of X one has

$$h_m(T^N,\beta) \geq H(T^N, (\mathfrak{U})_0^{N-1}) - p(\beta|\alpha)$$

where the last number was defined to be

$$p(\beta|\alpha) := \max_{B\in\beta} \text{card } \{A \in \alpha | \bar{A} \cap \bar{B} \neq \emptyset\}.$$

Using these facts one can show

<u>(20.11) Theorem:</u> If T is locally h-expansive then

$$\mathfrak{M}_{max}(T) \neq \emptyset.$$

<u>Proof:</u> Denote by $(\mathfrak{U}_n)_{n\in\mathbb{N}}$ the sequence of open covers given by the assumption of local h-expansiveness. Let $\epsilon_n > 0$ $(n \in \mathbb{N})$ such that $\sum_{n=1}^{\infty} \epsilon_n < \infty$. There exist **integers** $N_n \in \mathbb{N}$ $(n \in \mathbb{N})$ such that

$$N((\mathfrak{U}_n)_0^{N_n-1}) \leq \exp\{N_n (H(T,\mathfrak{U}_n) + \epsilon_n)\},$$

$$\lim_{n\to\infty} \frac{N_{n-1}}{N_n} = 0,$$

and N_{n-1} divides N_n $(n \geq 2)$.

By induction we shall construct a sequence m_n of

measures in $\mathfrak{M}_T(X)$ which is upper semicontinuous in entropy.

1) First we construct $\overline{m}_n \in \mathfrak{M}_{T^{N_n}}(X)$ and then set

$$m_n := \frac{1}{N_n} \sum_{k=0}^{N_n-1} T^k \overline{m}_n .$$

Starting with n = 1 we know from proposition (18.6) that there exists an open cover $\mathfrak{B}_1' \geq (\mathfrak{U}_1)_0^{N_1-1}$ such that $p(\mathfrak{B}_1') \leq N_1 \text{ card } \mathfrak{U}_1 + 1$. Write $(\mathfrak{U}_1)_0^{N_1-1} = (U_1,\dots,U_s)$ and define for $1 \leq i \leq s$

$$V_i := \bigcup \{V \in \mathfrak{B}_1' \mid V \subset U_i,\ V \not\subset U_j \quad \forall\, j < i\}.$$

Let $\mathfrak{B}_1 := (V_i \mid 1 \leq i \leq s)$. Clearly

$$\text{card } \mathfrak{B}_1 \leq \text{card } (\mathfrak{U}_1)_0^{N_1-1} ,$$

$$p(\mathfrak{B}_1) \leq p(\mathfrak{B}_1') \leq N_1 \text{ card } \mathfrak{U}_1 + 1$$

and

$$\mathfrak{B}_1 \geq (\mathfrak{U}_1)_0^{N_1-1} .$$

By lemma (16.6) there is a Borel partition α_1 finer than $\mathfrak{B}_1$ satisfying $p(\alpha_1) \leq N_1 \text{ card } \mathfrak{U}_1 + 1$ and $\text{card } \alpha_1 \leq \text{card } \mathfrak{B}_1$.

Hence by (18.7) there exists a measure $\overline{m}_1 \in \mathfrak{M}_{T^{N_1}}(X)$ such that

$$h_{\overline{m}_1}(T^{N_1},\beta) \geq H(T^{N_1}, (\mathfrak{U})_0^{N_1-1}) - \log p(\beta \mid \alpha_1)$$

for every Borel partition β of X.

Assume now that we have found for every $1 \leq k \leq n$

i) an open cover $\mathfrak{B}_k$ such that - setting $N_o = 1$ and $\mathfrak{B}_o = \mathfrak{U}_1$ -

$$\mathfrak{B}_k \geq \bigvee_{i=o}^{\frac{N_k}{N_{k-1}}-1} T^{-iN_{k-1}} \mathfrak{B}_{k-1} \vee \mathfrak{U}'_k =: \tilde{\mathfrak{U}}_k \geq (\mathfrak{U}_1 \vee \ldots \vee \mathfrak{U}_k)_o^{N_k-1}$$

- where $\mathfrak{U}'_k$ denotes a minimal subcover of $(\mathfrak{U}_k)_o^{N_k-1}$ for $k \geq 2$ and where $\mathfrak{U}'_1 = \mathfrak{U}_1$ -,

$$p(\mathfrak{B}_k) \leq \text{card } \mathfrak{U}'_k \left[\frac{N_k}{N_{k-1}} \prod_{j=1}^{k-1} (\text{card } \mathfrak{U}'_j)^{\frac{N_{k-1}}{N_j}} + 1\right]$$

and

$$\text{card } \mathfrak{B}_k \leq \text{card } \tilde{\mathfrak{U}}_k \leq (\text{card } \mathfrak{B}_{k-1})^{\frac{N_k}{N_{k-1}}} \text{ card } \mathfrak{U}'_k .$$

ii) a Borel partition α_k which is finer than $\mathfrak{B}_k$ such that card $\alpha_k \leq$ card $\mathfrak{B}_k$ and

$$p(\alpha_k) \leq \text{card } \mathfrak{U}'_k \left[\frac{N_k}{N_{k-1}} \prod_{j=1}^{k-1} (\text{card } \mathfrak{U}'_j)^{\frac{N_{k-1}}{N_j}} + 1\right].$$

iii) a measure $\bar{m}_k \in \mathfrak{M}_{T^{N_k}}(X)$ such that for every Borel partition β of X

$$h_{\bar{m}_k}(T^{N_k},\beta) \geq H(T^{N_k},(\mathfrak{U}_1 \vee \ldots \vee \mathfrak{U}_k)_o^{N_k-1}) - \log p(\beta \mid \alpha_k).$$

In order to construct $\mathfrak{B}_{n+1}$, α_{n+1} and $\bar{m}_{n+1}$ note that by (18.6) there is an open cover

$$\mathfrak{B}'_{n+1} \geq \bigvee_{i=o}^{\frac{N_{n+1}}{N_n}-1} T^{-iN_n}\mathfrak{B}_n \vee \mathfrak{U}'_{n+1} =: \tilde{\mathfrak{U}}_{n+1},$$

where $\mathfrak{U}'_{n+1}$ denotes a minimal subcover of $(\mathfrak{U}_{n+1})_o^{N_{n+1}-1}$, such that

$$\begin{aligned}
p(\mathfrak{B}'_{n+1}) &\leq (\frac{N_{n+1}}{N_n} \text{ card } \mathfrak{B}_n + 1) \text{ card } \mathfrak{U}'_{n+1} \\
&\leq (\frac{N_{n+1}}{N_n} \text{ card } \tilde{\mathfrak{U}}_n + 1) \text{ card } \mathfrak{U}'_{n+1} \\
&\leq [\frac{N_{n+1}}{N_n} (\text{card } \mathfrak{B}_{n-1})^{\frac{N_n}{N_{n-1}}} \text{ card } \mathfrak{U}'_n + 1] \text{ card } \mathfrak{U}'_{n+1} \\
&\leq \ldots \\
&\leq [\frac{N_{n+1}}{N_n} \prod_{j=1}^{n} (\text{card } \mathfrak{U}'_j)^{\frac{N_n}{N_j}} + 1] \text{ card } \mathfrak{U}'_{n+1}.
\end{aligned}$$

Write $\tilde{\mathfrak{U}}_{n+1} = (U_1, \ldots, U_s)$ and define for $1 \leq i \leq s$

$$V_i := \bigcup \{V \in \mathfrak{B}'_{n+1} \mid V \subset U_i;\ V \not\subset U_j \quad \forall\, j < i\}.$$

Let $\mathfrak{B}_{n+1} := (V_i \mid 1 \leq i \leq s)$. Clearly

$$p(\mathfrak{B}_{n+1}) \leq p(\mathfrak{B}'_{n+1}) \leq \text{card } \mathfrak{U}'_{n+1} [\frac{N_{n+1}}{N_n} \prod_{j=1}^{n} (\text{card } \mathfrak{U}'_j)^{\frac{N_n}{N_j}} + 1],$$

$$\text{card } \mathfrak{B}_{n+1} \leq \text{card } \tilde{\mathfrak{U}}_{n+1}$$

and

$$\mathfrak{B}_{n+1} \geq \tilde{\mathfrak{U}}_{n+1}.$$

Note that

$$\mathfrak{V}_{n+1} \geq \tilde{\mathfrak{U}}_{n+1} \geq \bigvee_{i=o}^{\frac{N_{n+1}}{N_n}-1} T^{-iN_n} \bigvee_{j=o}^{N_n-1} T^{-j}(\mathfrak{U}_1 \vee \dots \vee \mathfrak{U}_n) \vee (\mathfrak{U}_{n+1})_o^{N_{n+1}-1}$$

$$= (\mathfrak{U}_1 \vee \dots \vee \mathfrak{U}_{n+1})_o^{N_{n+1}-1} .$$

Using (16.6) again choose any Borel partition α_{n+1} finer than $\mathfrak{V}_{n+1}$ satisfying $p(\alpha_{n+1}) \leq p(\mathfrak{V}_{n+1})$ and card $\alpha_{n+1} \leq$ card $\mathfrak{V}_{n+1}$.

In order to complete the induction one only has to note that by proposition (18.7) $\bar{m}_{n+1}$ exists satisfying iii) for $k = n + 1$.

2) We may assume that the measures m_n defined by

$$m_n := \frac{1}{N_n} \sum_{j=o}^{N_n-1} T^j \bar{m}_n$$

converge weakly to a measure $m \in \mathfrak{M}_T(X)$.

We shall construct a sequence β_n $(n \in \mathbb{N})$ of m-continuity Borel partitions satisfying

$$\lim_{k\to\infty} h_m(T,\beta_k) = h_{top}(T) .$$

Using (16.6) for every $n \in \mathbb{N}$ there exists an m-continuity Borel partition β_n such that

$$\max_{B\in\beta_n} \text{card } \{V\in\mathfrak{V}_n | V \cap \bar{B} \neq \emptyset\} \leq p(\mathfrak{V}_n).$$

By induction over $k \geq 0$ we show for every fixed $n \in \mathbb{N}$:

$$\max_{B\in(\beta_n)_o^{N_{n+k}-1}} \text{card } \{V \in \mathfrak{V}_{n+k} | V \cap \bar{B} \neq \emptyset\} \leq$$

$$(p(\mathfrak{B}_n))^{N_{n+k}N_n^{-1}} \exp\{N_{n+k} \sum_{i=n+1}^{n+k} (H(T,\mathfrak{U}_i) + \varepsilon_i)\}.$$

For abbreviation call this last number $p_{n,k}$ $(k \geq 0)$. First of all for $k = 0$ the formula is trivial by definition of β_n. Suppose it is true for $k - 1$.

Let $B = \bigcap_{j=0}^{\frac{N_{n+k}}{N_{n+k-1}}} T^{-jN_{n+k-1}} B_{i_j} \in (\beta_n)_0^{N_{n+k}-1}$ where

$B_{i_j} \in (\beta_n)_0^{N_{n+k-1}-1}$ $(0 \leq j \leq \frac{N_{n+k}}{N_{n+k-1}} - 1)$. Since

$$\operatorname{card}\{V \in \mathfrak{B}_{n+k-1} | V \cap B_{i_j} \neq \emptyset\} \leq p_{n,k-1} \quad \text{for every } j,$$

it follows that

$$\operatorname{card}\{V \in T^{-jN_{n+k-1}} \mathfrak{B}_{n+k-1} | V \cap T^{-jN_{n+k-1}} \overline{B_{i_j}} \neq \emptyset\} \leq p_{n,k-1}.$$

Therefore we get

$$\operatorname{card}\{V \in \bigvee_{j=0}^{\frac{N_{n+k}}{N_{n+k-1}}} T^{-jN_{n+k-1}} \mathfrak{B}_{n+k-1} | V \cap \bar{B} \neq \emptyset\} \leq (p_{n,k-1})^{\frac{N_{n+k}}{N_{n+k-1}}}.$$

Since $\mathfrak{U}'_{n+k}$ was chosen to be a minimal subcover of $(\mathfrak{U}_{n+k})_0^{N_{n+k}-1}$ we obtain:

$$\operatorname{card}\{U \in \tilde{\mathfrak{U}}_{n+k} | U \cap \bar{B} \neq \emptyset\} \leq (p_{n,k-1})^{\frac{N_{n+k}}{N_{n+k-1}}} \exp\{N_{n+k}(H(T,\mathfrak{U}_{n+k})+\varepsilon_{n+k})\}.$$

Since by construction for every $V \in \mathfrak{B}_{n+k}$ there is $U_V \in \tilde{\mathfrak{U}}_{n+k}$ such that $V \subset U_V$ and since this correspondence is injective we also must have

$$\operatorname{card}\{V \in \mathfrak{B}_{n+k} | V \cap \overline{B} \neq \emptyset\} \leq$$
$$(p_{n,k-1})^{N_{n+k} N_{n+k-1}^{-1}} \exp\{N_{n+k}(H(T,\mathfrak{U}_{n+k}) + \epsilon_{n+k})\},$$

hence

$$\max_{B \in (\beta_n)_o^{N_{n+k}-1}} \operatorname{card}\{V \in \mathfrak{B}_{n+k} | V \cap \overline{B} \neq \emptyset\} \leq$$
$$\leq p(\mathfrak{B}_n)^{\frac{N_{n+k}}{N_n}} \exp\{N_{n+k} \sum_{i=n+1}^{n+k} (H(T,\mathfrak{U}_i) + \epsilon_i)\}.$$

Clearly these formulas imply for every $\alpha_{n+k}(k \geq 0,\ n \in \mathbb{N})$ that

$$p((\beta_n)_o^{N_{n+k}-1} | \alpha_{n+k}) \leq$$
$$\leq (p(\mathfrak{B}_n))^{\frac{N_{n+k}}{N_n}} \exp\{N_{n+k} \sum_{i=n+1}^{n+k} (H(T,\mathfrak{U}_i) + \epsilon_i)\},$$

since α_{n+k} is finer than $\mathfrak{B}_{n+k}$ and $\operatorname{card} \alpha_{n+k} = \operatorname{card} \mathfrak{B}_{n+k}$. Using proposition (18.7) we get

$$h_{\overline{m}_{n+k}}(T^{N_{n+k}}, (\beta_n)_o^{N_{n+k}-1}) \geq N_{n+k} H(T, \mathfrak{U}_1 \vee \ldots \vee \mathfrak{U}_{n+k}) -$$
$$- \log p((\beta_n)_o^{N_{n+k}-1} | \alpha_{n+k});$$

Clearly for m_{n+k} it follows that (cf. (18.5))

$$h_{m_{n+k}}(T,\beta_n) \geq H(T,\mathfrak{U}_1 \vee \ldots \vee \mathfrak{U}_{n+k}) -$$

$$- \frac{1}{N_{n+k}} \log p((\beta_n)_o^{N_{n+k}-1} \mid \alpha_{n+k})$$

$$\geq H(T,\mathfrak{U}_1 \vee \ldots \vee \mathfrak{U}_{n+k}) - \frac{1}{N_n} \log p(\mathfrak{B}_n) -$$

$$- \sum_{i=n+1}^{n+k} (H(T,\mathfrak{U}_i) + \varepsilon_i).$$

This finishes the construction of the m-continuity partitions $\beta_n (n \in \mathbb{N})$.

We shall show now that

$$\lim_{n\to\infty} h_m(T,\beta_n) = h_{top}(T).$$

Let $\varepsilon > 0$ be given.
By assumption on N_m and ε_m there is an $n \in \mathbb{N}$ such that

$$H(T,\mathfrak{U}_n) + \varepsilon_n + \frac{1}{N_n} (1 + \log \frac{N_n}{N_{n-1}}) +$$

$$+ \frac{N_{n-1}}{N_n} \sum_{j=1}^{n-1} \frac{1}{N_j} (H(T,\mathfrak{U}_j) + \varepsilon_j) +$$

$$+ \sum_{i=n+1}^{\infty} (H(T,\mathfrak{U}_i) + \varepsilon_i) < \varepsilon.$$

Pick any $L \in \mathbb{N}$ satisfying

$$h_m(T,\beta_n) \geq \frac{1}{L} H_m((\beta_n)_o^{L-1}) - \varepsilon.$$

Since β_n is an m-continuity partition there is a $k_o \in \mathbb{N}$, $k_o \geq n$ such that for every $k \geq k_o$

$$\frac{1}{L} \mid H_m((\beta_n)_o^{L-1}) - H_{m_k}((\beta_n)_o^{L-1}) \mid < \epsilon$$

holds. Therefore it follows that

$$h_m(T,\beta_n) \geq \frac{1}{L} H_{m_k}((\beta_n)_o^{L-1}) - 2\epsilon$$

$$\geq h_{m_k}(T,\beta_n) - 2\epsilon .$$

Using what we have shown for the β_k's we obtain for $k \geq k_o$:

$$2\epsilon + h_m(T,\beta_n) \geq H(T,\mathfrak{U}_1 \vee \ldots \vee \mathfrak{U}_k) - \frac{1}{N_n} \log p(\mathfrak{B}_n) - \sum_{i=n+1}^{k} (H(T,\mathfrak{U}_i) + \epsilon_i)$$

$$\geq H(T,\mathfrak{U}_1 \vee \ldots \vee \mathfrak{U}_k) - \sum_{i=n+1}^{k} (H(T,\mathfrak{U}_i) + \epsilon_i)$$

$$- \frac{1}{N_n} \log[\operatorname{card} \mathfrak{U}'_n (\frac{N_n}{N_{n-1}} \prod_{j=1}^{n-1} (\operatorname{card} \mathfrak{U}'_j)^{\frac{N_{n-1}}{N_j}} + 1)]$$

$$\geq H(T,\mathfrak{U}_1 \vee \ldots \vee \mathfrak{U}_k) - (H(T,\mathfrak{U}_n) + \epsilon_n) - \frac{1}{N_n} (1 + \log \frac{N_n}{N_{n-1}})$$

$$- \frac{1}{N_n} \sum_{j=1}^{n-1} \frac{N_{n-1}}{N_j} (H(T,\mathfrak{U}_j) + \epsilon_j) - \sum_{i=n+1}^{k} (H(T,\mathfrak{U}_i) + \epsilon_i)$$

$$\geq H(T,\mathfrak{U}_1 \vee \ldots \vee \mathfrak{U}_k) - \epsilon .$$

Now if $H(T,\mathfrak{U}_1 \vee \ldots \vee \mathfrak{U}_k) \geq h_{top}(T) - \epsilon$ (for k large enough) we get

$$h_m(T,\beta_n) \geq h_{top}(T) - 4\epsilon .$$

Letting $\epsilon \to 0$ it follows that

$$\lim_{n\to\infty} h_m(T,\beta_n) \geq h_{top}(T).$$

□

The converse of the last theorem holds as well:

(20.12) Theorem [46]: If $\mathfrak{M}_{max}(T) \neq \emptyset$ and $h_{top}(T) < \infty$, then T is locally h-expansive.

Proof: Since $h_{top}(T) < \infty$ we may pick an ergodic measure $m \in \mathfrak{M}_{max}(T)$ (see (19.4)). If m is periodic there is nothing left to show. Hence we assume that m is aperiodic.

For the proof let $\epsilon_n > 0$ $(n \in \mathbb{N})$ such that $\sum_{n=1}^{\infty} \epsilon_n < \infty$. Choose any sequence $\alpha_n (n \in \mathbb{N})$ of m-measurable partitions such that $h_{top}(T) - h_m(T,\alpha_n) < \epsilon_n$.

1. Definition of $\mathfrak{U}_1$:

Pick $p_1 \in \mathbb{N}$ sufficiently large and such that $p_1^{-1}\log p_1 < \epsilon_1$ and the theorem of Shannon, McMillan and Breiman (13.4) applies for ϵ_1 and $\delta > 0$ sufficiently small: There are atoms $B_i \in (\alpha_1)_0^{p_1-1}$ $(1\leq i\leq s)$ such that

$$\exp(-p_1(h_m(T,\alpha_1)+\epsilon_1)) \leq m(B_i) \leq \exp(-p_1(h_m(T,\alpha_1)-\epsilon_1))$$

and $m(\bigcup_{i=1}^{s} B_i) > 1 - \delta$.

Using (1.18) and the regularity of m it is easy to obtain an open (T,p_1,δ) Rohlin set V_1.

If $\delta > 0$ is chosen small enough then for

$$\beta_1 := (B_i \cap V_1,\ X \setminus \bigcup_{i=1}^{s} (B_i \cap V_1) \mid 1 \leq i \leq s)$$

one has $h_m(T,\beta_1) \geq h_m(T,\alpha_1) - \varepsilon_1$.

Approximating every set $B_i \cap V_1$ from inside by a compact set one obtains a partition

$$\gamma_1 = (K_i, X \setminus \bigcup_{i=1}^{s} K_i \mid 1 \leq i \leq s)$$

with $h_m(T,\gamma_1) \geq h_m(T,\alpha_1) - 2\varepsilon_1$. Since the K_i are compact subsets of V_1 there exists an $\varepsilon > 0$ and an open set $\tilde{V}_1 \supset X \setminus V_1$ such that for $1 \leq i \neq j \leq s$ $\overline{B_\varepsilon(K_i)} \subset V_1$, $\overline{B_\varepsilon(K_i) \cap B_\varepsilon(K_j)} = \emptyset$, $\overline{B_\varepsilon(K_i)} \cap \overline{\tilde{V}_1} = \emptyset$ and $m(\tilde{V}_1 \cup \bigcup_{i=1}^{s} B_\varepsilon(K_i)) > 1 - p_1^{-2}$. Choose any open set B_o satisfying $m(\operatorname{bd} B_o) = 0$,

$$X \setminus (\tilde{V}_1 \cup \bigcup_{i=1}^{s} B_\varepsilon(K_i)) \subset B_o \subset \bar{B}_o \subset V_1 \setminus \bigcup_{i=1}^{s} K_i$$

and $m(B_o) < p_1^{-2}$.

Setting $\mathfrak{U}_1 := (B_\varepsilon(K_i),\ B_o, \tilde{V}_1 \mid 1 \leq i \leq s)$ it follows easily that for $m \in \mathbb{N}$

$$2^{mp_1^{-1}+1}\ N((\mathfrak{U}_1)_o^{m-1}) \geq \operatorname{card}\ (\gamma_1)_o^{m-1},$$

since an element of $(\gamma_1)_o^{p_1-1}$ lies in at most 2 elements of $(\mathfrak{U}_1)_o^{p_1-1}$. Therefore

$$\begin{aligned} H(T,\mathfrak{U}_1) &\geq h_m(T,\gamma_1) - p_1^{-1} \log 2 \\ &\geq h_m(T,\alpha_1) - 3\varepsilon_1. \end{aligned}$$

On the other hand

$$N(\mathfrak{U}_1) \leq \exp\{p_1(h_m(T,\alpha_1) + \varepsilon_1)\} + 2$$

and because V_1 is a (T,p_1,δ)-Rohlin set it follows that

$$N((\mathfrak{U}_1)_o^{p_1-1}) \leq p_1(\exp\{p_1(h_m(T,\alpha_1) + \varepsilon_1\} + 2) + 1.$$

Therefore - if p_1 is chosen large enough -

$$H(T,\mathfrak{U}_1) \leq h_m(T,\alpha_1) + 3\varepsilon_1.$$

2. Induction hypothesis:

Assume that the open covers $\mathfrak{U}_1,\dots,\mathfrak{U}_{n-1}$ are constructed, such that the following is given:

(a) $\sum_{k=1}^{n-1} H(T,\mathfrak{U}_k) \leq h_{top}(T) + 7\sum_{k=1}^{n-2}\varepsilon_k + 3\varepsilon_{n-1}$

$H(T,\mathfrak{U}_1 \vee\dots\vee \mathfrak{U}_{n-1}) \geq h_m(T,\alpha_{n-1}) - 3\varepsilon_{n-1}$

(b) a measurable partition $\gamma_{n-1} = (C_o,C_1,\dots,C_q)$ and an open (T,p_{n-1},δ')-Rohlin V_{n-1} set such that

(i) C_i is a compact subset of V_{n-1} for $1 \leq i \leq q$

(ii) $h_m(T,\gamma_{n-1}) \geq h_m(T,\alpha_{n-1}) - 2\varepsilon_{n-1}$

(iii) $q \leq \exp\{p_{n-1}(h_{top}(T) + \varepsilon_{n-1})\}$

(c) an open cover $\mathfrak{B}_{n-1} = (\tilde{V}_{n-1},U_o',\dots,U_q')$ and a $P \in \mathbb{N}$ satisfying

(i) $C_k \subset U_k' \subset V_{n-1}$ for $1\leq k\leq q$, $U_o' \subset C_o \cap V_{n-1}$, $X \setminus V_{n-1} \subset \tilde{V}_{n-1}$, $m(U_o') \leq p_{n-1}^{-2}$ and $m(\mathrm{bd}\, U_o') = 0$.

(ii) $\overline{U_k'} \cap \overline{U_j'} = \emptyset$ $(1\leq k\neq j\leq q)$, $\overline{U_k'} \cap \overline{\tilde{V}_{n-1}} = \emptyset$ $(1\leq k\leq q)$

(iii) $(\mathfrak{U}_1\vee\dots\vee\mathfrak{U}_{n-1})_o^P > \mathfrak{U}_{n-1}$

Note that these conditions are fulfilled for n=2 with

$\mathfrak{B}_1 = \mathfrak{U}_1$.

3. Definition of $\mathfrak{U}_n$:

The first part is similar to 1., i.e. the construction of $\mathfrak{B}_1$.

(a) Let $\alpha \supset \gamma_{n-1} \vee \alpha_n$ be so fine such that for every $A \in \alpha$ and $W \in \mathfrak{B}_{n-1}$ either $A \cap W = \emptyset$ or $A \subset W$. Pick p_n sufficiently large and such that $p_n^{-1} \log 2p_n < \epsilon_n$ and the ergodic theorem and the theorem of Shannon, McMillan and Breiman apply for ϵ_n and $\delta > 0$ sufficiently small in the following manner:

(i) there are atoms $B_i \in (\alpha)_o^{p_n-1}$ $(1 \leq i \leq s)$ such that

$$\exp\{-p_n(h_m(T,\alpha)+\epsilon_n)\} \leq m(B_i) \leq \exp\{-p_n(h_m(T,\alpha)-\epsilon_n)\}$$

and $m(\bigcup_{i=1}^{s} B_i) > 1 - \delta$.

(ii) there are atoms $D_j \in (\gamma_{n-1})_o^{p_n-1}$ $(1 \leq j \leq r)$ such that

$$\exp\{-p_n(h_m(T,\gamma_{n-1}) + \epsilon_n)\} \leq m(D_j) \leq \exp\{-p_n(h_m(T,\gamma_{n-1})-\epsilon_n)\}$$

and $m(\bigcup_{j=1}^{r} D_j) > 1 - \delta$.

(iii) for $Z := \{x \in X \mid |p_n^{-1} \sum_{t=o}^{p_n-1} 1_{U_o'} T^t x - m(U_o')| < \delta\}$

one has $m(Z) > 1 - \delta$.

Now pick an open (T,p_n,δ) Rohlin set V_n and define

$$W := V_n \cap \bigcup_{i=1}^{s} B_i \cap \bigcup_{j=1}^{r} D_j \cap Z.$$

If $\beta_n := (B_i \cap W, X \setminus \bigcup_{i=1}^{s} (B_i \cap W) | 1 \leq i \leq s)$ then for sufficiently small δ we have $h_m(T,\beta_n) \geq h_m(T,\alpha) - \epsilon_n$, and approximating every set $B_i \cap W$ from inside by a compact set K_i it is possible to find a partition

$$\gamma_n := (K_i, X \setminus \bigcup_{i=1}^{s} K_i \mid 1 \leq i \leq s)$$

such that $h_m(T,\gamma_n) \geq h_m(T,\alpha) - 2\varepsilon_n \geq h_m(T,\alpha_n) - 2\varepsilon_n$.

Note that 2.(b) is fulfilled for n.

As in 1. there exist an $\varepsilon > 0$ and an open set $\tilde{V}_n$ such that for $1 \leq i \neq j \leq s$ $\overline{B_\varepsilon(K_i)} \cap \overline{B_\varepsilon(K_j)} = \emptyset$, $\overline{B_\varepsilon(K_i)} \subset V_n$, $\overline{B_\varepsilon(K_i)} \cap \overline{\tilde{V}_n} = \emptyset$, $\tilde{V}_n \supset X \setminus V_n$ and $m(\tilde{V}_n \cup \bigcup_{i=1}^{s} B_\varepsilon(K_i)) > 1-p_n^{-2}$. Again choose any open set B_o satisfying $m(\text{bd } B_o) = 0$,

$$X \setminus (\tilde{V}_n \cup \bigcup_{i=1}^{s} B_\varepsilon(K_i)) \subset B_o \subset \overline{B_o} \subset V_n \setminus \bigcup_{i=1}^{s} K_i$$

and $m(B_o) < p_n^{-2}$. We may of course assume that $B_\varepsilon(K_i) \subset Z$ for every $1 \leq i \leq s$. Define $\mathfrak{B}_n := (B_\varepsilon(K_i), B_o, \tilde{V}_n \mid 1 \leq i \leq s)$.

Note that 2. (c), (i) and (ii) are satisfied for n. Similarly to 1. one concludes

$$H(T,\mathfrak{B}_n) \geq h_m(T,\alpha_n) - 3\varepsilon_n.$$

(b) In order to obtain $\mathfrak{U}_n$ we have to match together certain elements of $\mathfrak{B}_n$. To prepare this coding we show the following:

Let

$$S_o := \exp\{p_n(h_m(T,\alpha) - h_m(T,\gamma_{n-1}) + 2\varepsilon_n + \varepsilon_{n-1})\},$$

$$S_1 := (p_{n-1}+2)^{p_n p_{n-1}^{-1}} \exp\{p_n p_{n-1}^{-1}(h_{top}(T) + \varepsilon_{n-1})\},$$

$$S_2 := S_1(S_1-1) \text{ and } S = S_2 S_o.$$

In the situation described in 2.(a) there exists a map

$$\varphi : \{B_\varepsilon(K_i) \mid 1 \leq i \leq s\} \longrightarrow \{1,\ldots,S\}$$

such that for every $V \in (\mathfrak{B}_{n-1})_o^{p_n-1}$ φ is injective on the set $\{B_\varepsilon(K_i) \cap V \neq \emptyset\}$.

Starting the construction of φ note that for every D_j $(1 \leq j \leq r)$

$$\operatorname{card}\{1 \leq i \leq s \mid B_i \cap D_j \neq \emptyset\} \leq \exp\{p_n(h_m(T,\alpha) - h_m(T,\gamma_{n-1}) + 2\varepsilon_n)\}$$

holds. If p_{n-1} is large enough it follows for $V \in (\mathfrak{B}_{n-1})_o^{p_n-1}$ that

$$\operatorname{card}\{1 \leq i \leq s \mid B_i \cap \overline{V} \neq \emptyset\} \leq$$
$$\leq 2^{p_n p_{n-1}^{-1}+1} \exp\{p_n(h_m(T,\alpha) - h_m(T,\gamma_{n-1}) + 2\varepsilon_n)\} = S_o.$$

If $\varepsilon > o$ is small enough the same inequality holds replacing B_i by $B_\varepsilon(K_i)$.

Step 1: Let $\mathfrak{C}_1 := \{V \in (\mathfrak{B}_{n-1})_o^{p_n-1} \mid V \cap B_\varepsilon(K_i) \neq \emptyset$ for some $i\}$ and for $V \in \mathfrak{C}_1$ set $\mathfrak{C}_1(V) = \{V' \in \mathfrak{C}_1 \mid V \cap V' \neq \emptyset\}$. It is not hard to see that for $V \in \mathfrak{C}_1$

$$\operatorname{card} \mathfrak{C}_1(V) \leq \sum_{k=o}^{p_n p_{n-1}^{-1}} \binom{p_n p_{n-1}^{-1}}{k} (p_{n-1}+1)^k (\operatorname{card} \mathfrak{B}_{n-1})^{t_V}$$

where $t_V = \operatorname{card}\{0 \leq j \leq p_n-1 \mid T^j V \subset U_o'\}$ (see 2.(c)).

Since for every $1 \leq i \leq s$ $B_\varepsilon(K_i) \subset Z$ and $V \cap B_\varepsilon(K_i) \neq \emptyset$ it follows that $t_V \leq p_n(m(U_o') + \delta) \leq p_n p_{n-1}^{-2}$ if δ is sufficiently small. Therefore one obtains $\operatorname{card} \mathfrak{C}_1(V) \leq S_1$ (use 2(b), (iii)).

We make use of the notation $V \sim V'$ if there exists a $V'' \in \mathfrak{C}_1$ such that $V \cap V'' \neq \emptyset$ and $V'' \cap V' \neq \emptyset$. Clearly for $V \in \mathfrak{C}_1$ one has $\operatorname{card}\{V' \in \mathfrak{C}_1 \mid V \sim V'\} \leq S_2$.

There exists a set $\mathfrak{J}_1 \subset \mathfrak{C}_1$ of maximal cardinality such that for any two $V, V' \in \mathfrak{J}_1$ $V \sim V'$ does not hold. Define

$$\mathfrak{D}_1 := \{B_\varepsilon(K_i) \mid \exists\, V \in \mathfrak{J}_1 : V \cap B_\varepsilon(K_i) \neq \emptyset\}$$

and $\varphi : \mathfrak{D}_1 \longrightarrow \{1,\dots,S_o\}$ such that for every $V \in (\mathfrak{B}_{n-1})_o^{p_n-1}$, $\varphi \mid \{B_\varepsilon(K_i) \cap V \neq \emptyset\}$ is injective. (This is possible since every B_i lies inside of V or is disjoint, and hence we may assume that $\varepsilon > 0$ is so small that

$$B_\varepsilon(K_i) \subset \bigcup \{W \in (\mathfrak{B}_{n-1})_o^{p_n-1} \mid W \cap V \neq \emptyset\}$$

if $B_i \subset V$.)

Step 2: Define $\mathfrak{C}_2 := \mathfrak{C}_1 \setminus \mathfrak{J}_1$ and $\mathfrak{D}_2' = \{B_\varepsilon(K_i) \mid 1 \leq i \leq s\} \setminus \mathfrak{D}_1$. Since $\mathfrak{J}_1$ is of maximal cardinality for $V \in \mathfrak{C}_2$ we must have that

$$\text{card}\,\{V' \in \mathfrak{C}_2 \mid V \sim V'\} \leq S_2 - 1.$$

Select $\mathfrak{J}_2 \subset \mathfrak{C}_2$ of maximal cardinality such that for any two $V, V' \in \mathfrak{J}_2$ $V \sim V'$ does not hold.

Denote by $\mathfrak{D}_2$ the set of all elements of $\mathfrak{D}_2'$ which intersect some member of $\mathfrak{J}_2$. Define

$$\varphi : \mathfrak{D}_2 \longrightarrow \{S_o + 1,\dots,2S_o\}$$

such that for every $V \in (\mathfrak{B}_{n-1})_o^{p_n-1}$ φ is injective on $\{B_\varepsilon(K_i) \in \mathfrak{D}_2 \mid B_\varepsilon(K_i) \cap V \neq \emptyset\}$. Repeating this procedure finitely many times one obtains a map φ which is injective on every set $\{B_\varepsilon(K_i) \cap V \neq \emptyset\}$ $(V \in (\mathfrak{B}_{n-1})_o^{p_n-1})$.
Since $\text{card}\,\{V' \in \mathfrak{C}_t \mid V \sim V'\} \leq S_2 - t + 1$ for every $V \in \mathfrak{C}_t$ we need at most S_2 steps for the construction of φ, hence φ is a map into $\{1,\dots,S\}$.

(c) We proceed with the definition of $\mathfrak{U}_n$.
Define $\mathfrak{U}_n := (B_o, \tilde{V}_n, U_k \mid 1 \leq k \leq S)$ where

$$U_k := \bigcup \{B_\varepsilon(K_i) \mid \varphi(B_\varepsilon(K_i)) = k\}$$

for $k = 1,\ldots,S$.

Since $N((\mathfrak{U}_n)_0^{p_n-1}) \leq p_n(S + 2) + 1$ it follows that $(S \leq S_1^2\, S_0!)$

$$\begin{aligned} H(T,\mathfrak{U}_n) &\leq p_n^{-1} \log\{p_n(S+2) + 1\} \\ &\leq p_n^{-1} \log 2\, p_n + 2\, p_{n-1}^{-1} \log(p_{n-1}+2) \\ &\quad + 2p_{n-1}^{-1}(h_{top}(T) + \epsilon_{n-1}) + h_{top}(T) - h_m(T,\gamma_{n-1}) \\ &\qquad + 2\epsilon_n + \epsilon_{n-1} \\ &\leq 3\, \epsilon_n + 4\, \epsilon_{n-1}, \end{aligned}$$

if p_{n-1} is sufficiently large. This shows the first part of 2.(a).

On the other hand it is easy to see that

$$(\mathfrak{B}_{n-1})_0^{p_n-1} \vee \mathfrak{U}_n \geq \mathfrak{B}_n$$

and therefore one gets $(\mathfrak{U}_1 \vee\ldots\vee \mathfrak{U}_n)_0^{P+p_n} > \mathfrak{B}_n$ and $H(T,\mathfrak{U}_1 \vee\ldots\vee \mathfrak{U}_n) \geq h_m(T,\alpha_n) - 3\, \epsilon_n$.

This shows the second part of 2.(a) and 2.(c) (iii) with $P + p_n$.

4. The statement of the theorem follows from

$$\sum_{k=1}^{\infty} H(T,\mathfrak{U}_k) \leq 7 \sum_{k=1}^{\infty} \epsilon_k + h_{top}(T) < \infty$$

and $\lim\limits_{n\to\infty} H(T,\mathfrak{U}_1 \vee\ldots\vee \mathfrak{U}_n) \geq \lim\limits_{n\to\infty} (h_m(T,\alpha_n) - 3\, \epsilon_n) =$

$= h_m(T) = h_{top}(T)$. □

Corollary: The following conditions are equivalent:

1) T is locally h-expansive.

2) $h_{top}(T) < \infty$ and $\mathfrak{M}_{max}(T) \neq \emptyset$.

Proof: It is easy to see that $h_{top}(T) < \infty$, if T is locally h-expansive, hence the corollary follows from the previous theorems. □

We finish this section discussing the previous theorems. Recall from section 19 the definition of a T-homogeneous measure m. The defining property holds uniformly on X, hence in view of theorem (20.12) it is not surprising that a homeomorphism T admitting a T-homogeneous measure is asymptotically h-expansive (see [138]). Thus we get together with (20.9) a new proof of (19.7). The simplest example of an asymptotically h-expansive but not h-expansive dynamical system is the following: Let (X_n, T_n) be expansive dynamical systems satisfying $\sum_{n=1}^{\infty} h_{top}(T_n) < \infty$. Let

$X = \prod_{n=1}^{\infty} X_n$ with the product topology and

$T((x_n)_{n\in\mathbb{N}}) := (T_n(x_n))_{n\in\mathbb{N}}$. It is easy to see that this example does it.

The simplest example of a locally but not asymptotically h-expansive system is the one given after (20.9). (To see this use the last corollary.)

Though the names "locally h-expansive" and "asymptotically h-expansive" are similar there is an important difference between the two notions: For the definition of the second one -intuitively- one has to use a sequence $\mathfrak{U}_n (n \in \mathbb{N})$ of open covers which separates all points of X under T (that is $h^*(T)$ is approximately $\lim_{n\to\infty} \lim_{m\to\infty} h(T,\mathfrak{U}_m|\mathfrak{U}_n))$. The condition of separating points has not been used for defining local h-expansiveness. One can understand local h-expansiveness as follows: There is a sequence $\mathfrak{U}_n$ $(n \in \mathbb{N})$ of open covers such that $\lim_n \lim_m h(T,\mathfrak{U}_m|\mathfrak{U}_n) = 0$ and $\lim_{n\to\infty} H(T,\mathfrak{U}_n) = h_{top}(T)$, where the last condition replaces the separation axiom.

21. The Specification Property

The following definition, which is due to Bowen [21], has turned out to be very useful in spite of its rather complicated appearance. Let (X,T) be a top. dynamical system and d a metric on X. Assume card $X > 1$.

(21.1) Definition [21]: (X,T) is said to satisfy the specification property if the following holds: for any $\epsilon > 0$ there exists an integer $M(\epsilon)$ such that for any $k \geq 2$, for any k points $x_1,\dots,x_k \in X$, for any integers

$a_1 \leq b_1 < a_2 \leq b_2 < \dots < a_k \leq b_k$ with $a_i - b_{i-1} \geq M(\epsilon)$

for $2 \leq i \leq k$ and for any integer p with $p \geq M(\epsilon) + b_k - a_1$, there exists a point $x \in X$ with $T^p x = x$ such that

$$d(T^n x, T^n x_i) \leq \epsilon \quad \text{for } a_i \leq n \leq b_i,\ 1 \leq i \leq k.$$

T is said to satisfy the weak specification property if the above condition holds for $k = 2$.

It is easy to see that this definition does not depend on the choice of the metric d.

The weak specification means that whenever there are two "pieces of orbits" $\{T^n x_1 | a_1 \leq n \leq b_1\}$ and $\{T^n x_2 | a_2 \leq n \leq b_2\}$, they may be approximated up to ϵ by one periodic orbit - the orbit of x - provided that the time for "switching" from the first piece of orbit to the second (namely a_2-b_1) and the time for "switching back" (namely $p - (b_2-a_1)$) are larger than $M(\epsilon)$, this number $M(\epsilon)$ being independent of the pieces of orbit, and in particular independent of their length. The stronger form of the specification property requires that such an approximation is possible for any number k of pieces of orbits, $M(\epsilon)$ being independent of k.

This seems to be a very strong condition, but it is satisfied by many examples. Most of them will be described later (see sections 23 and 24). Here we mention only

(21.2) Proposition: The shift on any compact metric state space satisfies the specification property. So do subshifts of finite type which are topologically mixing.

This can be seen directly from (17.6).

The following propositions are also easy to check:

(21.3) Proposition: If (X,T) satisfies the specification property, then the periodic points are dense and T is topologically mixing.

(21.4) Proposition:

(a) If T has the specification property, then T^k has the specification property, for any $k \neq 0$;

(b) the product of two systems with the specification property has the specification property;

(c) the factor of a system with the specification property has the specification property.

(21.5) Proposition [15]: If (X,T) satisfies the specification property then so does its extension $(\mathfrak{M}(X),T)$.

Proof: Let $\varepsilon > 0$ be given and $M(\frac{\varepsilon}{2})$ be as in the definition (21.1). Let $\mu_1,\dots,\mu_k \in \mathfrak{M}(X)$ be given, as well as integers $a_1 \leq b_1 < a_2 \leq b_2 < \dots < a_k \leq b_k$ and p with

$$a_i - b_{i-1} \geq M(\tfrac{\varepsilon}{2})$$

and

$$p \geq M(\varepsilon) + b_k - a_1.$$

Let $\bar{d}$ be the Prohorov metric on $\mathfrak{M}(X)$ (cf. the remark after (2.8)). Since $T : \mathfrak{M}(X) \longrightarrow \mathfrak{M}(X)$ is uniformly continuous, there exists an $\eta > 0$ such that $d(\mu,\nu) < \eta$ implies $d(T^j\mu, T^j\nu) < \frac{\varepsilon}{2}$ for $a_1 \leq j \leq b_k$. For some integer $n > 0$ there exist $\nu_i \in \mathfrak{M}_n(X)$ such that $d(\mu_i,\nu_i) < \eta$ for $1 \leq i \leq k$ (cf. (2.14)). Write

$$\nu_i = \frac{1}{n}\sum_{l=1}^{n} \delta(x_l^i) \qquad \text{for } i = 1,\dots,k.$$

Since $T : X \longrightarrow X$ satisfies the specification property, there exist a $z_l \in X$ with $T^p z_l = z_l$ and

$$d(T^j z_l, T^j x_l^i) < \frac{\varepsilon}{2} \qquad \text{for } a_i \leq j \leq b_i,\ i = 1,\dots,k \text{ and } l = 1,\dots,n.$$

Let $\rho = \frac{1}{n}\sum_{l=1}^{n} \delta(z_l)$. Obviously $T^p\rho = \rho$. Also

$$\bar{d}(T^j\rho, T^j\nu_i) = \bar{d}\Big(\frac{1}{n}\sum_{l=1}^{n} \delta(T^j z_l), \frac{1}{n}\sum_{l=1}^{n} \delta(T^j x_l^i)\Big) < \frac{\varepsilon}{2}$$

and hence

$$\bar{d}(T^j\rho, T^j\mu_i) < \varepsilon$$

for $a_i \leq j \leq b_i$, $i = 1,\ldots,k$. Hence $T : \mathfrak{M}(X) \longrightarrow \mathfrak{M}(X)$ satisfies the specification property. □

<u>(21.6) Proposition</u> [21]: If (X,T) satisfies the specification property, it has positive topological entropy.

<u>Proof:</u> Let $x,y \in X$ and $\varepsilon > 0$ be such that $d(x,y) > 3\varepsilon$. Let $M = M(\varepsilon)$ be given as in definition (21.1). For any n-tuple $(z_1,\ldots,z_n)$ with $z_i \in \{x,y\}$ there exists a $z \in X$ such that $d(T^{iM}z, z_i) \leq \varepsilon$ for $1 \leq i \leq n$.

Clearly to distinct n-tuples there correspond different z. Hence there are at least 2^n points which are (nM,ε)-separated. This implies

$$h_{top}(T) \geq \limsup \frac{1}{nM} \log 2^n = \frac{1}{M} \log 2 > 0 \,. \qquad \square$$

The specification property allows to investigate the space $\mathfrak{M}_T(X)$ in greater detail.

<u>(21.7) Definition:</u> A measure $\mu \in \mathfrak{M}_T(X)$ whose support is contained in one closed (i.e. periodic) orbit is said to be a <u>CO-measure</u>.

Thus to any point x with (minimal) period p there corresponds a unique CO-measure μ_x which has mass $\frac{1}{p}$ at each of the points $x, Tx, \ldots, T^{p-1}x$; and every CO-measure is of this form. We denote the set of these measures by $P(p)$. (Remark that the notation μ_x agrees with (4.6)).

Obviously CO-measures are ergodic. They are not weakly mixing, except if they are concentrated on one fixed point.

(21.8) Proposition [168]: If (X,T) satisfies the specification property, and if $l \in \mathbb{N}$, then $\bigcup_{p \geq l} P(p)$ is dense in $\mathfrak{M}_T(X)$.

Proof: Let $l \in \mathbb{N}$ and $\mu \in \mathfrak{M}_T(X)$ be given. Any neighborhood of μ contains a set of the form

$$W(\mu) = \{\nu \in \mathfrak{M}_T(X) \mid |\int f d\mu - \int f d\nu| < \epsilon \quad \text{for all } f \in F\}$$

where F is some finite subset of $C(X)$. We may assume $\|f\| \leq 1$ for all $f \in F$.

Thus one has to show that $W(\mu)$ contains an element of $P(p)$, for some $p \geq l$.

If N is large enough, then for any $f \in F$ and any x in $Q = Q_T(X)$ (the set of quasiregular points, see (4.6)) one has

$$\left|\frac{1}{N}\sum_{i=0}^{N-1} f(T^i x) - f^*(x)\right| < \frac{\epsilon}{4}$$

where $f^*(x)$ is the time average of f (see (4.1)). By (4.3) one has $\int f \, d\mu = \int f^* \, d\mu$. Since $\mu(Q) = 1$ (see (4.7)) this implies

$$\int_Q f^* d\mu = \int f d\mu.$$

Let $Q_1,\ldots,Q_k$ be a partition of Q into nonempty Borel sets such that $f^*|Q_j$ has oscillation $< \frac{\epsilon}{4}$, for all $f \in F$. Choose $x_j \in Q_j$. Clearly

$$\left|\int_Q f^* d\mu - \sum_{j=1}^{k} \mu(Q_j)\, f^*(x_j)\right| < \frac{\epsilon}{4}$$

and hence

$$\left|\int f d\mu - \sum_{j=1}^{k} \mu(Q_j) \left(\frac{1}{N}\sum_{i=0}^{N-1} f(T^i x_j)\right)\right| < \frac{\epsilon}{2}$$

for all $f \in F$ if N is large enough.

Let $m_1,\ldots,m_k$ be positive integers such that $\Sigma m_j = m \geq 1$ and $\frac{m_j}{m}$ is sufficiently close to $\mu(Q_j)$. Then

$$\left| \int f d\mu - \sum_{j=1}^{k} \frac{m_j}{m} \left(\frac{1}{N} \sum_{i=0}^{N-1} f(T^i x_j) \right) \right| < \frac{\varepsilon}{2}.$$

First we give an idea of the remainder of the proof. Let $\{ u_n \mid 1 \leq n \leq mN \} \subset X$ be a sequence of points defined by letting $\{ u_n \}$ run

m_1 times through $\{x_1, Tx_1, \ldots, T^{N-1}x_1\}$,

then m_2 times through $\{x_2, Tx_2, \ldots, T^{N-1}x_2\}$, etc.,

finally m_k times through $\{x_k, T_{x_k}, \ldots, T^{N-1}x_k\}$. Then one gets

$$\left| \int f d\mu - \frac{1}{mN} \sum_{n=1}^{mN} f(u_n) \right| \leq \frac{\varepsilon}{2}$$

for all $f \in F$.

Choose $\delta > 0$ such that for all $f \in F$ one has $|f(x) - f(y)| < \frac{\varepsilon}{4}$ whenever $d(x,y) < \delta$. Suppose there were a periodic point $z \in X$, of period mN, such that $d(T^n z, u_n) < \delta$ for all n. This would imply

$$(*) \qquad \left| \int f \, d\mu - \frac{1}{mN} \sum_{i=1}^{mN} f(T^i z) \right| \leq \frac{3\varepsilon}{4} .$$

Such a periodic point will not exist, in general. But by the specification property, there exists a point $y \in X$ which almost behaves that way. Indeed, let $M = M(\delta)$ be given as in (21.1) and define

$$a_t = t(N+M-1) \qquad \text{and} \qquad b_t = t(N+M-1)+N-1$$

for $t = 0,1,\ldots,m_1 + m_2 + \ldots + m_k - 1$. Let $\overline{m}_0 = 0$, $\overline{m}_1 = m_1$, $\overline{m}_2 = m_1 + m_2, \ldots, \overline{m}_k = m_1 + \ldots + m_k$. Define $y_t = x_j$ if $\overline{m}_{j-1} \leq t < \overline{m}_j$ for $j = 1,\ldots,k$. By the specification property, there exists a $y \in X$ with period $p = m(N+M-1) > 1$ such that

$$d(T^j y, T^j y_t) \leq \delta \quad \text{for} \quad a_t \leq j \leq b_t, \quad t = 0,\ldots,m-1.$$

For the CO-measure μ_y one has

$$\int f \, d\mu_y = \frac{1}{p} \sum_{i=1}^{p} f(T^i y) .$$

Let $A = \mathbb{Z} \cap \bigcup_{t=0}^{m-1} [a_t, b_t]$. A has mN elements. Just as in (*) one obtains

$$(**) \qquad \left| \int f \, d\mu - \frac{1}{mN} \sum_{i \in A} f(T^i y) \right| \leq \frac{3\varepsilon}{4}.$$

But $\sum_{i=1}^{p} f(T^i y)$ is just the sum $\sum_{i \in A} f(T^i y)$, plus mM - 1 extra terms. Since $\|f\| \leq 1$, the two sums therefore differ by at most mM - 1. This implies

$$(***) \qquad \left| \frac{1}{p} \sum_{i=1}^{p} f(T^i y) - \frac{1}{mN} \sum_{i \in A} f(T^i y) \right| \leq 2(M-1)N^{-1}.$$

If N is large enough one has $\frac{2(M-1)}{N} < \frac{\varepsilon}{4}$. Thus (**) and (***) imply

$$\left| \int f \, d\mu - \frac{1}{p} \sum_{i=1}^{p} f(T^i y) \right| \leq \varepsilon$$

for all $f \in F$, i.e. $\mu_y \in W(\mu)$. □

Together with (5.7) this implies

(21.9) Proposition [168]: If (X,T) satisfies the specification property, then the set of ergodic measures is a dense G_δ subset of $\mathfrak{M}_T(X)$.

(21.10) Proposition [168]: If (X,T) satisfies the specification property then the set of nonatomic invariant measures is a dense G_δ in $\mathfrak{M}_T(X)$.

Proof: For $r \in \mathbb{N}$ let K_r denote the set

$$\{\mu \in \mathfrak{M}_T(X) \mid \exists x \in X \text{ with } \mu(\{x\}) \geq \frac{1}{r}\}.$$

(a) K_r is closed. This can be seen just as in (2.16).

(b) K_r is nowhere dense. Indeed, suppose $V \subset K_r$ is an open set and $\mu \in V$. By (21.8) there exists a CO-measure $\mu_y \in V$ concentrated on an orbit of period > r. But then μ_y cannot belong to K_r, a contradiction.

The set of T-invariant measures with atoms is just $\bigcup_{r=1}^{\infty} K_r$, and therefore of first category. Since $\mathfrak{M}_T(X)$ is compact, the proposition follows. □

(21.11) Proposition: Let (X,T) be a top. dyn. system. Then the set of T-invariant measures with support X is either empty or a dense G_δ in $\mathfrak{M}_T(X)$.

Proof: Let there be a $\mu \in \mathfrak{M}_T(X)$ with Supp $\mu = X$. Let U be a nonempty open set in X and denote by D(U) the set $\{\mu \in \mathfrak{M}_T(X) \mid \mu(U) = 0\}$.

(a) D(U) is closed. This follows from (2.7)

(B) D(U) is nowhere dense: for $\nu \in D(U)$, the measure $(1+\epsilon)\nu + \epsilon\mu$ does not belong to D(U), but converges to ν as $\epsilon \downarrow 0$.

Now let U_i be a countable basis for the open sets in X. $\bigcup_{i=1}^{\infty} D(U_i)$ is of first category in $\mathfrak{M}_T(X)$. Its complement is the set of all T-invariant measures with support X. □

(21.12) Proposition: If the periodic points of (X,T) are dense (for example, if (X,T) satisfies the specification property), then a dense G_δ subset of T-invariant measures has support X.

Proof: It is enough to construct a $\mu \in \mathfrak{M}_T(X)$ with Supp μ = X. If $\{x_i\}$ is a dense sequence of periodic points in X, then $\sum 2^{-i} \mu_{x_i}$ obviously is such a μ. □

(21.13) Proposition: If (X,T) satisfies the specification property, then the set of strongly mixing measures is of first category in $\mathfrak{M}_T(X)$.

Proof: (Parthasarathy [151].) Let F_1, F_2 be two disjoint compact subsets of X and $\epsilon \in (0, \frac{1}{2})$. Write

$$W = \{\mu \in \mathfrak{M}_T(X) \mid \mu(F_1) \geq \epsilon,\ \mu(F_2) \geq \epsilon\}.$$

W is closed by (2.7). Let

$$S(F_1,F_2,\epsilon) = \{\mu \in W \mid \mu \text{ is strongly mixing}\}.$$

We claim that $S(F_1,F_2,\epsilon)$ is of first category. Let U_n be a sequence of open sets decreasing to F_1. Since

$$\lim_{k\to\infty} \mu(U_n \cap T^{-k} U_n) = \mu^2(U_n)$$

for $\mu \in S(F_1,F_2,\epsilon)$, there is a rational $r \in [0,\mu^2(F_1) + \epsilon^2]$ such that for some U_n

$$\lim \mu(U_n \cap T^{-k} U_n) \leq r.$$

Hence $S(F_1,F_2,\epsilon)$ is contained in

$$\{\mu \in W \mid \exists n \in \mathbb{Z},\ \exists \text{ rational } r \in (0,1) \text{ such that}$$

$$\limsup_k \mu(U_n \cap T^{-k} U_n) \leq r \quad \text{and } r \leq \mu^2(F_1) + \epsilon^2\}.$$

This set is just

$$\bigcup_{n\in\mathbb{Z}} \ \bigcup_{r\in\mathbb{Q}\cap(0,1)} \ \bigcup_{m\in\mathbb{Z}} E(n,r,m)$$

where

$$E(n,r,m) = \bigcap_{k=m}^{\infty} \{\mu \in W \mid \mu(U_n \cap T^{-k} U_n) \leq r,\ \ r \leq \mu^2(F_1) + \epsilon^2\}.$$

By (2.7) the sets $E(n,r,m)$ are closed. Thus $S(F_1,F_2,\epsilon)$ is contained in a countable union of closed sets. It remains to show that the complements of $E(n,r,m)$ are dense.

By (21.7) it is enough to show that $\bigcup_{k\geq m} P(k)$ lies in the complement of $E(n,r,m)$, or even that $P(k)$ is in the complement of

$$E_k = \{\mu \in W \mid \mu(U_n \cap T^{-k} U_n) \leq r,\ \ r \leq \mu^2(F_1) + \epsilon^2\}.$$

Let x be a point of period k. If $\mu_x(F_1) < \epsilon$ or $\mu_x(F_2) < \epsilon$, then

μ_x is not in W, and hence not in E_k. Otherwise recall that F_1 and F_2 are disjoint, hence

$$\varepsilon \leq \mu_x(F_1) \leq 1 - \varepsilon.$$

Also

$$\mu_x(U_n \cap T^{-k} U_n) = \mu_x(U_n).$$

Therefore it is enough to prove that

$$(*) \qquad \mu_x(U_n) \geq \mu_x^2(F_1) + \varepsilon^2$$

But $U_n \supset F_1$ and $\varepsilon < \frac{1}{2}$, so

$$\mu_x(U_n) - \mu_x^2(F_1) \geq \mu_x(F_1) - \mu_x^2(F_1) \geq \varepsilon(1 - \varepsilon) > \varepsilon^2 .$$

Thus (*) is proved, and $S(F_1,F_2,\varepsilon)$ is of first category.

Now let $\{x_n\}$ be a dense sequence in X, and consider all closed spheres centered at these points and having rational radii. This is a countable class, call it $\mathfrak{C}$.

If μ is strongly mixing, then it belongs to the set

$$\bigcup_{\varepsilon \in \mathbb{Q} \cap (0,\frac{1}{2})} \bigcup_{\substack{F_1,F_2 \in \mathfrak{C} \\ F_1 \cap F_2 = \emptyset}} S(F_1,F_2,\varepsilon)$$

which is of first category, or else it is concentrated on one fixed point. The set of these point measures is also of first category, hence the proposition is proved. □

For the following recall the definition (3.7) and the fact (3.8) that the set $V_T(\mu)$ of accumulation points of time averages $\mu^N = \frac{1}{N} \sum_{j=0}^{N-1} T^j \mu$ is a nonempty, closed and connected subset of $\mathfrak{M}_T(X)$.

(21.14) Proposition [172]: Let (X,T) satisfy the specification property, and let $V \subset \mathfrak{M}_T(X)$ be a nonempty closed **connected set.** Then there exists an $x \in X$ such that $V_T(x) = V$. Moreover, the set of such x is dense.

Proof: Since V is closed and connected, there exists a sequence of closed balls B_n in $\mathfrak{M}(X)$ with radii ϵ_n (in some metric $\bar{d}$ compatible with the weak topology) such that the following holds:

(a) $B_n \cap B_{n+1} \cap V \neq \emptyset$

(b) $\bigcap_{N=1}^{\infty} \bigcup_{n \geq N} B_n = V$

(c) $\epsilon_n \longrightarrow 0$

By (21.7) we may also assume that the center of B_n is a CO-measure γ_n. The support of γ_n is the orbit of some periodic point $x_n \in X$ whose period is p_n.

Let $x_o \in X$ be given and U_o the open ball of radius δ around x_o. We have to show that there exists an $x \in U_o$ such that $V_T(x) = V$.

Let $M_n = M(2^{-n}\delta)$ be defined as in the definition of specification (21.1). Define

$$
\begin{aligned}
a_o &= 0 & b_o &= 0 \\
a_1 &= b_o + M_1 & b_1 &= a_1 + 2(a_1 + M_2)p_1 \\
a_2 &= b_1 + M_2 & b_2 &= a_2 + 2^2(a_2 + M_3)p_2 \\
&\ldots\ldots & &\ldots\ldots \\
a_n &= b_{n-1} + M_n & b_n &= a_n + 2^n(a_n + M_{n+1})p_n \\
&\ldots\ldots\ldots & &\ldots\ldots
\end{aligned}
$$

Note that as $n \longrightarrow \infty$, b_n and a_{n+1} become much larger than a_n, M_{n+1} and p_n . Now define a sequence $z_n \in X$ inductively. Let

$z_o = x_o$. By specification, there exists a z_1 such that

$$d(T^j z_o, T^j z_1) < 2^{-1}\delta \qquad \text{for } j = a_o = b_o = 0$$

and

$$d(T^j x_1, T^j z_1) < 2^{-1}\delta \qquad \text{for } a_1 \leq j \leq b_1.$$

There exists a z_2 such that

$$d(T^j z_1, T^j z_2) < 2^{-2}\delta \qquad \text{for } a_o \leq j \leq b_1$$

and

$$d(T^j x_2, T^j z_2) < 2^{-2}\delta \qquad \text{for } a_2 \leq j \leq b_2.$$

In general there exists a z_n such that

$$d(T^j z_{n-1}, T^j z_n) < 2^{-n}\delta \qquad \text{for } a_o \leq j \leq b_{n-1}$$

$$d(T^j x_n, T^j z_n) < 2^{-n}\delta \qquad \text{for } a_n \leq j \leq b_n.$$

It is easy to check that for $m > n$

$$d(T^j x_n, T^j z_m) < 2^{-n+1}\delta \qquad \text{for } a_n \leq j \leq b_n.$$

Since $d(z_{n-1}, z_n) < 2^{-n}\delta$, the sequence z_n converges to some point $x \in U_o$, and one has

$$(*) \qquad d(T^j x_n, T^j x) \leq 2^{-n+1}\delta \qquad \text{for } a_n \leq j \leq b_n.$$

Remark that if A is a finite subset of $\mathbb{N}$,

$$(**) \qquad \left| \frac{1}{\text{card } A} \sum_{j \in A} f(T^j x) - \frac{1}{\max A+1} \sum_{j=o}^{\max A} f(T^j x) \right| \leq$$

$$\leq 2(\text{card } A)^{-1}(\max A + 1 - \text{card } A) \, \|f\|$$

for any $x \in X$ and $f \in C(X)$.

<u>Claim a:</u> $V \subset V_T(x)$.

Let $\mu \in V$ be given. By (b) and (c) there exists a sequence

$n_k \uparrow \infty$ such that $\gamma_{n_k} \longrightarrow \mu$. Let $f \in C(X)$ be given with $\|f\| \leq 1$, and denote by $\omega_f(\epsilon)$ the oscillation

$$\max \{ |f(y) - f(z)| \mid d(y,z) \leq \epsilon \}.$$

Let ρ_n denote the measure $\delta(x)^{b_n}$. Thus

$$\int f d\rho_n = \frac{1}{b_n} \sum_{j=o}^{b_n - 1} f(T^j x) .$$

Also

$$\int f \, d\gamma_n = \frac{1}{b_n - a_n} \sum_{j=a_n}^{b_n - 1} f(T^j x_n) .$$

So by (*)

$$\left| \int f \, d\gamma_n - \frac{1}{b_n - a_n} \sum_{j=a_n}^{b_n - 1} f(T^j x) \right| \leq \omega_f(2^{-n+1} \delta)$$

Since $\|f\| \leq 1$, (**) implies, with $A = [a_n, b_n[\cap \mathbb{Z}$,

$$\left| \frac{1}{b_n - a_n} \sum_{j=a_n}^{b_n - 1} f(T^j x) - \frac{1}{b_n} \sum_{j=o}^{b_n - 1} f(T^j x) \right| \leq \frac{2a_n}{b_n - a_n} .$$

Since $\omega_f(2^{-n+1}\delta) \to 0$ and $\dfrac{2a_n}{b_n - a_n} \longrightarrow 0$, this shows that

$$\left| \int f \, d\rho_n - \int f \, d\gamma_n \right| \longrightarrow 0$$

Hence $\rho_{n_k} \longrightarrow \mu$ and thus $\mu \in V_T(x)$.

<u>Claim b:</u> $V_T(x) \subset V$.

Let $\mu \in V_T(x)$ be given. Write $\mu(n)$ for the measure $\delta(x)^n$. There exists a sequence $n_k \uparrow \infty$ such that $\mu(n_k) \longrightarrow \mu$. Let $\epsilon > 0$ and $f \in C(X)$ with $\|f\| \leq 1$ be given.

For fixed n_k let $i = i(n_k)$ be the largest integer such that $b_{i-1} \leq n_k$. Let n_k (and hence i) be so large that

$$\omega_f(2^{-i+1}) < \omega_f(2^{-i+2}) < \frac{\varepsilon}{4} .$$

We may of course assume that n_k, if it is larger than a_i, is of the form $a_i + mp_i$, with $m(= m_k)$ an integer > 0. In this case

$$\int f \, d\gamma_i = \frac{1}{n_k - a_i} \sum_{j=a_i}^{n_k-1} f(T^j x_i) .$$

Also

$$\int f \, d\gamma_{i-1} = \frac{1}{b_{i-1} - a_{i-1}} \sum_{j=a_{i-1}}^{b_{i-1}-1} f(T^j x_{i-1}).$$

Hence

$$\left| \int f \, d\gamma_i - \frac{1}{n_k - a_i} \sum_{j=a_i}^{n_k-1} f(T^j x) \right| < \frac{\varepsilon}{4}$$

(if $n_k > a_i$) and

$$\left| \int f \, d\gamma_{i-1} - \frac{1}{b_{i-1} - a_{i-1}} \sum_{j=a_{i-1}}^{b_{i-1}-1} f(T^j x) \right| < \frac{\varepsilon}{4}.$$

Write

$$\alpha = (b_{i-1} - a_{i-1})(b_{i-1} - a_{i-1} + n_k - a_i)^{-1}$$

if $n_k > a_i$, and 1 otherwise. Set

$$\rho(n_k) = \alpha \, \gamma_{i-1} + (1 - \alpha) \, \gamma_i .$$

One has

$$\int f \, d\mu(n_k) = \frac{1}{n_k} \sum_{j=o}^{n_k-1} f(T^j x).$$

Using (**) again, with $A = ([a_{i-1}, b_{i-1}) \cup [a_i, n_k)) \cap \mathbb{Z}$, one obtains

$$\left| \int f \, d\mu(n_k) - \frac{1}{b_{i-1}-a_{i-1}+n_k-a_i} \left(\sum_{j=a_{i-1}}^{b_{i-1}-1} f(T^j x) + \sum_{j=a_i}^{n_k-1} f(T^j x) \right) \right|$$

$$\leq 2(a_i+a_{i-1}-b_{i-1})\,(b_{i-1}-a_{i-1}+n_k-a_i)^{-1}$$

$$\leq 2\left(\frac{a_{i-1}}{b_{i-1}-a_{i-1}} + \frac{a_i - b_{i-1}}{b_{i-1}-a_{i-1}} \right) < \frac{\varepsilon}{2}$$

provided n_k (and i) are large enough. Hence

$$\left| \int f \, d\mu(n_k) - \left[\alpha \frac{1}{b_{i-1}-a_{i-1}} \sum_{j=a_{i-1}}^{b_{i-1}-1} f(T^j x) + \right.\right.$$

$$\left.\left. + (1-\alpha) \frac{1}{n_k - a_i} \sum_{j=a_i}^{n_k-1} f(T^j x) \right] \right| < \frac{\varepsilon}{2}$$

and so

$$\left| \int f \, d\mu(n_x) - \int f \, d\rho(n_k) \right| \leq \varepsilon \quad \text{for k large enough.}$$

Thus $\rho(n_k)$ has the same limit as $\mu(n_k)$, namely μ. On the other hand the limit of $\rho(n_k)$ has to be in V, since

$$\bar{d}\,(\gamma_{i-1}, V) \leq \varepsilon_{i-1},$$

$$\bar{d}\,(\gamma_i, V) \leq \varepsilon_i,$$

$$\bar{d}\,(\gamma_{i-1}, \gamma_i) \leq \varepsilon_{i-1} + \varepsilon_i$$

and $\varepsilon_i \downarrow 0$. Hence $\mu \in V$. □

(21.15) Corollary: If (X,T) satisfies the specification property, then every T-invariant measure has **generic** points.

By a slight modification of the proof above, one gets

(21.16) Proposition [172]: Let (X,T) satisfy the specification property, and let $V \subset \mathfrak{M}_T(X)$ be nonempty, closed and connected. Then the set

$$\{x \in X \mid V_{T^n}(x) = V \quad \text{for } n = 1,2,\ldots\}$$

is dense in X.

(21.17) Definition: A point $x \in X$ is said to have maximal oscillation if $V_T(x) = \mathfrak{M}_T(X)$.

(21.18) Proposition: If (X,T) satisfies the specification property, then the set of points having maximal oscillation is residual in X.

Proof: There exist open balls V_i, U_i in $\mathfrak{M}(X)$ such that

(a) $V_i \subset \overline{V_i} \subset U_i$;

(b) $\operatorname{diam} U_i \longrightarrow 0$;

(c) $V_i \cap \mathfrak{M}_T(X) \neq \emptyset$;

(d) each point of $\mathfrak{M}_T(X)$ lies in infinitely many V_i.

Put $P(U_i) = \{x \in X \mid V_T(x) \cap U_i \neq \emptyset\}$.

It is easy to see that the set of points with maximal oscillation is just $\bigcap P(U_i)$. Now

$$P(U_i) \supset \{x \in X \mid \forall N_o \in \mathbb{N}, \exists N > N_o \text{ with } \delta(x)^N \in V_i\}$$

$$= \bigcap_{N_o=1}^{\infty} \bigcup_{N>N_o} \{x \in X \mid \delta(x)^N \in V_i\}.$$

Since $x \longrightarrow \delta(x)^N$ is continuous (for fixed N), the sets $\bigcup_{N\geq N_o} \{x \in X \mid \delta(x)^N \in V_i\}$ are open. Since $V_i \cap \mathfrak{M}_T(X) \neq \emptyset$, these sets are also dense, as shown by (21.15). Hence $\bigcap P(U_i)$ contains a dense G_δ-set. □

(21.19) Corollary: If (X,T) satisfies the specification property then the set $Q_T(X)$ of quasiregular points is of first category.

This also follows from (4.9).

(21.20) Proposition [174]: If (X,T) satisfies the specification property and $V \subset \mathfrak{M}_T(X)$ is nonempty, closed and connected, then the set

$$\{\mu \in \mathfrak{M}(X) \mid V_T(\mu) = V\}$$

is dense in $\mathfrak{M}(X)$.

Proof: By (21.14) there exists an $x \in X$ such that $V_T(x) = V$. Let $\mu_o \in \mathfrak{M}(X)$ and $\epsilon > 0$ be given. By (21.5) $T: \mathfrak{M}(X) \longrightarrow \mathfrak{M}(X)$ satisfies the specification property. Let $M_i = M(2^{-i}\epsilon)$ be given as in definition (21.1), but with respect to the extension T on $\mathfrak{M}(X)$. There exists a $\mu_1 \in \mathfrak{M}(X)$ such that $\bar{d}(\mu_o, \mu_1) < 2^{-1}\epsilon$ and $\bar{d}(T^n\mu_o, T^n\mu_1) < 2^{-1}\epsilon$ for $B_1 \leq n \leq E_1$, where $B_1 = M_1$ and $E_1 = 2(B_1 + M_2)$. There exists a $\mu_2 \in \mathfrak{M}(X)$ such that $\bar{d}(T^n\mu_1, T^n\mu_2) < 2^{-2}\epsilon$ for $0 \leq n \leq E_1$ and $\bar{d}(T^n\delta(x), T^n\mu_2) < 2^{-2}\epsilon$ for $B_2 \leq n \leq E_2$ where $B_2 = E_1 + M_2$ and $E_2 = 2^2(B_2+M_3)$. By induction, one constructs for $i > 2$ a $\mu_i \in \mathfrak{M}(X)$ such that $\bar{d}(T^n\mu_{i-1}, T^n\mu_i) < 2^{-i}\epsilon$ for $0 \leq n \leq E_{i-1}$ and $\bar{d}(T^n\delta(x), T^n\mu_i) < 2^{-i}\epsilon$ for $B_i \leq n \leq E_i$, where $B_i = E_{i-1} + M_i$ and $E_i = 2^i(B_i + M_{i+1})$. Clearly μ_i converges to some $\mu \in \mathfrak{M}(X)$ with $\bar{d}(\mu, \mu_o) < \epsilon$. Since $\bar{d}(T^n\delta(x), T^n\mu) < 2^{-i+1}\epsilon$ for $B_i \leq n \leq E_i$, and $\bigcup_{i=1}^{\infty} [B_i, E_i]$ is of density one in $\mathbb{N}$, one also gets $V_T(\mu) = V_T(x)$ and hence $V_T(\mu) = V$. □

(21.21) Proposition: If (X,T) satisfies the specification property then the set of measures with maximal oscillation, i.e. the set $\{\mu \in \mathfrak{M}(X) \mid V_T(\mu) = \mathfrak{M}_T(X)\}$, is residual in $\mathfrak{M}(X)$.

Proof: This follows as in (21.18). □

We mention that all the results in this section are also valid for systems satisfying the weak specification property. On the other hand, they need not hold for transformations which are topologically mixing and have periodic points dense. In [189] B.Weiss constructed a nontrivial subshift with these properties, but having topological entropy 0. For this subshift the ergodic measures are not dense in $\mathfrak{M}_T(X)$, and not every invariant measure has generic points.

22. Specification and Expansiveness

Let (X,T) be a top.dynamical system which is expansive and satisfies the specification property. (The nonwandering parts of Axiom A diffeomorphisms are built up by such transformations.) The present section deals mainly with the construction of the unique measure of maximal entropy for such systems, which is due to Bowen. It closely follows his proofs in [21] and [27].

First we recall some notation. δ^* is some expansive constant, $P^n(T)$ is the set of periodic points of period n, $\mathrm{Per}_n(T) = \mathrm{card}\, P^n(T)$, $s_n(\epsilon)$ is the maximal cardinality of an (n,ϵ)-separated set in X and
$B^n_\epsilon(x) = \{ y \in X \mid d(T^j x, T^j y) \leq \epsilon \text{ for } 0 \leq j < n \}$.
Clearly $P^n(T) \neq \emptyset$ for n large enough. We shall write h for $h_{top}(T)$.

(22.1) Definition: Denote by μ_n the measure

$$\frac{1}{\mathrm{Per}_n(T)} \sum_{x \in P^n(T)} \delta(x)$$

and by μ_B some limit point of the sequence μ_n (which exists by (2.8)). Thus there is an increasing sequence $S = \{n_k\}$ of integers such that

$$\mu_B = \lim \mu_{n_k}.$$

Clearly μ_n and μ_B are T-invariant. We shall show that μ_B is the unique measure of maximal entropy (and hence independent of the sequence S). μ_B is called the Bowen measure.

(22.2) Lemma: If ϵ and δ are small enough, there exists a constant $C_{\delta,\epsilon}$ such that

$$s_n(\delta) \leq C_{\delta,\epsilon}\; s_n(\epsilon) \qquad \text{for all} \quad n \geq 0.$$

<u>Proof:</u> Let $\epsilon < \frac{1}{2}\,\delta^*$. By the corollary to (16.9) there is an $N = N(\delta)$ such that $d(T^j x, T^j y) \leq 2\epsilon$ for $-N \leq j \leq N$ implies $d(x,y) \leq \delta$.

Choose $\beta > 0$ so small that $d(x,y) < \beta$ implies $d(T^j x, T^j y) \leq \delta$ for $-N \leq j \leq N$. Let $K = K(\beta)$ be the maximal number of β-separated points in $X \times X$ (with the maximum metric).

Let $F \subset X$ be maximal (n,ϵ)-separated and $E \subset X$ maximal (n,δ)-separated.

For $x \in E$ there exists a $z(x) \in F$ with

$$d(T^j x, T^j z(x)) \leq \epsilon \qquad \text{for } 0 \leq j < n.$$

(Otherwise $F \cup \{x\}$ would be (n,ϵ)-separated, a contradiction to the maximality of F.)

For $z \in F$ write

$$E_z = \{x \in E \mid z(x) = z\}.$$

One has $\operatorname{card} E_z \leq K$. Indeed, if $x,y \in E_z$ then

$$d(T^j x, T^j y) \leq 2\epsilon \qquad \text{for} \quad 0 \leq j < n$$

and hence

$$d(T^j x, T^j y) \leq \delta \qquad \text{for} \quad N \leq j < n - N.$$

But $\{x,y\}$ is (n,δ)-separated, so that one has $d(x,y) > \beta$ or $d(T^n x, T^n y) > \beta$. Hence the set $\{(x, T^n x) \mid x \in E_z\} \subset X \times X$ is β-separated (i.e. $(1,\beta)$-separated). Thus

$$s_n(\delta) = \operatorname{card} E \leq K \operatorname{card} F = K\, s_n(\epsilon)$$

as was to be shown. (Note that we used only the expansiveness of (X,T)). □

(22.3) Lemma: If ε is small enough, there exists a constant D_ε such that

$$s_{n_1+\ldots+n_k}(\varepsilon) \leq \prod_{i=1}^{k} D_\varepsilon \cdot s_{n_i}(\varepsilon)$$

whenever $n_1,\ldots,n_k \geq 1$.

Proof: Let $E \subset X$ be $(n_1 + \ldots + n_k,\varepsilon)$-separated and $F_i \subset X$ maximal $(n_i, \frac{\varepsilon}{2})$-separated, for $1 \leq i \leq k$.

For $x \in E$ choose $\underline{z}(x) = (z_1(x),\ldots,z_k(x)) \in F_1\times\ldots\times F_k$ such that

$$d(T^{n_1+\ldots+n_{i-1}+j}x,\ T^j z_i(x)) \leq \frac{\varepsilon}{2} \quad \text{for} \quad 0 < j < n$$

(such $z_i(x)$ exist by the maximality of F_i). The map $\underline{z} : E \longrightarrow F_1 \times\ldots\times F_k$ is injective since E is $(n_1+\ldots+n_k,\varepsilon)$-separated. Hence

$$\text{card } E \leq \prod_{i=1}^{k} \text{card } F_i = \prod_{i=1}^{k} s_{n_i}(\tfrac{\varepsilon}{2}).$$

By lemma (22.2)

$$s_{n_i}(\tfrac{\varepsilon}{2}) \leq C_{\frac{\varepsilon}{2},\varepsilon}\ s_{n_i}(\varepsilon)$$

and hence the lemma follows with $D_\varepsilon := C_{\frac{\varepsilon}{2},\varepsilon}$. □

(22.4) Lemma: If ε is small enough, there exists a constant E_ε with

$$s_{n_1+\ldots+n_k}(\varepsilon) \geq \prod_{i=1}^{k} E_\varepsilon\ s_{n_i}(\varepsilon).$$

Proof: Choose a maximal $(n_i, 3\epsilon)$ separated set $E_i \subset X$ for $1 \leq i \leq k$. Let $M = M(\epsilon)$ be as in the definition (21.1) of the specification property. Write
$a_i = n_1+\ldots+n_{i-1} + (i-1)M$, $m = n_1+\ldots+n_k + (k-1)M$.

By the specification property, for each $\underline{z} = (z_1,\ldots,z_k) \in E_1 \times\ldots\times E_k$ there exists an $x = x(\underline{z})$ with $d(T^{a_i+j}x, T^j z_i) \leq \epsilon$ for $0 \leq j < n_i$, $1 \leq i \leq k$.

Clearly the set

$$E = \{x(\underline{z}) \mid \underline{z} \in E_1 \times\ldots\times E_k)\}$$

is (m,ϵ)-separated. Hence

$$\prod_{i=1}^{k} s_{n_i}(3\epsilon) \leq s_m(\epsilon).$$

On the other hand, lemma (22.3) implies that

$$s_m(\epsilon) \leq D_\epsilon^k \, s_{n_1+\ldots+n_k}(\epsilon) \, (s_M(\epsilon))^{k-1}$$

and lemma (22.2) says that

$$C_{\epsilon,3\epsilon}^{-1} \, s_{n_i}(\epsilon) \leq s_{n_i}(3\epsilon).$$

Putting $E_\epsilon = [C_{\epsilon,3\epsilon} \cdot D_\epsilon \cdot s_M(\epsilon)]^{-1}$ one obtains the result. □

(22.5) Lemma: For small ϵ one has

$$D_\epsilon^{-1} \cdot e^{hn} \leq s_n(\epsilon) \leq E_\epsilon^{-1} \cdot e^{hn}$$

for all $n \geq 1$. $(h = h_{top}(T))$

Proof: Suppose $s_n(\varepsilon) < D_\varepsilon^{-1} e^{hn}$ for some n. By (22.3) one has

$$s_{kn}(\varepsilon) < [D_\varepsilon \cdot s_n(\varepsilon)]^k$$

and hence

$$s_{kn}(\varepsilon) \leq e^{hnk}.$$ By corollary 2 of (16.9)

$$h = \lim_{k\to\infty} \frac{1}{kn} \log s_{kn}(\varepsilon).$$ Thus

$$h < \frac{1}{n} \log e^{hn} = h ,$$

a contradiction. Similarly for the other inequality. □

(22.6) Lemma: There are constants $0 < D < E$ such that for n large enough

$$De^{hn} \leq Per_n(T) \leq Ee^{hn}.$$

Proof: Clearly if $\varepsilon < \delta^*$ then $P^n(T)$ is (n,ε)-separated. Thus $Per_n(T) \leq s_n(\varepsilon)$ and therefore by (22.5)

$$Per_n(T) \leq Ee^{hn}$$

with $E = E_\varepsilon^{-1}$.

Now let $M = M(\varepsilon)$ be given as in the definition of the specification property, and let $E \subset X$ be $(n - M, 3\varepsilon)$-separated set, with $n > M$. By the specification property, for each $z \in E$ there is an $x(z) \in P^n(T)$ with

$$d(T^j z, T^j x(z)) \leq \varepsilon \quad \text{for} \quad 0 \leq j < n - M.$$

Clearly $z \neq z'$ implies $x(z) \neq x(z')$. Hence

$$\mathrm{Per}_n(T) \geq s_{n-M}(3\epsilon) \geq D_{3\epsilon}^{-1}\, e^{h(n-M)} = De^{hn}$$

with $D = [D_{3\epsilon}\, e^{hM}]^{-1}$. □

<u>(22.7) Theorem:</u> If (X,T) satisfies expansiveness and the specification property, then

$$h_{top}(T) = \lim \frac{1}{n} \log \mathrm{Per}_n(T)$$

<u>Proof:</u> This follows immediately from (22.6). □

<u>(22.8) Lemma:</u> For ϵ small enough there exists an $A_\epsilon > 0$ such that

$$\mu_B(B_\epsilon^n(x)) \geq A_\epsilon e^{-nh}$$

for all $x \in X$ and $n \geq 1$.

<u>Proof:</u> Let $E \subset X$ be a maximal $(m,3\epsilon)$-separated set. Write $r = n + m + 2M$, where $M = M(\epsilon)$ is given as in the definition of the specification property. For each $z \in E$ there exists by the specification property an $x(z) \in P^r(T)$ with

$$d(T^j x(z), T^j x) \leq \epsilon \quad \text{for } 0 \leq j < n$$

and

$$d(T^{n+M+j} x(z), T^j z) \leq \epsilon \text{ for } 0 \leq j < m.$$

Clearly if z and z' are two distinct points of E then $x(z) \neq x(z')$. Hence

$$\mathrm{card}\ (B_\epsilon^n(x) \cap P^r(T)) \geq \mathrm{card}\ E = s_m(3\epsilon)\ .$$

This implies

$$\mu_r(B_\epsilon^n(x)) = \frac{1}{\mathrm{Per}_r(T)} \ \mathrm{card}\ (B_\epsilon^n(x) \cap P^r(T))$$

$$\geq \frac{s_m(3\epsilon)}{\mathrm{Per}_r(T)}$$

$$\geq \frac{e^{-hr}}{E} \ \frac{e^{hm}}{D_{3\epsilon}} = A_\epsilon \ e^{-hn}$$

with $A_\epsilon := [E\ D_{3\epsilon}\ e^{2Mh}]^{-1}$.

Remark that $B_\epsilon^n(x)$ is closed. Letting $r \longrightarrow \infty$ (through the sequence S) one obtains by (2.6)

$$\mu_B(B_\epsilon^n(x)) \geq \limsup_{r \in S} \mu_r(B_\epsilon^n(x)) \geq A_\epsilon \ e^{-hn}. \qquad \square$$

<u>(22.9) Lemma:</u> For ϵ small enough there exist a $\delta = \delta(\epsilon)$ and a constant B_ϵ such that

$$\mu_B(B_\delta^n(y)) \leq B_\epsilon \ e^{-nh}$$

for all $y \in X$ and $n \geq 1$.

<u>Proof:</u> Let ϵ be given and $M = M(\epsilon)$ as in the definition of the specification property. Choose $\delta = \delta(\epsilon) < \frac{\epsilon}{6}$ so small that $d(x,y) < 6\delta$ implies $d(T^j x, T^j y) \leq \epsilon$ for $-M \leq j \leq M$. Write $r = n + m + 2M$. By expansiveness, any two distinct points x,z in $P^r(T) \cap B_{3\delta}^n(y)$ are (r,ϵ)-separated. Since $d(T^j x, T^j z) \leq \epsilon$ for $-M \leq j \leq M + n$, this implies that $T^{n+M}x$ and $T^{n+M}z$ are (m,ϵ)-separated. Hence

$$\mathrm{card}(B_{3\delta}^n(y) \cap P^r(T)) \leq s_m(\epsilon).$$

Thus by (22.5) and (22.6)

$$\mu_r(B^n_{3\delta}(y)) = \frac{1}{\mathrm{Per}_r(T)} \text{ card } (B^n_{3\delta}(y) \cap P^r(T)) \leq \frac{1}{\mathrm{Per}_r(T)} s_m(\epsilon)$$

$$\leq D^{-1} e^{-hr} E_\epsilon^{-1} e^{hm} = B_\epsilon e^{-hn}$$

with $B_\epsilon = [D \cdot E_\epsilon \cdot e^{2hM}]^{-1}$.

Let V be the open set $\{x \in X | d(T^j x, T^j y) < 2\delta \text{ for } 0 \leq j < n\}$. Clearly $B^n_\delta(y) \subset V \subset B^n_{3\delta}(y)$. Thus

$$\mu_r(V) \leq B_\epsilon \cdot e^{-hn}$$

and hence by (2.7), letting $r \to \infty$ through the sequence S,

$$\mu_B(B^n_\delta(y)) \leq \mu_B(V) \leq \liminf \mu_r(V) \leq B_\epsilon \cdot e^{-hn}. \qquad \square$$

<u>(22.10) Theorem:</u> The Bowen measure has maximal entropy.

<u>Proof:</u> Lemmas (22.8) and (22.9) show that μ_B is T-homogeneous (cf. (19.6)). By (19.7) such measures have maximal entropy. □

<u>(22.17) Proposition:</u> μ_B has support X.

<u>Proof:</u> This is a corollary from (22.8). □

<u>(22.12) Lemma:</u> μ_B is ergodic.

<u>Proof:</u> Let $A, B \subset X$ be compact sets and U, V be μ_B-continuity 2δ-neighborhoods of A, B for some arbitrary small suitable $\delta > 0$. Set $\tilde{U} = B_\delta(A)$ and $\tilde{V} = B_\delta(B)$. Choose $\epsilon < 2^{-1}\delta*$. By the corollary to (16.9) there exists an $N = N(\delta)$ such that $d(T^j x, T^j y) < \epsilon$ for $-N \leq j \leq N$ implies $d(x,y) < \delta$.

Assume $n > N$ and let s, t be positive integers. Let $M = M(\epsilon)$ be given as in the definition of the specification property, let $E_s \subset X$ be $(s, 2\epsilon)$-separated and $E_t \subset X$ be $(t, 2\epsilon)$-separated. Let

$$I_i = [a_i, b_i) \cap \mathbb{Z} \qquad \text{for } i = 1,2,3,4$$

be intervals of integers with

$$a_1 = -n \qquad\qquad b_1 = n$$

$$a_2 = b_1 + M \qquad\qquad b_2 = a_2 + s$$

$$a_3 = b_2 + M = n + s + 2M \qquad b_3 = a_3 + 2n$$

$$a_4 = b_3 + M \qquad\qquad b_4 = a_4 + t$$

Let $m = b_4 - a_1 + M = t + s + 4n + 4M$.

Assume $\underline{z} = (z_1, z_2, z_3, z_4)$ is such that

$$z_1 \in P^{2n}(T) \cap T^{-n}\tilde{U}$$

$$z_2 \in E_s$$

$$z_3 \in P^{2n}(T) \cap T^{-n}\tilde{V}$$

$$z_4 \in E_t.$$

By the specification property there exists a point $x = x(\underline{z})$ with $T^m x = x$ such that

$$d(T^{a_i+j}x, T^j z_i) < \epsilon \quad \text{for } 0 \leq j < b_i - a_i,\ i = 1,2,3,4.$$

Since $n > N$ and $T^n z_1 \in \tilde{U}$, it is easy to see that $x \in U$. Similarly $T^{s+2M+2n}(x) \in V$. With $r = s+2(M+n)$ one has therefore

$$x \in P^m(T) \cap U \cap T^{-r}V.$$

Clearly if $\underline{z} \neq \underline{z}'$ then $x(\underline{z}) \neq x(\underline{z}')$. Indeed if $x(\underline{z}) = x(\underline{z}')$, then $z_1 \neq z_1'$ would imply $d(T^j z_1, T^j z_1') < 2\epsilon < \delta^*$

for all $j \in \mathbb{Z}$, a contradiction to expansiveness. $z_2 \neq z_2'$ would imply $d(T^j z_2, T^j z_2') < 2\varepsilon$ for $0 \le j < s$, a contradiction to the definition of E_s. Similarly for z_3 and z_4. There are more than

$$\mathrm{Per}_{2n}(T)\mu_{2n}(\tilde{U})\ s_s(2\varepsilon)\ \mathrm{Per}_{2n}(T)\mu_{2n}(\tilde{V})\ s_t(2\varepsilon)$$

different $\underline{z}$'s. Hence by (22.5) and (22.6)

$$\begin{aligned}\mu_m(U \cap T^{-r}V) &= \frac{1}{\mathrm{Per}_m(T)}\ \mathrm{card}(P^m(T) \cap U \cap T^{-r}V)\\
&\ge \frac{1}{\mathrm{Per}_m(T)}\left(\mathrm{Per}_{2n}(T)\right)^2 s_s(2\varepsilon)\ s_t(2\varepsilon)\ \mu_{2n}(\tilde{U})\ \mu_{2n}(\tilde{V})\\
&\ge \frac{e^{-hm}}{E}\left(\frac{De^{h2n}}{D_{2\varepsilon}}\right)^2 e^{hs}\ e^{ht}\ \mu_{2n}(\tilde{U})\ \mu_{2n}(\tilde{V})\\
&= R\ \mu_{2n}(\tilde{U})\ \mu_{2n}(\tilde{V})\end{aligned}$$

with

$$R = \frac{e^{-4Mh}}{E}\left(\frac{D}{D_{2\varepsilon}}\right)^2 > 0.$$

Since t is chosen independent of $\delta,\varepsilon,n,s,U,V$ (and hence independent of r,N,M,R) we obtain by (2.7) letting $t \longrightarrow \infty$ such that m runs through S while n and $r \ge 2(n+M)$ are fixed

$$\mu_B(U \cap T^{-r}V) = \lim_m \mu_m(U \cap T^{-r}V) \ge R\ \mu_{2n}(\tilde{U})\ \mu_{2n}(\tilde{V}).$$

But $n > N$ is independent of δ and ε (and hence of N,M,R, U and V). Therefore if $n \longrightarrow \infty$ such that 2n runs through S (we may of course assume that S consists of even numbers) then $r \longrightarrow \infty$ also and we get using (2.7) again

$$\liminf_r \mu_B(U \cap T^{-r}V) \geq \liminf_n R\, \mu_{2n}(\tilde{U})\, \mu_{2n}(\tilde{V}) \geq$$

$$\geq R\, \mu_B(\tilde{U})\, \mu_B(\tilde{V}) \geq R\, \mu_B(A)\, \mu_B(B).$$

Since $\mu_B(A \cap T^{-r}B) \geq \mu_B(U \cap T^{-r}V) - \mu_B(U \setminus A) - \mu_B(V \setminus B)$, it follows that

$$\liminf \mu_B(A \cap T^{-r}B) \geq R\, \mu_B(A)\, \mu_B(B) - \mu_B(U \setminus A) - \mu_B(V \setminus B).$$

We note that ϵ and hence M and R are chosen independent of δ. Moreover if $\delta \longrightarrow 0$ then $\mu_B(U \setminus A) \longrightarrow 0$ and $\mu_B(V \setminus B) \longrightarrow 0$. Therefore

$$\liminf_r \mu_B(A \cap T^{-r}B) \geq R\, \mu_B(A)\, \mu_B(B) .$$

(We remark that R is independent of A and B!)

Now let $P, Q \subset X$ be such that $\mu_B(P) > 0$, $\mu_B(Q) > 0$. By regularity, there are compact sets $A \subset P$, $B \subset Q$ with $\mu_B(A) > 0$, $\mu_B(B) > 0$. Hence $\liminf_r \mu_B(P \cap T^{-r}Q) > 0$, and thus the system (X, T, μ_B) is ergodic. □

<u>(22.13) Lemma:</u> If $a_1, \ldots, a_m$ are reals ≥ 0 such that $0 < a_1 + \ldots + a_m = s \leq 1$, then

$$-\Sigma\, a_i \log a_i \leq s \log m - s \log s \leq s \log m + \frac{1}{2}$$

<u>Proof:</u> This is a trivial consequence of Jensen's inequality. □

<u>(22.14) Lemma:</u> Let ϵ be an exp.const., $n \in \mathbb{Z}$ and α a partition of X such that diam $T^jA < \epsilon$ for $A \in \alpha$ and $0 \leq j < n$. Then

$$\frac{1}{n} H_\mu(\alpha) \geq h_\mu(T).$$

<u>Proof:</u> Since T is expansive, α is an m.t. generator for T^n.

Hence

$$H_\mu(\alpha) \geq h_\mu(T^n,\alpha) = h_\mu(T^n) = nh_\mu(T). \qquad \square$$

(22.15) Theorem: If (X,T) satisfies expansiveness and the specification property, then (X,T) is intrinsically ergodic.

Proof: Assume $\nu \in \mathfrak{M}_T(X)$ is such that $h_\nu(T) = h$. We have to show that $\nu = \mu_B$. Since entropy is affine, we may suppose that ν is ergodic, and hence by (5.4) singular with respect to μ_B. Then there exists a T-invariant set $C \subset X$ such that $\mu_B(C) = 0$ and $\nu(C) = 1$.

Fix some ϵ small enough and let $E \subset X$ be a maximal $(n,2\epsilon)$-separated set. Thus E is also $(n,2\epsilon)$-spanning. Hence

$$X \subset \bigcup_{x\in E} B^n_{2\epsilon}(x),$$

and if $x,y \in E$ are distinct,

$$B^n_\epsilon(x) \cap B^n_\epsilon(y) = \emptyset.$$

For each x there exists a Borel set A_x with

$$B^n_\epsilon(x) \subset A_x \subset B^n_{2\epsilon}(x)$$

such that $\alpha_n = (A_x | x \in E)$ is a Borel partition (cf. (16.6)). Now expansiveness implies

$$\operatorname{diam} T^{[\frac{n}{2}]} B^n_{2\epsilon}(x) \longrightarrow 0$$

and hence

$$\operatorname{diam} T^{[\frac{n}{2}]} \alpha_n \longrightarrow 0.$$

Thus the sequence $T^{[\frac{n}{2}]}\alpha_n$ generates.

It is easy to see that there exist sets C_n which are unions of atoms of α_n such that

$$\nu(T^{[\frac{n}{2}]} C_n \Delta C) \longrightarrow 0$$

and

$$\mu_B(T^{[\frac{n}{2}]} C_n \Delta C) \longrightarrow 0 .$$

Since C is T-invariant this implies

$$\nu(C_n \Delta C) \longrightarrow 0 \quad \text{and} \quad \mu_B(C_n \Delta C) \longrightarrow 0.$$

By (22.14)

$$h = h_\nu(T) \leq \frac{1}{n} H_\nu(\alpha_n) .$$

Hence

$$\begin{aligned} nh &\leq -\sum_{A_x \in \alpha_n} \nu(A_x) \log \nu(A_x) \\ &= -\sum_{A_x \subset C_n} \nu(A_x) \log \nu(A_x) + \sum_{A_x \cap C_n = \emptyset} \nu(A_x) \log \nu(A_x) . \end{aligned}$$

Applying (22.13) to both sums

$$nh \leq \nu(C_n) \log \operatorname{card}\{A_x | A_x \subset C_n\} + \nu(C_n^c) \log \operatorname{card}\{A_x | A_x \cap C_n = \emptyset\} + 1 .$$

Hence

$$-1 \leq \nu(C_n)[\log \operatorname{card}\{A_x | A_x \subset C_n\} - nh] + \nu(C_n^c)[\log \operatorname{card}\{A_x | A_x \cap C_n = \emptyset\} - nh] .$$

But by (22.8)

$$e^{-nh} \leq A_\epsilon^{-1} \mu_B(B_\epsilon^n(x)) \leq A_\epsilon^{-1} \mu_B(A_x)$$

and so

$$-1 \leq \nu(C_n) \log A_\epsilon^{-1} \mu_B(C_n) + \nu(C_n^c) \log A_\epsilon^{-1} \mu_B(C_n^c).$$

If $n \to \infty$ one has $\mu_B(C_n) \to 0$ and $\nu(C_n^c) \to 0$, hence the right hand side of this inequality tends to $-\infty$, a contradiction. □

<u>(22.16) Proposition</u> [161, 168]: The set of $\mu \in \mathfrak{M}_T(X)$ with $h_\mu(T) = 0$ is a residual set in $\mathfrak{M}_T(X)$.

<u>Proof:</u> The density of this set follows from the fact that CO-measures are dense (21.8) and have zero entropy. Since T is expansive, $\mu \longmapsto h_\mu(T)$ is upper semicontinuous (see (16.7)). Hence

$$\{\mu \in \mathfrak{M}_T(X) \mid h_\mu(T) < \tfrac{1}{n}\}$$

is open, and $\{\mu \in \mathfrak{M}_T(X) \mid h_\mu(T) = 0\}$ is a G_δ. □

23. Basic Sets for Axiom A

The last twelve years have seen some remarkable progress in the ergodic theory of differentiable dynamical systems, and in particular of structurally stable systems. Many of these results are based on topological properties like specification and expansiveness. In this section we shall sketch some of the differentiable background. Some elementary knowledge about manifolds will be assumed (see, for example, Lang's book: Differentiable manifolds, Wiley and Sons, 1962).

Let M be a connected differentiable manifold.

For $x \in M$ we denote by T_xM the tangent space at x and by TM the tangent bundle $\bigcup_{x \in M} T_xM$. A differentiable map $\Phi : M \longrightarrow M$ induces a map from T_xM to $T_{\Phi(x)}M$ and thereby a vector bundle homomorphism $D\Phi : TM \longrightarrow TM$ called the tangent map of Φ.

If for any $x \in M$ there exists a bilinear symmetric positive definite form on T_xM which depends smoothly on x, then this defines a Riemannian structure on TM. The corresponding norm on T_xM will be denoted by $||\cdot||$. It induces a metric d on M. Riemannian structures exist on every manifold.

(23.1) Proposition [22]: Let $\varphi : M \longrightarrow M$ be a differentiable map. Then

$$h_d(\varphi) \leq \max \{0, m \log a\}$$

where m is the dimension of M and $a = \sup\{||D\varphi | T_xM|| \mid x \in M\}$.

We refer to Bowen [22] for a proof. In the next section we shall need this result for $M = R^m$ and φ linear: in this case it follows easily from (14.20).

(23.2) Theorem of Kushnirenko [129]: Let M be a compact manifold and $\varphi : M \longrightarrow M$ differentiable. Then $h_{top}(\varphi)$ is finite.

Proof: This is a corollary of (23.1). □

Two differentiable maps $\varphi_1, \varphi_2 : M \longrightarrow N$ are said to be C^k-close $(k = 1,\ldots,\infty)$ if in some local coordinate system their derivatives up to order k are close. This induces the C^k-topology in the space of differentiable maps $M \longrightarrow N$, and in particular in the space of diffeomorphisms from M onto itself.

For the remainder of this section we consider only compact manifolds. In [136] Misiurewicz showed that $\varphi \longrightarrow h_{top}(\varphi)$ is not continuous with respect to the C^k-topology $(k=1,\ldots,\infty)$. We only give a sketch of his construction:

(23.3) Example: a) Let $\mathbb{T}^1 = \mathbb{R}|\mathbb{Z}$ be the one-dimensional torus. For $n = 1,2,\ldots$ define

$$\tilde{\Psi}_n(x) = \begin{cases} 2x & \text{if } x \in [0,\frac{1}{4n}] \\ x + (4n)^{-1} & \text{if } x \in [\frac{1}{4n}, 1-\frac{2}{4n}] \\ \frac{1}{2}(1+x) & \text{if } x \in [1-\frac{2}{4n}, 1). \end{cases}$$

This is a well-defined homeomorphism with 0 as unique fixed point. By 'smoothing the edges' of $\tilde{\Psi}_n$ one obtains a diffeomorphism Ψ_n having 0 as unique fixed point, and indeed even as unique nonwandering point. Write Ψ_∞ for the identity map on $\mathbb{T}^1$. By smoothing well enough one can make the Ψ_n converge to Ψ_∞ in any C^k-topology.

(b) Let $\{\varphi_t\}_{t\in\mathbb{R}}$ be a one parameter group of diffeomorphisms on some compact manifold N such that $h_{top}(\varphi_1) > 0$. (This can be obtained by "suspending" a diffeomorphism with positive entropy, see [180]).

(c) Let $f : \mathbb{T}^1 \longrightarrow [0,1]$ be differentiable with $f(0) = 0$ and $f(\frac{1}{2}) = 1$. Write $M = N \times \mathbb{T}^1$ and define $\Phi_n : M \longrightarrow M$ by

$$\Phi_n(x,y) = (\varphi_{f(y)}(x), \Psi_n(y))$$

for $n = 1,2,\ldots,\infty$. Now each Φ_n is a diffeomorphism, and Φ_n converges to Φ_∞ in C^k. But

$$h_{top}(\Phi_\infty) \geqq h_{top}(\Phi_\infty | N \times \{\tfrac{1}{2}\}) = h_{top}(\varphi_1) > 0$$

and

$$h_{top}(\Phi_n) = h_{top}(\Phi_n | \Omega_{\Phi_n}) = 0$$

since the nonwandering set Ω_{Φ_n} is in $N \times \{0\}$ and $\Phi_n | N \times \{0\}$ is the identity.

Actually Misiurewicz showed in [137] that $\varphi \longrightarrow h_{top}(\varphi)$ need not even be upper semicontinuous.

(23.4) Definition: A differentiable map $\Phi : M \longrightarrow M$ is said to be C^k-structurally stable $(k = 1,2,\ldots,\infty)$ if there exists a C^k-neighborhood $U(\Phi)$ of Φ such that any $\Phi' \in U(\Phi)$ is topologically conjugate to Φ.

This means that the topological structure of the orbits of Φ doesn't change if Φ is slightly perturbed (in the C^k-topology). This property is obviously very desirable for applications.

In order to describe some structurally stable transformations we need a few more notions.

(23.5) Definition: A fixed point x of a diffeomorphism $\Phi : M \longrightarrow M$ is said to be hyperbolic if no eigenvalue of the tangent map $D\Phi : T_xM \longrightarrow T_xM$ lies on the complex unit circle. If all eigenvalues have modulus < 1, x is called a sink. If all eigenvalues have modulus > 1, x is called a source; otherwise x is called a saddle. A periodic point x is said to be hyperbolic if it is a hyperbolic fixed point of Φ^p, for some $p \in \mathbb{N}$.

(23.6) Definition: Let T be a homeomorphism of a compact space X with metric d. The set

$$W^+(x) = \{y \in X \mid d(T^n x, T^n y) \longrightarrow 0 \quad \text{for } n \longrightarrow +\infty\}$$

is called the in-set of the point $x \in X$, and the set

$$W^-(x) = \{y \in X \mid d(T^n x, T^n y) \longrightarrow 0 \quad \text{for } n \longrightarrow -\infty\}$$

is called its out-set.

(23.7) Definition: A diffeomorphism $\Phi : M \longrightarrow M$ is said to be a Morse-Smale diffeomorphism if the following conditions are satisfied:

(a) the nonwandering set $\Omega_\Phi(M)$ is finite, and hence consists only of periodic points;

(b) all periodic points are hyperbolic;

(c) for $x, y \in \Omega_\Phi(M)$ the sets $W^+(x)$ and $W^-(y)$ are transversal.

This last condition means that if $W^+(x) \cap W^-(y)$ contains a point z, then the tangents from z to $W^+(x)$ and $W^-(y)$ (which can be shown to exist) span the tangent space $T_x M$.

A simple example of a Morse-Smale system is the map defined on the two-sphere (regarded as the Riemann sphere $\mathbb{C} \cup \{\infty\}$) by $\Phi(z) = 2z$. Here $\Omega_\Phi(M)$ consists of two fixed points, the north pole which is a sink and the south pole which is a source.

<u>(23.8) Theorem:</u> Morse-Smale diffeomorphisms are structurally stable.

For one dimensional manifolds all structurally stable diffeomorphisms are Morse-Smale. Once this was conjectured to hold for higher dimensions, too. Fortunately, for ergodic theorists, it turned out that in all dimensions ≥ 2 there exist structurally stable diffeomorphisms with more interesting nonwandering sets. In particular, it was shown that hyperbolic local automorphisms (which will be described in section 24) are structurally stable. This (and geodesic flows on compact manifolds with negative curvature) suggested the definition of Anosov systems:

<u>(23.9) Definition:</u> A diffeomorphism $\phi : M \longrightarrow M$ is said to be an <u>Anosov diffeomorphism</u> if there exists a splitting of the tangent bundle into a continuous sum of subbundles

$$TM = E^{+} \oplus E^{-}$$

such that the tangent map $D\phi$ preserves the splitting and there exist constants $c > 0$, $\lambda > 1$ satisfying

$$||D\phi^{n}(v)|| \leq c\lambda^{-n}||\, v\, || \qquad \text{for} \quad v \in E^{+}$$

$$||D\phi^{n}(v)|| \geq c\lambda^{n}||\, v\, || \qquad \text{for} \quad v \in E^{-}$$

for $n = 1,2,\ldots$.

This means that ϕ contracts along the direction of E^{+} and expands along the direction of E^{-}. This condition is independent of the choice of the Riemannian metric $||\cdot||$.

<u>(23.10) Theorem</u> [8]: Anosov diffeomorphisms are structurally stable.

In order to find a unifying description of Morse-Smale diffeomorphisms, Anosov diffeomorphisms and other, more complicated examples of structurally stable diffeomorphisms, Smale was led to a condition which he called Axiom A.

(23.11) Definition: A diffeomorphism $\Phi : M \longrightarrow M$ is said to satisfy Axiom A if the following two conditions are satisfied:

(a) the periodic points are dense in $\Omega_\Phi(M) = \Omega$.

(b) Ω is hyperbolic, i.e. there exists a splitting of the restriction of the tangent bundle to Ω, $T_\Omega M = \bigcup_{x\in\Omega} T_x M$, into a continuous sum of $D\Phi$-invariant subbundles

$$T_\Omega M = E^+ \oplus E^-$$

such that there exist constants $c > 0$, $\lambda > 1$ satisfying

$$\|D\Phi^n(v)\| \leq c\lambda^{-n} \|v\| \quad \text{for} \quad v \in E^+$$

$$\|D\Phi^n(v)\| \geq c\lambda^{n} \|v\| \quad \text{for} \quad v \in E^-$$

for $n = 1,2,\dots$.

It is conjectured that (b) implies (a).

(23.12) Definition: An Axiom A diffeomorphism Φ is said to satisfy the transversality condition if for $x,y \in \Omega_\Phi(M)$ the sets $W^+(x)$ and $W^-(y)$ are transversal (cf. (23.7)).

Anosov diffeomorphisms and Morse-Smale diffeomorphisms are examples of Axiom A diffeomorphisms satisfying the transversality condition.

(23.13) Theorem of Robbin [157]: If the diffeomorphism $\Phi : M \longrightarrow M$ satisfies Axiom A and the transversality condition, then it is C^2-structurally stable.

The converse is conjectured. It holds if structural

stability is replaced by a slightly stronger stability condition (cf. [157]).

For the remainder of this section let Φ be an Axiom A diffeomorphism and Ω its nonwandering set. Ω can be quite complicated, but the restriction of Φ to Ω has a very strong local property (local product structure) which in some sense satisfies all the needs of ergodic theory. We refer to [31] for the proofs of the following consequences of this local product structure.

(23.14) Smale's spectral decomposition theorem: Ω is a finite union of disjoint closed Φ-invariant sets Ω_j $(1 \leq j \leq s)$ such that $\Phi|\Omega_j$ is topologically transitive.

Such sets Ω_j are called basic sets.

Note that if Ω is finite (i.e. if Φ is Morse-Smale) then this theorem gives the decomposition of Ω into distinct periodic orbits.

(23.15) Bowen's decomposition theorem: Any basic set Ω_j is a finite union of disjoint closed sets Ω_j^k $(1 \leq k \leq m = m(j))$ such that $\Phi\,\Omega_j^k = \Omega_j^{k+1}$ (and $\Phi\Omega_j^m = \Omega_j^1$) and such that $\Phi^m : \Omega_j^k \longrightarrow \Omega_j^k$ is topologically mixing.

If Ω_j is finite, i.e. if it consists of a single periodic orbit, then this theorem gives the decomposition into distinct points.

(23.16) Definition: A transformation of the form $\Phi^m : \Omega_j^k \to \Omega_j^k$ as described in (23.15) is called an elementary part of the Axiom A diffeomorphism Φ.

In order to study invariant measures for Φ, it is enough to concentrate on elementary parts. Indeed, any invariant measure is concentrated on Ω (cf. 6.16)). Hence by (23.14), any Φ-invariant μ can be written in the form $a_1\mu_1+\ldots+a_s\mu_s$

with $\mu_j \in \mathfrak{M}_\Phi(\Omega_j)$ and $a_j \geq 0$, $\Sigma\, a_j = 1$. Therefore it is enough to consider invariant measures concentrated on basic sets.

(23.17) Proposition: Let Ω_j be a basic set and $\Omega_j^1 \cup \ldots \cup \Omega_j^m$ the decomposition described in (23.15). For $\mu \in \mathfrak{M}(\Omega_j)$ define $\sigma\mu \in \mathfrak{M}(\Omega_j^1)$ by

$$\sigma\mu(A) = \mu(A \cup \Phi A \cup \ldots \cup \Phi^{m-1} A)$$

for $A \in \mathfrak{B}(\Omega_j^1)$. Then σ is a homeomorphism from $\mathfrak{M}_\Phi(\Omega_j)$ onto $\mathfrak{M}_{\Phi^m}(\Omega_j^1)$.

This is easy to check. Remark that for $\nu \in \mathfrak{M}_{\Phi^m}(\Omega_j^1)$ one has

$$\sigma^{-1}\nu = \frac{1}{m}(\nu + \Phi\nu + \ldots + \Phi^{m-1}\nu) \in \mathfrak{M}_\Phi(\Omega_j).$$

A point $x \in \Omega_j^1$ is generic (resp. quasigeneric) for $\mu \in \mathfrak{M}_\Phi(\Omega_j)$ with respect to Φ iff it is generic (resp. quasigeneric) for $\sigma\mu$ with respect to Φ^m. By (14.18) and (14.21) one has

$$h_{top}(\Phi^m | \Omega_j^1) = m\, h_{top}(\Phi | \Omega_j)$$

and by (18.5) and (10.12.b)

$$h_{\sigma\mu}(\Phi^m | \Omega_j^1) = m\, h_\mu(\Phi | \Omega_j).$$

Also, σ preserves the properties of being ergodic, nonatomic, positive on all open sets. But if $m \neq 1$ then a measure in $\mathfrak{M}_\Phi(\Omega_j)$ never can be weakly mixing.

Let T be a continuous transformation of a compact space X with metric d.

<u>(23.18) Definition:</u> A sequence $\{x_i\}_{i=a}^{b}$ is called a <u>δ-pseudo-orbit for T</u> if $d(Tx_i, x_{i+1}) < \delta$ for $a \leq i < b$ ($a = -\infty$ and $b = +\infty$ are permitted). A point $x \in X$ ε-traces the δ-pseudo-orbit $\{x_i\}_{i=a}^{b}$ if $d(T^i x, x_i) \leq \varepsilon$ for $a \leq i \leq b$. (X,T) has the <u>tracing property</u> if for any $\varepsilon > 0$ there is some $\delta > 0$ such that every δ-pseudo-orbit is ε-traced by some $x \in X$. (This property is independent of the choice of the metric d).

From now on assume that (X,T) is an elementary part of some Axiom A diffeomorphism.

<u>(23.19) Proposition:</u> (X,T) is expansive and has the tracing property.

For a proof, we refer to [31]. An easy consequence is

<u>(23.20) Proposition</u> [21]: (X,T) has the specification property.

<u>Proof:</u> Let $\varepsilon > 0$ be given. We may assume $\varepsilon < \frac{1}{2}\delta^*$, where δ^* is some expansive constant for T. Choose $\delta = \delta(\varepsilon)$ as in the definition of the tracing property. Cover X by a finite family $\mathfrak{U}$ of δ-balls. Since T is topologically mixing (by (23.15)), for any two $U_i, U_j \in \mathfrak{U}$ there exists an integer $M_{ij} > 0$ such that $T^n U_i \cap U_j \neq \emptyset$ for $n > M_{ij}$. Let M (= M(ε)) be the largest of the M_{ij}.

Let $x_1, \ldots, x_k$ be points in X and $a_1 \leq b_1 < \ldots < a_k \leq b_k$ be integers with $a_j - b_{j-1} > M$ for $j = 2, \ldots, k$.

Let p be an integer such that $p - (b_k - a_1) > M$. Put $a_{k+1} = p + a_1$ and $x_{k+1} = x_1$.

For $z \in X$ denote by U(z) some $U \in \mathfrak{U}$ with $z \in U$. For $j = 1, \ldots, k$ there exists an $y_j \in U(T^{b_j} x_j)$ such that $T^{a_{j+1} - b_j} y_j \in U(T^{a_{j+1}} x_{j+1})$. Now consider the δ-pseudo-orbit

$\{z_i\}_{i=-\infty}^{+\infty}$ defined as follows

(1) $z_i = T^i x_j$ for $a_j \leq i < b_j$;

(2) $z_i = T^{i-b_j} y_j$ for $b_j \leq i < a_{j+1}$;

(3) $z_{i+p} = z_i$ for $i \in \mathbb{Z}$.

It is easy to check that this is indeed a δ-pseudo-orbit. Thus there is some $x \in X$ which ϵ-traces it. This x is obviously periodic: indeed, $T^p x$ is also ϵ-tracing $\{z_i\}$. Hence $d(T^{p+n}x, T^n x) < 2\epsilon \leq \delta^*$ for all $n \in \mathbb{Z}$ and so $x = T^p x$ by expansiveness. Clearly

$$d(T^n x, T^n x_j) \leq \epsilon \quad \text{for } a_j \leq n \leq b_j,\ j = 1,\dots,k. \quad \square$$

24. Automorphisms of the Torus

Let $\mathbb{T}^m = \mathbb{R}^m | \mathbb{Z}^m$ denote the m-dimensional torus. Thus the elements of $\mathbb{T}^m$ are m-tuples of reals mod 1. $\mathbb{T}^m$ is a compact abelian group with respect to addition mod 1. The Haar measure ω is just Lebesgue measure.

If $A = (a_{ij})$ is an element of $GL(m,\mathbb{Z})$, i.e. an $m \times m$-matrix with integer entries and determinant ± 1, then A induces a linear automorphism A_R of $\mathbb{R}^m$ sending $\mathbb{Z}^m$ onto $\mathbb{Z}^m$ and hence an automorphism A_T of $\mathbb{T}^m$. Conversely every continuous automorphism of $\mathbb{T}^m$ is of this form.

Clearly A_T and A_R are differentiable and preserve the Lebesgue measure ω.

(24.1) Proposition: The m.t.dynamical system $(\mathbb{T}^m, A_T, \omega)$ is ergodic iff A has no eigenvalues which are roots of unity.

For a proof we refer to [80] or [185].

For the computation of the topological entropy of $(\mathbb{T}^m, A_T)$ we follow the proof by Bowen [22] and Walters [185].

Let d denote a metric on $\mathbb{R}^m$ given by some norm (any two such metrics are uniformly equivalent). Let $L : \mathbb{R}^m \longrightarrow \mathbb{R}^m$ be linear. L is uniformly continuous with respect to d.

(24.2) Lemma: Let λ be an eigenvalue of L with maximal absolute value. Then

$$h_d(L) \leq \max \{0, m \log |\lambda|\}.$$

Proof: By (14.20)

$$h_d(L^n) \leq \max \{0, m \log ||L^n||\}.$$

By (14.17)

$$h_d(L) \leq \max\{0, m \log \|L^n\|^{\frac{1}{n}}\}$$

By the spectral theorem $\|L^n\|^{\frac{1}{n}} \to |\lambda|$. □

<u>(24.3) Lemma:</u> $h_d(L) \geq \log |\det L|$.

<u>Proof:</u> We may of course assume $\det L \neq 0$. For any $x \in \mathbb{R}^m$,

$$B^n_\varepsilon(x) = x + B^n_\varepsilon(0)$$

(cf. the example of (19.7)). Let $K \subset \mathbb{R}^m$ be compact with $\omega(K) > 0$ and F (n,ε)-spanning K.

$$K \subset \bigcup_{x \in F} B^n_x(\varepsilon) = \bigcup_{x \in F} \{x + B^n_\varepsilon(0)\}$$

and so

$$\omega(K) \leq \omega(B^n_\varepsilon(0)) \cdot r_n(\varepsilon, K).$$

This implies

$$\bar{r}(\varepsilon,K) = \limsup \frac{1}{n} \log r_n(\varepsilon,K) \geq - \limsup \frac{1}{n} \log \omega(B^n_\varepsilon(0)).$$

For any Borel set $E \subset \mathbb{R}^m$,

$$\omega(L(E)) = \frac{\omega(E)}{|\det L|}$$

and so

$$\omega(B^n_\varepsilon(0)) \leq \omega(L^{n-1} B_\varepsilon(0)) = \frac{\omega(B_\varepsilon(0))}{|\det L|^{n-1}} .$$

This implies

$$\bar{r}(\varepsilon,K) \geq \log |\det L|.$$

Hence

$$h_d(L) \geq \log |\det L|. \qquad \square$$

<u>(24.4) Lemma:</u> Let $\lambda_1,\ldots,\lambda_m$ be the eigenvalues of L. Then

$$h_d(L) = \sum_{|\lambda_i|>1} \log |\lambda_j|.$$

<u>Proof:</u> Let $\{\alpha_1,\ldots,\alpha_s\}$ be the set of absolute values of $\lambda_1,\ldots,\lambda_m$. Let E_j be the subspace corresponding to the eigenvalues with absolute value α_j. By the Jordan Decomposition Theorem

$$\mathbb{R}^m = E_1 \oplus \ldots \oplus E_s.$$

One has $L(E_j) \subset E_j$. Put $L_j = L|E_j$. By (24.2)

$$h_d(L_j) \leq \max \{0, \dim E_j \cdot \log \alpha_j\}.$$

Since $L = L_1 \oplus \ldots \oplus L_s$, one has

$$h_d(L) \leq \sum_{j=1}^{s} h_d(L_j)$$

by (14.23) and (14.13). Hence by (24.2)

$$h_d(L) \leq \Sigma \dim E_j \log \alpha_j = \sum_{|\lambda_i|>1} \log |\lambda_i|.$$

Now let V be the product of the subspaces E_j with $\alpha_j > 1$ and L_V the restriction of L to V. Obviously

$$h_d(L) \geq h_d(L_V).$$

By (24.3) $\quad h_d(L_V) \geq \log |\det L_V| = \sum_{|\lambda_i|>1} \log |\lambda_i|. \qquad \square$

(24.5) Theorem: Let $A \in GL(m,\mathbb{Z})$ and let $\lambda_1,\dots,\lambda_m$ be the eigenvalues of A. Then

$$h_{top}(A_T) = \sum_{|\lambda_i|>1} \log |\lambda_i| .$$

Proof [18]: By (24.4) it is enough to show that $h_{top}(A_T) = h_d(A_R)$. Let $\pi : \mathbb{R}^m \longrightarrow \mathbb{T}^m$ denote the projecting map $x \longmapsto x(\mathrm{mod}\ 1)$. Obviously $\pi \circ A_R = A_T \circ \pi$. Let d_R (resp. d_T) be the Euclidean metric on $\mathbb{R}^m$ (resp. $\mathbb{T}^m$). For some $\delta > 0$, π is an isometry from $B_\delta(x)$ onto $B_\delta(\pi(x))$ for all $x \in \mathbb{R}^m$. Let $\epsilon < \delta$ be such that $d_R(x,y) < \epsilon$ implies $d_R(A_R\, x, A_R\, y) < \delta$.

Let $K \subset \mathbb{R}^m$ be compact with diam $K < \delta$, and let $E \subset K$ be (n,ϵ)-separated. Then $\pi(E)$ is (n,ϵ)-separated. Indeed, let $x,y \in E$ be such that $x \neq y$. Then $\pi(x) \neq \pi(y)$. There is some $0 \leq j < n$ such that $d_R(A_R^i\, x, A_R^i\, y) \leq \epsilon$ for $i \leq j$ and $d_R(A_R^{j+1}\, x, A_R^{j+1} y) > \epsilon$. But $d_R(A_R^{j+1}\, x, A_R^{j+1} y) < \delta$ and thus $d_T(A_T^{j+1}\pi(x), A_T^{j+1}\pi(y)) = d_R(A_R^{j+1}x, A_R^{j+1}y) > \epsilon$. Thus $\pi(E)$ is (n,ϵ)-separated. Hence

$$s_n(\epsilon,K) \leq s_n(\epsilon,\pi(K)).$$

If, conversely, $E' \subset \pi(K)$ is (n,ϵ)-separated, then $\pi^{-1}(E') \cap K$ is obviously also (n,ϵ)-separated. Thus

$$s_n(\epsilon,K) = s_n(\epsilon,\pi(K))$$

and

$$h_{d_R}(A_R,K) = h_{d_T}(A_T,\pi(K)).$$

Letting K vary through the compact sets of $\mathbb{R}^m$ of diameter $< \delta$ is the same as letting $\pi(K)$ run through the compact sets of $\mathbb{T}^m$ of diameter $< \delta$. Using (14.21(a)) this implies the result.

$\square$

Corollary: For all $g \in \mathbb{T}^m$,

$$h_\omega(A_T) = h_{top}(A_T) = h_{top}(A_T \circ R_g) = h_\omega(A_T \circ R_g)$$

where R_g is the translation $x \mapsto g + x$ in $\mathbb{T}^m$. This follows by (19.7).

If A has no eigenvalues which are roots of unity, then A has some eigenvalues outside of the complex unit circle. (This is a consequence of a theorem of Kronecker: see for example VIII. 200 in Polya-Szegö: Aufgaben und Lehrsätze der Analysis, Springer 1925). Using (24.1) one obtains

Corollary: If (T^m, A_T, ω) is ergodic, then $h_\omega(A_T) > 0$.

Actually much more is true:

(24.6) Theorem of Katznelson [103]: If $(\mathbb{T}^m, A_T, \omega)$ is ergodic, then it is m.t. conjugate to a Bernoulli shift.

It follows that any two ergodic automorphisms of the torus with the same entropy are m.t. conjugate. For the case $m = 2$ this was proved by Adler and Weiss in [6].

(24.7) Proposition: If the automorphism $(\mathbb{T}^m, A_T, \omega)$ is ergodic, then the periodic points are exactly those with rational coordinates. Hence $\Omega = \mathbb{T}^m$.

Proof: (a) Let $\underline{x} \in \mathbb{T}^m$ have rational coefficients and let r be the lowest common denominator of the coefficients. Thus

$$\underline{x} = \left(\frac{x_1}{r}, \ldots, \frac{x_m}{r}\right)$$

When the x_i are in $\mathbb{Z}_r$, the set of integers mod r. A induces a map A_r from the finite set $(\mathbb{Z}_r)^m$ into itself, which is easily seen to be an automorphism. Thus there exists a k such that $A_r^k\, r\underline{x} = r\underline{x}$ (mod r). Hence $A_T^k\, \underline{x} = \underline{x}$.

(b) Since A has no roots of unity as eigenvalues, $\det(A^k - \mathrm{Id}) \neq 0$ for all $k \neq 0$. So if $\underline{x} \in \mathbb{T}^m$ has period k, i.e. if $(A^k - \mathrm{Id})\underline{x} \in \mathbb{Z}^m$, then $\underline{x}$ has rational coordinates. □

(24.8) Definition: A matrix $A \in GL(m,\mathbb{Z})$ and the corresponding automorphism $A_T : \mathbb{T}^m \longrightarrow \mathbb{T}^m$ are said to be hyperbolic if A has no eigenvalues of absolute value 1.

Since $\det A = \pm 1$, this implies that some eigenvalues of A are inside the unit circle, and some outside. The former are called contracting, the latter expanding. $\mathbb{R}^m$ can be split into a direct sum $E^+ \oplus E^-$, where E^+ is the eigenspace corresponding to the contracting eigenvalues, E^- corresponding to the expanding ones. It is then easy to see that $A_T : \mathbb{T}^m \longrightarrow \mathbb{T}^m$ is an Anosov diffeomorphism (see (23.9)). Hence it is structurally stable (23.10) and satisfies expansiveness (23.19) and the specification property (23.20). Also $(\mathbb{T}^m, A_T, \omega)$ is obviously ergodic.

If $m < 4$ and $(\mathbb{T}^m, A_T, \omega)$ is ergodic then A is hyperbolic. For $m \geq 4$ there exist ergodic nonhyperbolic automorphisms (see [184] for an example).

$A, B \in GL(m,\mathbb{Z})$ are said to be rationally dependent if $A^k = B^l$ for some integers $k, l \neq 0$. In [173] it is shown that if A and $B \in GL(2,\mathbb{Z})$ have the same eigenvectors, they are rationally dependent. From this follows

(24.9) Proposition [173]: Let $A, B \in GL(2,\mathbb{Z})$ such that $(\mathbb{T}^2, A_T, \omega)$ and $(\mathbb{T}^2, B_T, \omega)$ are ergodic. Then (i) and (ii) are equivalent:

(i) A and B are rationally dependent;

(ii) $G_{A_T}(\omega) = G_{B_T}(\omega)$, where

$$G_{A_T}(\omega) = \{x \in \mathbb{T}^2 \mid \frac{1}{N} \sum_{j=1}^{N-1} \delta(A_T^j x) \longrightarrow \omega\}.$$

Proof: (i)$\Longrightarrow$(ii) is easy. Now suppose A and B are rationally independent. Let $\bar{x}_A$ (resp. $\bar{x}_B$) be the contracting eigenvectors of A (resp. B). Choose $\mu \in \mathfrak{M}_{B_T}(\mathbb{T}^2)$ with $\mu \neq \omega$ and $y_\mu \in G_{B_T}(\mu)$, $y_\omega \in G_{A_T}(\omega)$. Since $\bar{x}_A$ and $\bar{x}_B$ are independent, there exists a $y \in \mathbb{T}^2$ such that for some constants λ_μ, λ_ω one has:

$$y = y_\mu + \lambda_\mu \bar{x}_B \qquad \text{and}$$

$$y = y_\omega + \lambda_\omega \bar{x}_A$$

Since $\bar{x}_A$ is contracting, it is easy to see that $d(A_T^n y, A_T^n y_\omega) \to 0$ and hence that $y \in G_{A_T}(\omega)$. Similarly, $y \in G_{B_T}(\mu)$. But $G_{B_T}(\mu) \cap G_{B_T}(\omega) = \emptyset$. □

We refer to [163] and [164] for much deeper results of this type concerning endomorphisms of $\mathbb{T}^m$.

<u>(24.10) Proposition:</u> Let $A,B \in GL(2,\mathbb{Z})$ such that $(\mathbb{T}^2, A_T, \omega)$ and $(\mathbb{T}^2, B_T, \omega)$ are ergodic. Then $Q_{A_T}(\mathbb{T}^2) = Q_{B_T}(\mathbb{T}^2)$ iff $A = B$.

For the proof, we refer to [173]. Note that if $(\mathbb{T}^2, A_T, \omega)$ is not ergodic, then $Q_{A_T}(\mathbb{T}^2) = \mathbb{T}^2$.

In [6], [184] and [185] one can find more on automorphisms of the torus. In [22] Bowen proved an analogue of (24.5) for endomorphisms of Lie groups.

25. More on Subshifts of Finite Type

The basic properties of f.t.subshifts were proved in section 17. We are going to study further properties of these subshifts in the present section taking into account what has been done in sections 15,17-19 and 21-24. The first part of this section deals with canonical coordinates and subshifts, the second one with topological conjugacy between subshifts and their associated transition matrices, and the third one with sofic systems.

Let (X,T) denote a topological dynamical system with metric d. For every $\delta > 0$ and every $x \in X$ define the <u>δ-in-set</u> of x to be the set

$$W_\delta^+(x) := \{y \in X \mid d(T^n x, T^n y) \leq \delta \quad \text{for all} \quad n \geq 0\}$$

and the <u>δ-out-set</u> of x to be

$$W_\delta^-(x) := \{y \in X \mid d(T^n x, T^n y) \leq \delta \quad \text{for all} \quad n \leq 0\}.$$

(Compare definition (23.6).)

<u>(25.1) Definition</u> [18]: A topological dynamical system (X,T) has <u>canonical coordinates</u> if there exists a metric d such that:

For every $\delta > 0$ there exists an $\varepsilon > 0$ such that for every $x,y \in X$ with $d(x,y) < \varepsilon$ one has

$$W_\delta^+(x) \cap W_\delta^-(y) \neq \emptyset.$$

Note that definition (25.1) is independent of the metric used. This is not true in the following

<u>(25.2) Definition</u> [18]: A topological dynamical system (X,T) has got <u>hyperbolic canonical coordinates</u> if (X,T)

has canonical coordinates and if there exist constants $\delta^* > 0$, $0 < \lambda < 1$ and $c \geq 1$ such that for every $x \in X$ the following two conditions are fulfilled:

a) Every $y \in W^+_{\delta^*}(x)$ satisfies for $n \geq 0$

$$d(T^n x, T^n y) \leq c\lambda^n d(x,y)$$

b) Every $z \in W^-_{\delta^*}(x)$ satisfies for $n \leq 0$

$$d(T^n x, T^n z) \leq c\lambda^{-n} d(x,y).$$

Remark: It follows easily from [21] (see [196]) that a topological dynamical system (X,T) with hyperbolic canonical coordinates has a dense set of periodic points iff (X,T) is nonwandering.

If (X,T) has hyperbolic canonical coordinates and the set of periodic points is dense in X, then Smale's and Bowen's spectral decomposition theorems hold for (X,T) (see (23.14) and (23.15) and replace X for Ω and T for Φ).

Similarly (23.19) and (23.20) are valid also: If (X',T') is an elementary part in the spectral decomposition then it has got the tracing and specification property and it is expansive. Therefore chapter 22 applies to this situation.

An example of a system (X,T) with hyperbolic canonical coordinates one gets from Smale [180] and Hirsch, Pugh [200] (see [21]):

(25.3) Proposition: Let T be an Axiom A diffeomorphism and Ω its nonwandering set. Then $(\Omega, T|\Omega)$ has got hyperbolic canonical coordinates with respect to some metric.

The connection between hyperbolic canonical coordinates and f.t. subshifts is given by

(25.4) Proposition: Let $\Lambda \subset \{1,\ldots,s\}^{\mathbb{Z}}$ be an f.t.subshift and let d be the metric on Λ defined by

$$d(x,y) := \sum_{n\in\mathbb{Z}} 2^{-|n|} (1 - \delta_{x_n y_n})$$

where δ_{ij} denotes the Kronecker symbol and where $x = (x_n)_{n\in\mathbb{Z}}$, $y = (y_n)_{n\in\mathbb{Z}} \in \Lambda$. Then $(\Lambda,\sigma|\Lambda)$ has hyperbolic canonical coordinates with respect to d.

Proof: Let N be an order of Λ (see (17.4)).

1) Firstly we show that $(\Lambda,\sigma|\Lambda)$ has canonical coordinates. Let $\delta > 0$ be given. There exists an $M \in \mathbb{N}$ such that for $x \in \Lambda$

$$W_\delta^+(x) \supset \{z \in \Lambda \mid z_k = x_k \text{ for all } k \geq -M\}$$

and

$$W_\delta^-(x) \supset \{z \in \Lambda \mid z_k = x_k \text{ for all } k \leq M\}.$$

Choose $\varepsilon > 0$ such that $d(x,y) < \varepsilon$ implies $x_k = y_k$ for each $k \in \{-N - M,\ldots, N + M\}$. Note that (17.6) (3) is just the property of having canonical coordinates.

2) Secondly we show that $(\Lambda,\sigma|\Lambda)$ has hyperbolic canonical coordinates with respect to d.
Choose $\delta^* = \frac{1}{2}$, $\lambda = \frac{1}{2}$ and $c = 1$. Let $x \in \Lambda$. Note that

$$W_{\delta^*}^+(x) = \{y \in \Lambda \mid y_k = x_k \text{ for all } k \geq 0\}$$

and

$$W_{\delta^*}^-(x) = \{y \in \Lambda \mid y_k = x_k \text{ for all } k \leq 0\}.$$

Clearly for $y \in W_{\delta^*}^+(x)$ it follows that for $n \geq 0$

$$d(\sigma^n(x),\sigma^n(y)) = \sum_{k\in\mathbb{Z}} 2^{-|k|} (1 - \delta_{x_{k+n},y_{k+n}})$$

$$= 2^{-n} \sum_{k\geq 1} 2^{-k}(1 - \delta_{x_{-k},y_{-k}}) = 2^{-n} d(x,y).$$

Similarly one shows for $y \in W^-_{\delta^*}(x)$ and $n \leq 0$

$$d(\sigma^n(x), \sigma^n(y)) = 2^n d(x,y). \qquad \square$$

<u>(25.5) Proposition:</u> Let $\Lambda \subset \{1,\ldots,s\}^{\mathbb{Z}}$ be a subshift such that $(\Lambda,\sigma|\Lambda)$ has canonical coordinates. Then $(\Lambda,\sigma|\Lambda)$ is of finite type.

<u>Proof:</u> [21] Let $\epsilon = \frac{1}{2}$ and choose the corresponding δ as in the definition (25.1). Let M be so large that $d(x,y) < \delta$ if $x,y \in \Lambda$ with $x_n = y_n$ for $|n| \leq M$. Let $N = 2M + 2$. It is enough to show that $(a_1,\ldots,a_m)$ occurs in Λ if all its N-subblocks occur in Λ. This is trivial for $m \leq N$. If it has been shown for m, consider the (m+1)-block $(a_1,\ldots,a_{m+1})$ all of whose N-blocks occur in Λ. By induction hypothesis, $(a_1,\ldots,a_m)$ and $(a_2,\ldots,a_{m+1})$ occur in Λ. Hence there is an $x \in \Lambda$ with $x_{-M-2+i} = a_i$ for $i = 1,\ldots,m$, and a $y \in \Lambda$ with $y_{-M-2+i} = a_i$ for $i = 2,\ldots,m$. Thus x and y agree at the places from -M to M, and thus by canonical coordinates there is a $z \in \Lambda$ with $z_i = x_i$ for $i \leq 0$, $z_i = y_i$ for $i \geq 0$. Thus $(a_1,\ldots,a_{m+1})$ occurs in z, and hence in Λ. $\square$

<u>Corollary 1</u> ([18]): For subshifts with finite alphabets the finite type property is an invariant of topological conjugacy.

<u>Proof:</u> Since the property of having canonical coordinates is an invariant for conjugacy, the corollary follows from (25.4) and (25.5). $\square$

<u>Corollary 2:</u> Let (X,T) be 0-dimensional, expansive with canonical coordinates. Then (X,T) is topologically conjugate to some f.t.subshift.

Proof: (16.5) and (25.5). □

It is possible to characterize the f.t. subshifts by canonical coordinates. However, we saw that there are other transformations having canonical coordinates which are not f.t.subshifts ((25.3)). What is the connection between both examples? In the 0-dimensional and expansive case the answer is given by corollary 2 of (25.5) and by William's theorem in [193] (which is inspired by the famous horseshoe example of Smale in [180]).

(25.6) Proposition: Every topologically transitive f.t. subshift for which the periodic points are dense is topologically conjugate to (Ω,T) where T is a diffeomorphism from S^3 onto itself satisfying Samle's Axiom A and where Ω denotes some 0-dimensional basic set of (S^3,T).

We omit the proof of this proposition and start to investigate the connection between systems with hyperbolic canonical coordinates and f.t.subshifts in general.

In order to solve this problem we return to section 15. Because of expansiveness and in view of theorem (15.2) one can hope to find a finite topological generator for X such that the subshift associated by (15.2) is of finite type. Partitions of this type are called Markov partitions.

In order to prepare the definition of a Markov partition we introduce the following notation. Let Z, Z_1, Z_2 be topological spaces and let

$$\varphi : Z \longrightarrow Z_1 \times Z_2$$

be an isomorphism. Then for $x \in Z$ define

$$Z_1(x) := \{y \in Z \mid \varphi(y) \in Z_1 \times \{x_2\}\}$$

and

$$Z_2(x) := \{y \in Z \mid \varphi(y) \in \{x_1\} \times Z_2\}$$

where $\varphi(x) = (x_1, x_2)$.

<u>(25.7) Definition:</u> Let (X,T) be a topological dynamical system. A finite topological generator $\alpha = (A_1, \ldots, A_N)$ for X satisfying $A_i \subset \overline{\text{int } A_i}$ $(1 \leq i \leq N)$ is called a <u>Markov partition</u> for T if the following conditions are satisfied:

1) (Local product structure). For every $1 \leq i \leq N$ there exist compact spaces E_i and F_i and there exists a topological isomorphism

$$\varphi_i : \overline{A_i} \rightarrow E_i \times F_i$$

such that for every $x \in \text{int } A_i$ with $T(x) \in \text{int } A_j$ for some $1 \leq j \leq N$ one has

$$T(E_i(x) \cap \text{int } A_i) \supset E_j(T(x)) \cap \text{int } A_j$$

and

$$T^{-1}(F_j(T(x)) \cap \text{int } A_j) \supset F_i(x) \cap \text{int } A_i.$$

2) (Boundary condition). There exists a decomposition

$$X \setminus \bigcup_{i=1}^{N} \text{int } A_i = B^+ \cup B^-$$

(not necessarily disjoint sets B^+ and B^-) such that

$$T B^+ \subset B^+ \text{ and } T^{-1} B^- \subset B^-.$$

There are several variants of the definition of Markov partitions. The original definition was due to Sinai [176] and [177] for the case of Anosov diffeomorphism. In [19] and [20] Bowen constructed Markov partitions for Axiom A diffeomorphisms (see [31] for a very elegant exposition).

Other papers are [76] and [77] by Gurevič and [123] by Kryzewski , where he gives a definition for expansive diffeomorphisms. Actually the first example of a Markov partition may be found in Adler and Weiss [5].

The main difference between our definition and the other ones is condition 2) in (25.7) and the fact that the isomorphisms φ_i are defined on $\overline{A_i}$ instead of an open neighbourhood of $\overline{A_i}$. We state the definition in this way in order to make the ideas clear.

(25.8) Theorem [19]: Let (X,T) be a basic set of a topological dynamical system with hyperbolic canonical coordinates. Then (X,T) admits a Markov partition.

Remark: Bowen in [19], showed this theorem for (X,T) being a basic set of an Axiom A diffeomorphism. Using the facts proved in [21] this can be extended by showing that the map

$$(x,y) \mapsto W_\delta^+(x) \cap W_\delta^-(y)$$

for $x,y \in X$ with $d(x,y) < \varepsilon$ is continuous, if $\varepsilon > 0$ is small enough. For a detailed proof of this lemma see [196].

Let us suppose now that $\alpha = (A_1,\ldots,A_N)$ is a Markov partition for (X,T). Let

$$B := \{(i,j) \mid \text{int } A_i \cap T^{-1} \text{ int } A_j \neq \emptyset \quad (1 \leq i,\ j \leq N)\}$$

and define $\Lambda \subset \{1,\ldots,N\}^{\mathbf{Z}}$ to be the f.t.subshift of order 2 for which B is a defining system of blocks (see (17.6)).

Recall now theorem (15.2). Since α is a topological generator, the associated subshift Λ_α is well defined, $\Lambda_\alpha \subset \{1,\ldots,N\}^{\mathbf{Z}}$, and (X,T) is a factor of (Λ_α,σ).

In the situation just described (see [19]):

<u>(25.9) Proposition:</u> $\Lambda_\alpha = \Lambda$.

<u>Proof:</u> In view of (15.2) we have to show for every block $(i_o,\dots,i_{n-1})$ occuring in Λ that

$$\bigcap_{k=o}^{n-1} T^{-k} \operatorname{int} A_{i_k} \neq \emptyset.$$

By induction we shall prove:

$$\emptyset \neq U := \bigcap_{k=o}^{n-1} T^{-k} \operatorname{int} A_{i_k} \supset F_{i_o}(y) \cap \operatorname{int} A_{i_o}$$

for every $y \in U$.

For $n = 1$ $\operatorname{int} A_{i_o} \supset F_{i_o}(y) \cap \operatorname{int} A_{i_o}$ holds for every choice of $y \in \operatorname{int} A_{i_o}$ by the definition of a Markov partition.

Assume that the relation is proved for every block of length n occuring in Λ.

Let $(i_o,\dots,i_n)$ occur in Λ. Then

$$\emptyset \neq V := \bigcap_{k=o}^{n-1} T^{-k} \operatorname{int} A_{i_{k+1}} \supset \operatorname{int} A_{i_1} \cap F_{i_1}(z)$$

for every $z \in V$ by hypothesis.

Also

$$T \operatorname{int} A_{i_o} \cap \operatorname{int} A_{i_1} \neq \emptyset$$

is given by the definition of Λ. Let w be any point in this set. Then there is a unique point $v \in \operatorname{int} A_{i_1}$ such that

$$\{v\} = F_{i_1}(z) \cap E_{i_1}(w).$$

Because α is a Markov partition one has

$$T[E_{i_o}(T^{-1}(w)) \cap \operatorname{int} A_{i_o}] \supset E_{i_1}(w) \cap \operatorname{int} A_{i_1}$$

and therefore it follows that

$$T^{-1}(v) \in T^{-1}[\text{int } A_{i_1} \cap F_{i_1}(z)] \cap T^{-1}[E_{i_1}(w) \cap \text{int } A_{i_1}]$$

$$\subset T^{-1} V \cap E_{i_o}(T^{-1}(w)) \cap \text{int } A_{i_o}$$

$$\subset \text{int } A_{i_o} \cap T^{-1} V =: U.$$

Now let $y \in U$. Then $T(y) \in V$ and hence

$$T[F_{i_o}(y) \cap \text{int } A_{i_o}] \subset F_{i_1}(T(y)) \cap \text{int } A_{i_1} \subset V.$$

Therefore one obtains

$$F_{i_o}(y) \cap \text{int } A_{i_o} \subset T^{-1}V \cap F_{i_o}(y) \cap \text{int } A_{i_o} \subset U. \quad \square$$

For further use of Markov partitions we note that proposition (15.7) applies to the situation described for the last proposition, since X_α is residual. Thus we can transport some topological and measure theoretic properties from X to Λ_α and vice versa.

If $\alpha = (A_1,\dots,A_N)$ is a Markov partition for (X,T) then by theorem (15.2) and proposition (25.10) (X,T) is a factor of the associated f.t.subshift $\Lambda_\alpha \subset \{1,\dots,N\}^{\mathbb{Z}}$. Denote by μ_α the Parry measure on Λ_α (see (17.15)). Clearly - by definition - it is positive on open sets. Hence the transported measure $\pi\mu_\alpha$ is by (15.7) (4) positive on open sets as well. Since μ_α is ergodic, $\pi\mu_\alpha$ is also ergodic. Now apply (25.7) (2) to conclude that the Markov partition is a $\pi\mu_\alpha$-continuity partition. It follows that

$$h_{\pi\mu_\alpha}(T,\alpha) = h_{top}(\sigma|\Lambda_\alpha).$$

Now (14.22) and (18.4) yield

$$h_{top}(T) = h_{top}(\sigma | \Lambda_\alpha).$$

In addition let (X,T) - and, by (15.7), (Λ_α, σ) also - be topologically transitive. By (18.14) (Λ_α, σ) is intrinsically ergodic, the Parry measure μ_α being the unique measure in $\mathfrak{M}_\sigma(\Lambda_\alpha)$. What is said so far, especially $h_{top}(T) = h_{top}(\sigma | \Lambda_\alpha)$, and proposition (3.2) clearly shows that (X,T) is intrinsically ergodic. Thus the transported Parry measure $\pi\mu_\alpha$ equals the Bowen measure μ_B (see (22.1)) (if (X,T) satisfies the specification property and is expansive).

Furthermore one can apply (15.7) (2) and the corollary of (17.14): If (X,T) is in addition topologically mixing then (Λ_α, σ) is topologically mixing and μ_B is Bernoulli, since $\pi : \Lambda_\alpha \longrightarrow X$ is an m.t. isomorphism with respect to μ_α and $\pi\mu_\alpha = \mu_B$.

Now we shall turn towards the second problem mentioned in the beginning. In the first part we saw among others that being of finite type is a conjugacy invariant for subshifts. It is a natural (but highly nontrivial) problem to look for relations between the transition matrices corresponding to topological conjugate subshifts of finite type. B.Williams solved this problem in terms of an equivalence relation between matrices which is rather difficult to compute.

<ins>(25.10) Definition:</ins> Let A and B be two square matrices (not necessarily of the same order) with entries in $\mathbb{Z}_+$. A is said to be <ins>related</ins> to B ($A \underset{1}{\sim} B$) if A = RS and B = SR for some rectangular matrices R,S over $\mathbb{Z}_+$.

A and B are said to be <ins>strongly shift equivalent</ins> ($A \sim B$)

if there exists a chain $C_1,\dots,C_k$ such that

$A \underset{1}{\sim} C_1 \underset{1}{\sim} \cdots \underset{1}{\sim} C_k \underset{1}{\sim} B.$

(Remark that the relation $\underset{1}{\sim}$ is not transitive!)

(25.11) Proposition [193]: Let L and L' be two transition matrices (not necessarily on the same state space) and $\Lambda(L)$ and $\Lambda(L')$ their corresponding subshifts of finite type. Then $\Lambda(L)$ and $\Lambda(L')$ are topologically conjugate iff L and L' are strongly shift equivalent.

(25.12) Proposition [193]: Given a strong shift equivalence between the transition matrices L and L', there exists a finite procedure for constructing a conjugacy between $\Lambda(L)$ and $\Lambda(L')$.

Finally we turn towards the study of subshifts which are factors of f.t.subshifts. We saw in the first part that the factors of f.t.subshifts which arise from (15.2) have many of the interesting properties of the f.t.subshifts. This can be said also for the factors we want to describe now (see [190] and [40]).

Consider the subshift of finite type Λ defined by the transition matrix

$$\begin{bmatrix} 0 & 1 & 0 \\ 1 & 0 & 1 \\ 1 & 0 & 1 \end{bmatrix}$$

Let $F : \{1,2,3\} \to \{1,2\}$ be defined by $F(1) = F(2) = 1$, $F(3) = 2$. F induces a map $F_\infty : \{1,2,3\}^{\mathbb{Z}} \to \{1,2\}^{\mathbb{Z}}$ (see sect. 7). It is easy to see that the image of Λ under F_∞ is the subshift $\Lambda' \subset \{1,2\}^{\mathbb{Z}}$ defined by the condition that all blocks of ones which have maximal length have even length. Using (25.4) and (25.5) it is easy to see that Λ' is not of finite type. Thus the class of subshifts of finite type (and the class of transformations with canonical coordinates) are not closed under the operation of taking factors.

(25.13) Definition [190]: A subshift which can be displayed as a factor of a subshift of finite type is said to be a sofic subshift.

They can be characterized in several ways:

(25.14) Definition [190]: Let G be a finite semigroup, $e \in G$ such that $eg = ge = e$ for all $g \in G$, and $a(1),\dots,a(s) \in G \setminus \{e\}$ such that $e, a(1),\dots,a(s)$ generate G. Write

$$\Lambda_G = \{x \in S^{\mathbf{Z}} \mid a(x_n)a(x_{n+1})\dots a(x_m) \neq e \text{ for all } n < m\}$$

It is easy to see that Λ_G is a subshift. Any subshift of this form will be called a G-admissible subshift.

(25.15) Definition [190]: For any subshift Λ let $W(\Lambda)$ denote the set of all blocks occuring in Λ. For $w \in W(\Lambda)$ set

$$P(w) = \{w' \in W(\Lambda) \mid w'w \in W(\Lambda)\} \quad \text{(the predecessors of w)}$$

$$F(w) = \{w' \in W(\Lambda) \mid ww' \in W(\Lambda)\} \quad \text{(the followers of w)}$$

Λ is said to be P-finitary if $\{P(w) \mid w \in W(\Lambda)\}$ is finite, and F-finitary if $\{F(w) \mid w \in W(\Lambda)\}$ is finite.

(25.16) Proposition [190]: Let Λ be a subshift of $S^{\mathbf{Z}}$. The following conditions are equivalent:

(a) Λ is sofic;

(b) Λ is G-admissible, for some appropriate G;

(c) Λ is P-finitary;

(d) Λ is F-finitary.

In [40] Coven and Paul showed:

(25.17) Proposition: Every sofic subshift is a finite-to-one factor of some subshift of finite type with the same topological entropy.

(25.18) Proposition [40]: A sofic subshift Λ admits a unique measure of maximal entropy with support Λ iff it is top.transitive and has periodic points dense.

A similar result (together with another characterization of sofic systems) has been obtained recently by Fischer (see [210] and [211]) .

26. Preparations for Generator Theorems

Rohlin Sets

(26.1) Definition: Let (X,m,Ψ) be an m.t.dynamical system and $F \in \Sigma$. The (F,Ψ)-tower is the sequence of sets $(F_0, F_1, \ldots)$ where $F_i = \Psi^i F \setminus \bigcup_{j=0}^{i-1} \Psi^j F \ (i \geq 0)$. $\bigcup_{i \geq 0} F_i = \bigcup_{i \geq 0} \Psi^i F$ is Ψ-invariant (mod 0). If it has measure 1, F is called a sweep-out set, and in this case the (F,Ψ)-tower is a partition of X. If Ψ is ergodic and $m(F) > 0$, F always is a sweep-out set.

To represent the (F,Ψ)-tower, we use the following picture:

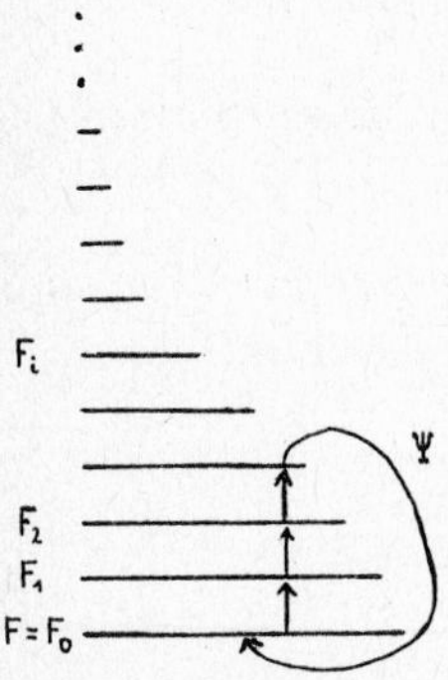

The levels are counted upwards, beginning from 0. The interval in the i-th level represents F_i, and its length is $m(F_i)$. The vertical ordering means: every point $x \in F_i$ moves vertically upwards by one step; if there is no corresponding point in F_{i+1}, Ψ brings x back to the basis. So we can say that $\Psi^{-1}F$ is the roof of the tower.

Rohlin sets are sets for which the lower levels of the tower have all the same size. Part of the following definition repeats (1.18).

(26.2) Definition: $F \in \Sigma$ is called a (Ψ,n)-Rohlin set if

$$\Psi^i F \cap F = \emptyset \quad (1 \leq i < n).$$

F is a (Ψ,n,ε)-Rohlin set if in addition

$$m\left(\bigcup_{i=0}^{n-1} \Psi^i F\right) > 1 - \varepsilon.$$

F is a uniform (Ψ,n)-Rohlin set (resp. uniform (Ψ,n,ε)-Rohlin set) if in addition to the other conditions

$$m_x(F) = \text{const} = m(F) \quad \text{a.e.},$$

i.e. F is independent of the σ-algebra of Ψ-invariant sets.

<u>(26.3) Lemma:</u> Let Ψ be an aperiodic transformation.

a) For every $n \in \mathbb{N}$ and $\varepsilon > 0$ there is a uniform (Ψ,n,ε)-Rohlin set $F \subset X$.

b) If $Q \in \Sigma$ with $m_x(Q) > 1 - \delta$ a.e. and $n \in \mathbb{N}$, $\varepsilon > 0$, then there is a uniform $(\Psi,n,\varepsilon+\delta)$-Rohlin set $F \subset Q$.

<u>Proof:</u>

a) By lemma (1.18) we obtain a $(\Psi,n,\varepsilon/4)$-Rohlin set G. Let $D = \{x \in \mathfrak{X} \mid m_x(G) \geq n^{-1}(1 - \frac{\varepsilon}{2})\}$. Then $m(D) \geq \frac{1}{2}$, because $m(\bigcup_{i=o}^{n-1} \Psi^i G) > 1-\frac{\varepsilon}{4}$. As a consequence of (13.1), we find a set $F_1 \subset D \cap G$ with $m_x(F_1) \doteq n^{-1}(1-\frac{\varepsilon}{2})$ for almost all $x \in D$. F_1 is a (Ψ_D,n,ε)-Rohlin set. In $\mathfrak{X} \setminus D$ we repeat the same procedure and exhaust $\mathfrak{X}$ in countably many steps.

b) Let $H \in \Sigma$ be a uniform $(\Psi,n,\frac{\varepsilon}{2})$-Rohlin set, and for $x \in \mathfrak{X}$

$$j(x) = \min\{i \geq 0 \mid m_x(Q \cap \Psi^i H) \geq n^{-1}(1 - \delta - \tfrac{\varepsilon}{2})\}.$$

Since $$m_x(Q \cap \bigcup_{i=o}^{n-1} \Psi^i H) > 1 - \delta - \tfrac{\varepsilon}{2}, \quad j(x) < n \quad \text{a.e.},$$

j(x) is invariant and measurable, and therefore the set

$$\overline{H} = Q \cap \bigcup_{i=o}^{n-1} \Psi^i H \cap \{x \mid j(x) = i\}$$

is measurable. As above, we can cut from $\overline{H}$ a set F for which

$$m_x(F) = n^{-1}(1 - \delta - \tfrac{\varepsilon}{2}) \quad \text{a.e.}$$

Of course, F is a $(\Psi,n,\varepsilon+\delta)$-Rohlin set. □

The following lemma (26.4) is the special case of lemma (26.5) for ergodic transformations; we formulate and prove it separately because this will make the idea clearer, and in the generator theorem for ergodic transformations only this variant is needed.

(26.4) Lemma: Let $(\mathfrak{X},\Sigma,m,\Psi)$ be ergodic and aperiodic, $A \in \Sigma$ with $0 < m(A) < 1$ and $0 < \varepsilon < 1$, $0 < \delta < 1$. Then for sufficiently large $q \in \mathbb{N}$ there is an $n > q\cdot m(A)$ such that for every set $Q \in \Sigma$ with $m(A \cap Q) \geq m(A)(1 - \delta)$ there exists a set $F \subset Q \cap A$ with the properties

F is a $(\Psi,q,\varepsilon+\delta)$-Rohlin set

F is a $(\Psi_A,n,\varepsilon+\delta)$-Rohlin set.

Proof: Ψ_A is ergodic. Let $N \in \mathbb{N}$ with $N \cdot m(A) \cdot \frac{\varepsilon}{4} > 1$ be so large that, if $q \geq N$ and

$$B_q = \{x \in A \mid \sum_{j=0}^{q-1} 1_A(\Psi^j x) \leq q \cdot m(A)(1 + \tfrac{\varepsilon}{4})\} ,$$

$$m(B_q) \geq m(A)(1 - \tfrac{\varepsilon}{4}).$$

Such an N exists by the ergodic theorem.

For $q \geq N$ let $n = \lceil q\, m(A)(1 + \frac{\varepsilon}{4})\rceil$. ($\lceil r \rceil$ is the smallest integer $\geq r$.) Then

$$q \cdot m(A) < n \leq q \cdot m(A) \cdot (1 + \tfrac{\varepsilon}{2}).$$

Since $m(B_q \cap Q) \geq m(A)(1 - \delta - \frac{\varepsilon}{4})$, there is a $(\Psi_A,n,\delta+\frac{\varepsilon}{2})$-Rohlin set $F \subset B_q \cap Q$ (lemma (26.3.b)). This F is also a (Ψ,q)-Rohlin set; for, if $x \in F$, then $x \in B_q$, and of the first $(q - 1)$ Ψ-images of x $(\Psi x,\ldots,\Psi^{q-1}x)$, at most $n - 1$ are in A. But since under the induced transformation Ψ_A x does not return to F in the first $n - 1$ steps, the assertion follows.

To estimate the size of F, we compute

$$m\Big(\bigcup_{j=o}^{q-1} \Psi^j F\Big) = q\cdot m(F) \geq m(A)\ (1 - \delta - \frac{\varepsilon}{2})\cdot \frac{q}{n} \geq$$

$$\geq (1 - \delta - \frac{\varepsilon}{2})\cdot (1 + \frac{\varepsilon}{2})^{-1} >$$

$$> (1 - \delta - \frac{\varepsilon}{2})\ (1 - \frac{\varepsilon}{2}) > 1 - (\delta + \varepsilon);$$

$$m_A\Big(\bigcup_{j=o}^{n-1} \Psi_A^j\ F\Big) = n\cdot m_A(F) \geq q \cdot m(F) \geq 1 - (\delta + \varepsilon). \quad \square$$

Remark: In general $\bigcup_{j=o}^{n-1} \Psi_A^j\ F \not\subset \bigcup_{j=o}^{q-1} \Psi^j\ F$.

For a stronger version of (26.4) in the case of a topological transformation see a4) in the proof of (28.3).

(26.5) Lemma: Let $(\mathfrak{X},\Sigma,m,\Psi)$ be aperiodic, $A \in \Sigma$ with $0 < m(A) < 1$, $m_x(A) = m(A)$ a.e., and $0 < \varepsilon < 1$, $0 < \delta < 1$. Then for sufficiently large $q \in \mathbb{N}$ there is an $n > q \cdot m(A)$ such that for every set $Q \in \Sigma$ with $m(Q \cap A) \geq m(A)\cdot(1 - \delta^2)$ there exist a Ψ-invariant set $\hat{X}$ and a set $F \subset Q \cap A \cap \hat{X}$ with the properties

$$m(\hat{X}) > 1 - \delta - \varepsilon$$

F is a uniform $(\Psi_{\hat{X}}, q, \varepsilon+\delta)$-Rohlin set,

F is a uniform $(\Psi_{\hat{X}\cap A}, n, \varepsilon+\delta)$-Rohlin set.

Proof: Let $N \in \mathbb{N}$ with $N \cdot m(A) \cdot \frac{\varepsilon}{4} > 1$ be so large that for $q \geq N$ and

$$B_q = \{x \in A \mid \sum_{j=o}^{q-1} 1_A(\Psi^j x) \leq q \cdot m(A)\cdot(1 + \frac{\varepsilon}{4})\}$$

$$m(B_q) \geq m(A)\cdot(1 - (\frac{\varepsilon}{2})^2).$$

For $q \geq N$ let $n = \lceil q \cdot m(A) \cdot (1 + \frac{\varepsilon}{4}) \rceil$; again

$$q \cdot m(A) < n \leq q \cdot m(A) \cdot (1 + \frac{\varepsilon}{2}).$$

Since $m(A \setminus (B_q \cap Q)) \leq m(A) \cdot (\delta^2 + (\frac{\varepsilon}{2})^2)$, we have

$$m(\{x \in \mathfrak{X} \mid m_x(A \setminus (B_q \cap Q)) \geq (\delta + \frac{\varepsilon}{4})\, m(A)\}) < \delta + \varepsilon$$

Set

$$\hat{X} = \{x \in \mathfrak{X} \mid m_x(B_q \cap Q) > (1 - \delta - \frac{\varepsilon}{4})\, m(A)\}.$$

Using lemma (26.3.b)), we find a uniform $(\Psi_{A \cap X}, n, \delta + \frac{\varepsilon}{2})$-Rohlin set $F \subset B_q \cap Q \cap \hat{X}$. The rest of the proof is as in lemma (26.4). □

Blocks and Subshifts

A block is a finite sequence of natural numbers: $P = (p_1, \ldots, p_n)$. (Sometimes we begin to count with 0 or some other number. The p_i will be taken only from a finite set of numbers, but this will be clear in the context.)

Bl is the set of all blocks.

For $P = (p_1, \ldots, p_n)$, $l(P) = n$ is the length of the block. Bl_r is the set of blocks of length r.

For $P = (p_1, \ldots, p_m) \in Bl_m$, $Q = (q_1, \ldots, q_n) \in Bl_n$ we write $PQ = P \cdot Q = (p_1, \ldots, p_m, q_1, \ldots, q_n) \in Bl_{m+n}$, the juxtaposition of P and Q.

If $l(Q) \geq l(P)$, we define

$$\mu_Q(P) = (l(Q) - l(P) + 1)^{-1} \cdot \operatorname{card}\{j \mid 1 \leq j \leq l(Q) - l(P) + 1,\ (q_j, \ldots, q_{j+l(P)-1}) = P\},$$

the relative frequency of (the occurence of) P in Q as a sub-block. μ_Q is a probability measure on Bl_m for every $m \leq l(Q)$.

We shall have to work a lot in some shift space $S^{\mathbb{Z}}$ with finite state space $S = \{1,\ldots,\text{card } S\}$. The number card S corresponds to the cardinality of a certain partition α of the m.t. dynamical system $(\mathfrak{X},m,\Psi)$. Points of $S^{\mathbb{Z}}$ will be denoted by Greek letters $\omega,\eta,\ldots$, subshifts of $S^{\mathbb{Z}}$ by capitals $M,K,\ldots$, the shift transformations on $M,K,\ldots$ always by σ.

For $\omega \in S^{\mathbb{Z}}$, $O(\omega) = \{\sigma^j\omega \mid j \in \mathbb{Z}\}$ is the <u>orbit</u> and $\overline{O(\omega)}$ the <u>orbit closure</u> of ω, and if $s \leq t \in \mathbb{Z}$:

$$\omega_{\langle s,t\rangle} = (\omega_s,\ldots,\omega_t) \in Bl_{t-s+1}$$

For a subshift K of $S^{\mathbb{Z}}$ and $r \in \mathbb{N}$,

$$Bl_r(K) = \{P \in Bl_r \mid {}_o[P] \cap K \neq \emptyset\} = \{\omega_{\langle 1,r\rangle} \mid \omega \in K\}$$

is the <u>set of r-blocks occuring in K</u>, and $Bl(K) = \bigcup_{r\in\mathbb{N}} Bl_r(K)$ the <u>set of all K-blocks</u>. For the <u>number of blocks</u> we write

$$\theta_r(K) = \text{card } Bl_r(K), \quad \theta(K) = \theta_1(K).$$

We shall often assume that we have some subshift K without specifying $S^{\mathbb{Z}}$. In such a case we use $\theta(K)$ as a bound for the number of states.

We write $h(K) = h_{top}(\sigma|K)$ and $h(\omega) = h(\overline{O(\omega)})$. With the notations above, the topological entropy of K is

$$h(K) = \lim r^{-1} \log \theta_r(K) \quad \text{(see (16.11))}.$$

If $\nu \in \mathfrak{M}_\sigma(S^{\mathbb{Z}})$ (or $\nu \in \mathfrak{M}_\sigma(K)$) we abbreviate

$$\nu(P) = \nu({}_o[P]) \quad (P \in Bl(S^{\mathbb{Z}})) \ ; \ h(\nu) = h_\nu(\sigma) \ .$$

If some measure in $\mathfrak{M}_\sigma(S^{\mathbb{Z}})$ can be considered to be natural, we denote it by the letter μ:

For a quasiregular $\omega \in S^{\mathbb{Z}}$, μ_ω is the measure for which it is generic (cf. (4.6)).

For a uniquely ergodic subshift $K \subset S^{\mathbb{Z}}$ and $\omega \in K$,

$\mu_K = \mu_\omega$ is the unique element of $\mathfrak{M}_\sigma(K)$ (cf. (5.14)).

If α is a partition for the m.t. dynamical system $(\mathfrak{X}, m, \Psi)$ with index set S, and m_x an ergodic component of m, then $\mu_\alpha = \Phi_\alpha m$ and $\mu_{x,\alpha} = \Phi_\alpha m_x$. ($\mu_{x,\alpha}$ is ergodic and defined for a.e. $x \in \mathfrak{X}$.)

If $M \subset S^{\mathbb{Z}}$ is a topologically transitive subshift of finite type, μ_M is the Parry measure on M (i.e. the unique measure with maximal entropy; cf. (17.15) and (19.14)).

The symbol $\prec$ will express all kinds of occurences of blocks:

<u>(26.6) Definition:</u>

a) If $P, Q \in Bl$, then

$P \prec Q$ if P is a subblock of Q, or $\mu_Q(P) > 0$.

If $P \in Bl$, $\omega \in S^{\mathbb{Z}}$, then

$P \prec \omega$ if P occurs in ω or $\omega \in \bigcup_{i=-\infty}^{+\infty} {}_i[P]$.

If $P \in Bl$ and $K \subset S^{\mathbb{Z}}$ is a subshift, then

$P \prec K$ if there is an $\omega \in K$ with $P \prec \omega$, or $P \in Bl(K)$.

We say that P "occurs" or "appears" in ω resp. K.

b) If $P \in Bl$ and $\omega \in S^{\mathbb{Z}}$ (resp. $K \subset S^{\mathbb{Z}}$ is a subshift), then $P \underset{d}{\prec} \omega$ (resp. $P \underset{d}{\prec} K$) if there exists $k \in \mathbb{N}$ such that the distances of the occurrences of P in ω (resp. in $\eta \in K$) are at most k, i.e.

$$O(\omega) \subset \bigcup_{i=o}^{k-1} {}_i[P] \quad (\text{resp. } K \subset \bigcup_{i=o}^{k-1} {}_i[P]).$$

(Because of the compactness of K, $P \underset{d}{\prec} K \Leftrightarrow \forall\, \eta \in K : P \underset{d}{\prec} \eta$.)

We say that P "<u>occurs densely</u>".

c) If $P \in Bl$, $\omega \in S^{\mathbb{Z}}$ (resp. $K \subset S^{\mathbb{Z}}$ is a subshift) and $\varepsilon > 0$, then $P \underset{\varepsilon\text{-reg}}{\prec} \omega$ (resp. $P \underset{\varepsilon\text{-reg}}{\prec} K$) if there is some $t > l(P)$ such that

$$Q_1, Q_2 \prec \omega \ (\text{resp. } Q_1, Q_2 \prec K),\ l(Q_i) = t \Rightarrow |\mu_{Q_1}(P) - \mu_{Q_2}(P)| < \varepsilon.$$

In this case we can even find some $s > l(P)$ such that

$Q_1, Q_2 \prec \omega$ (resp. $Q_1, Q_2 \prec K$), $l(Q_i) \geq s \Rightarrow |\mu_{Q_1}(P) - \mu_{Q_2}(P)| < \epsilon$,

and here it is allowed that $l(Q_1) \neq l(Q_2)$. (Note that $P \underset{\epsilon\text{-reg}}{\prec} \omega$ does not mean that also $P \prec \omega$, and $P \underset{\epsilon\text{-reg}}{\prec} \omega$, $P \prec \omega$ does not imply $P \underset{d}{\prec} \omega$.)

We say that P "<u>occurs ϵ-regularly</u>".

We call $\omega \in S^{\mathbf{Z}}$ <u>minimal</u>, if $\overline{O(\omega)}$ is minimal (see (6.12)), <u>uniquely ergodic</u>, if $\overline{O(\omega)}$ is uniquely ergodic (see (5.14)), and ω as well as $\overline{O(\omega)}$ <u>strictly ergodic</u>, if $\overline{O(\omega)}$ is minimal and uniquely ergodic. (Usually, minimal ω are called <u>almost periodic</u> and uniquely ergodic ω <u>strictly transitive</u>; but we do not want to have too many concepts.)

An easy consequence of the definition, resp. of (5.15.b)) is

<u>(26.7) Theorem:</u> The subshift $K \subset S^{\mathbf{Z}}$ is

minimal $\iff \forall\ P \prec K : P \underset{d}{\prec} K$

uniquely ergodic $\iff \forall\ P \prec K\ \ \forall\ \epsilon > 0 : P \underset{\epsilon\text{-reg}}{\prec} K.$

<u>Partitions and Generators</u>

Recall the definition of μ_α and $\mu_{x,\alpha}$ in the last paragraph.

<u>(26.8) Definition:</u> If α is a partition for $(\mathfrak{X}, m, \Psi)$ with index set S, then $\Lambda_\alpha \subset S^{\mathbf{Z}}$ is the support of the measure μ_α (see (2.9)). If $M \subset S^{\mathbf{Z}}$ is a subshift and $\Lambda_\alpha \subset M$ (i.e. $\mu_\alpha(M) = 1$), then α is called an <u>M-partition</u> (or an <u>M-generator</u> if it generates).

It is clear that for a.e. $x \in \mathfrak{X}$, $\mu_{x,a}(\Lambda_\alpha) = 1$.

We shall construct generators as limits of other generators. The following two lemmas deal with the behaviour of such limits.

(26.9) Lemma: If (α_i) is a sequence of partitions with index set S and $\alpha_i \longrightarrow \alpha$ (i.e. $\|\alpha_i,\alpha\| \to 0$), then $\mu_{\alpha_i} \longrightarrow \mu_\alpha$ weakly.

Proof: Indicator functions of cylinders form a total set in $C(S^{\mathbb{Z}})$, and

$$\sum_{P\in Bl_r} |\mu_{\alpha_i}(P) - \mu_\alpha(P)| \leq \|(\alpha_i)_o^{r-1}, (\alpha)_o^{r-1}\| \leq r\|\alpha_i,\alpha\|. \quad \square$$

(26.10) Lemma: Suppose we construct recursively a sequence of generators (α_i) with the same index set and a sequence of positive numbers (ϵ_i) such that $\|\alpha_i,\alpha_{i+1}\| < \epsilon_i$. Then, if in each step ϵ_i is chosen sufficiently small (depending on $\alpha_1,\dots,\alpha_i$), the limit partition $\alpha = \lim \alpha_i$ is a generator.

Proof: If $\Sigma\,\epsilon_i < \infty$, it is clear that $\lim \alpha_i$ exists. Put $\epsilon_o = 1$ and suppose that $\alpha_1,\dots,\alpha_i$ have been constructed ($i \geq 1$). We fix $t_i \in \mathbb{N}$ and $\delta_i > 0$ such that

$$(\alpha_i)_{-t_i}^{+t_i} \overset{2^{-i}}{\supset} \alpha_1$$

and

$$\|\rho,\ \alpha_i\| < \delta_i \Longrightarrow \|(\rho)_{-t_i}^{t_i}\ \| < 2^{-i}.$$

Let $\epsilon_i = 3^{-i}(\delta_i \wedge \epsilon_{i-1})$ ($\leq 3^{-i}\,\delta_k \ \ \forall\, k \leq i$). Now we construct α_{i+1} with $\|\alpha_i,\alpha_{i+1}\| < \epsilon_i$. Since $\sum_{j=i}^{\infty} \epsilon_j < \delta_i$, we get $\|\alpha,\alpha_i\| < \delta_i$ and therefore

$$(\alpha)_{-t_i}^{t_i} \overset{2^{-i+1}}{\supset} \alpha_1,$$

i.e. α is a generator. $\square$

The Entropy of Induced Transformations

Let $(\mathfrak{X},m,\Psi)$ be an m.t. dynamical system and $E \in \Sigma$ with

$m(E) > 0$. The dynamical system (E, m_E, Ψ_E) was defined in (1.16). The most interesting case is when E is a sweep-out set, i.e. $m(\bigcup_{i\in\mathbb{Z}} \Psi^i E) = 1$. In this case the σ-algebra $\Sigma|E$ and Ψ together contain full information about Σ (for $A \in \Sigma$ there are sets $A_i \in \Sigma|E$ such that $A = \dot{\bigcup_i} \Psi^i A_i$ mod 0). The formula of Abramov expresses in terms of entropy that all the information is conserved. We give here a very simple proof of the formula, due to H.Scheller. Scheller's complete proof covers the much more difficult case of a non-invertible transformation; it is published in Krengel [113]. Note that the proof does not depend on $\mathfrak{X}$ being a Lebesgue space.

We have to use conditional entropies with respect to σ-algebras: If α is a partition with $H_m(\alpha) < \infty$ and $\Sigma' \subset \Sigma$ a sub-σ-algebra, then

$$H_m(\alpha|\Sigma') = \inf \{H_m(\alpha|\beta) \mid \beta \subset \Sigma' \text{ is a partition}\}$$

and if $(\alpha_k)_{k\in\mathbb{N}}$ is a generating sequence for Σ', then

$$H_m(\alpha|\Sigma') = \lim \downarrow H_m(\alpha|\alpha_k) \qquad \text{(see [149]).}$$

To have shorter notations, a countable refinement of partitions is understood here to be the σ-algebra generated by all these partitions: thus

$$(\alpha)_1^\infty = \bigvee_{k\geq 1} \Sigma(\Psi^{-k}\alpha).$$

With this notation, we can write for α with $H_m(\alpha) < \infty$

$$h_m(\alpha,\Psi) = H_m(\alpha|(\alpha)_1^\infty)$$

We use only partitions out of the class

$$\mathfrak{B}(E) = \{\xi \mid \xi \text{ is a partition of } \mathfrak{X},\ \mathfrak{X}\setminus E \text{ is an atom of } \xi,\ H_m(\xi) < \infty\}.$$

For a partition $\xi \in \mathfrak{Z}(E)$ we write

$$\Psi_E^i \xi = (\mathfrak{X} \setminus E, \Psi_E^i \xi_E) \quad (\in \mathfrak{Z}(E));$$

$$(\Psi_E^{-i} \xi)_{i=s}^t = \bigvee_{s \leq i \leq t} \Psi_E^{-i} \xi \overset{o}{=} (\mathfrak{X} \setminus E, (\xi_E)_s^t);$$

$$(\Psi_E^{-i} \xi)_{i=1}^\infty = \bigvee_{i \geq 1} \Psi_E^{-i} \xi = \bigvee_{i \geq 1} (\mathfrak{X} \setminus E, (\xi_E)_1^i) \quad (\sigma\text{-algebras!})$$

An important element of $\mathfrak{Z}(E)$ is the <u>return time partition</u> ρ:

$\rho = (\mathfrak{X} \setminus E, R_1, R_2, R_3, \ldots)$ where

$$R_i = \{x \in E \mid \Psi^i x \in E \text{ for the first time}\} =$$

$$= E \cap [\Psi^{-i}E \setminus \bigcup_{j=o}^{i-1} \Psi^{-j}E].$$

All the sets $\Psi^j R_i$ $(i \geq 1, 0 \leq j < i)$ are disjoint, and

$$\bigcup_i \bigcup_{0 \leq j < i} \Psi^j R_i = \bigcup_{i \geq o} \Psi^i E \mod 0.$$

Hence $\sum_{i \geq 1} i\, m(R_i) = m(\bigcup_{i \geq 0} \Psi^i E) \leq 1$. (This is the recurrence theorem of Kac, see e.g. [94]). That $H_m(\rho) < \infty$ and therefore $\rho \in \mathfrak{Z}(E)$ is immediate from the following lemma (put $p_o = m(\mathfrak{X} \setminus E)$, $p_i = m(R_i)$ $(i \geq 1)$).

<u>(26.11) Lemma:</u> If $p = (p_i)_{i \geq o}$ is a probability vector with $\Sigma\, i\, p_i \leq 1$, then $H(p) < \infty$.

<u>Proof:</u> We rearrange the p_i in decreasing order. Then still $\Sigma\, i\, p_i \leq 1$. Let $i \geq 1$ and $p_i \geq a$. Then

$1 \geq \sum_{j=o}^{i} j\, p_j \geq \frac{1}{2} i(i+1)a$, therefore $a \leq 2 \cdot i^{-2}$, so that $p_i \leq 2 \cdot i^{-2}$. The lemma follows now from $\sum_{i \geq 1} i^{-2} \log i < \infty$. □

(26.12) Theorem (Abramov): Let $(\mathfrak{X},m,\Psi)$ be an m.t. dynamical system and $E \subset \mathfrak{X}$ a sweep-out set. Then

$$h_m(\Psi) = m(E)\cdot h_{m_E}(\Psi_E) .$$

Proof: We prove this in several small steps.

a) It is easy to see that $h_m(\Psi) = \sup\{h_m(\xi,\Psi) | \xi \in \mathfrak{Z}(E)\}$.

b) From now on $\Psi_E\, \rho \subset \xi \in \mathfrak{Z}(E)$. For such ξ, we have because of $R_i = \Psi^{-i}(\Psi_E R_i)$: $R_i \in (\xi)_1^\infty$, $R_i \in (\xi)_1^\infty \cap E$ and $E = \bigcup R_i \in (\xi)_1^\infty$.

c) $(\Psi_E^{-i}\, \xi)_{i=1}^\infty = (\xi)_1^\infty \cap E$ mod 0.

The direction $\subset$ is clear since $R_i \in (\xi)_1^\infty \cap E$, and if $A \subset E$:

$$\Psi_E^{-1} A = \bigcup_{i=1}^{\infty} \Psi_E^{-1} (\Psi_E R_i \cap A) = \bigcup_{i=1}^{\infty} R_i \cap \Psi^{-i} A.$$

For the other direction, let $n \in \mathbb{N}$ and $A \subset E$. Then

$$\Psi^{-n} A \cap E = \bigcup_{j=1}^{n} [\Psi_E^{-j} A \cap \bigcup_{i_1+\dots+i_j=n} R_{i_1} \cap \Psi_E^{-1} R_{i_2} \cap\dots\cap$$
$$\cap\dots\cap \Psi_E^{-(j-1)} R_{i_j}].$$

If A is a union of atoms of ξ_E, this implies $\Psi^{-n} A \cap E \in (\xi_E)_1^\infty$, and in particular:

$$\Psi^{-n} E \cap E \in (\xi_E)_1^\infty,$$

$$\Psi^{-n}(\mathfrak{X}\setminus E) \cap E \overset{0}{=} E \setminus (\Psi^{-n} E \cap E) \in (\xi_E)_1^\infty.$$

d) b,c) and the triviality of $\xi | \mathfrak{X}\setminus E$ give for $\Psi_E \rho \subset \xi \in \mathfrak{Z}(E)$:

$$h_m(\xi,\Psi) = H_m(\xi|(\xi)_1^\infty) = H_m(\xi|(\xi)_1^\infty \cap E) = H(\xi|(\Psi_E^{-i}\ \xi)_{i=1}^\infty)$$

$$= m(E)\ H_{m_E}(\xi_E|(\xi_E)_1^\infty) = m(E)\ h_{m_E}(\xi_E,\Psi_E).$$

With a) this implies immediately that $h_m(\Psi) = m_E\ h_{m_E}(\Psi_E)$. □

(26.13) Corollary: If Ψ is ergodic, (Q,R) a partition with $0 < m(Q) < 1$ and α a partition of $\mathfrak{X}$, then

$$h(\alpha,\Psi) \leq h(\alpha \cap Q,\Psi) + m(R)\ h(\alpha_R,\Psi_R).$$

Proof: Using the inequality

$$\left(\Sigma_\Psi(\alpha \vee (Q,R))\ |\ R\right) \subset \left(\Sigma_{\Psi_R}(\alpha_R) \vee [\Sigma_\Psi(\alpha \cap Q)|R]\right)$$

and twice theorem (26.12) we can compute

$$h(\alpha,\Psi) \leq h(\alpha \vee (Q,R),\Psi) = m(R)\ h(\Sigma_\Psi(\alpha \vee (Q,R))|R,\Psi_R)$$

$$\leq m(R)\ h(\Sigma_{\Psi_R}(\alpha_R),\Psi_R) + m(R)\ h(\Sigma_\Psi(\alpha \cap Q)|R,\ \Psi_R)$$

$$= m(R)\ h(\alpha_R,\Psi_R) + h(\alpha \cap Q,\Psi)\ .$$ □

Mixing Subshifts of Finite Type

We have to use many different subshifts of finite type, and we are not able to restrict ourselves to the standardized case of order 2, as it was possible in section 17 for the general theory.

For short we write "f.t.-subshift" for "subshift of finite type" and "m.f.t.-subshift" for "(topologically) mixing f.t.-subshift".

For the definition of an f.t.-subshift we always give the blocks of a certain length occuring in M, which is a possible description by (17.6.2).

(26.14) Definition: Let M be an f.t.-subshift.

a) If $P,Q \prec M$, an M-transition block from P to Q is a block $U \prec M$ such that $P \cdot U \cdot Q \prec M$. $L \in \mathbb{N}$ is a transition length from P to Q if there exist M-transition blocks $U \in Bl_L(M)$ and $U' \in Bl_{L+1}(M)$ from P to Q.

b) $L_0 \in \mathbb{N}$ is a transition length for M if for all $L \geq L_0$ and $P,Q \prec M$, L is a transition length from P to Q.

We recall (17.10) and the following remark in a form adapted to our applications.

(26.15) Lemma: Equivalent conditions for the f.t.-subshift M are

a) M is an m.f.t.-subshift;

b) there exists a transition length;

c) M is top. transitive and there are two blocks $P,Q \prec M$ (both at least as long as the order of M) such that there exists a transition length from P to Q.

Clearly two m.f.t.-subshifts always have a common transition length.

In the remainder of this section we prove the existence of some m.f.t.-subshifts which are important for the constructions in the subsequent theorems, especially for some coding procedures.

(26.16) Lemma: Let M be an m.f.t.-subshift, $K_1, K_2 \subset M$ disjoint subshifts with $K_1 \neq \emptyset$, and $\varepsilon > 0$.

a) There is an m.f.t.subshift M_1 with
$K_1 \subset M_1 \subset M$, $K_2 \cap M_1 = \emptyset$ and $h(M_1) < h(K_1) + \varepsilon$.

b) If C,C' are finite sets of M-blocks and K_1 has one or both of the properties

$$P \in C \Rightarrow P \underset{d}{\prec} K_1 \quad ; \quad P \in C' \Rightarrow P \underset{\varepsilon\text{-reg}}{\prec} K_1 ,$$

then the corresponding properties can be achieved for M_1.

Proof: a) Let r be so large that $Bl_r(K_1) \cap Bl_r(K_2) = \emptyset$, and that for $j \geq r$ $\theta_j(K_1) < \exp[j(h(K_1) + \frac{\varepsilon}{2})]$.
Let L be an M-transition length, choose $j > 3r + L$ so large that

$$(*) \qquad j^{-1}[\log 4(L+j) + (r + L + 1) \log \theta(M)] < \frac{\varepsilon}{2}$$

and take as the defining block system for M_1:

$$Bl_j(M_1) := \{P \in Bl_j(M) \mid P \prec QUR, \text{ where } Q,R \text{ run through } Bl_j(K_1) \text{ and }$$

$$U \in Bl_L(M) \cup Bl_{L+1}(M) \text{ is an M-transition block from Q to R}\}.$$

Then, of course, $M_1 \supset K_1$, and $M_1 \cap K_2 = \emptyset$ because $j > 3r + L$.
M_1 is mixing because of the variable length admitted for the U-blocks (see (26.15)). We count now the j-subblocks of blocks of the form QUR with $l(U) = L$: grouping them according to the place where they begin in QUR, **we see that their number is**

$$\leq \sum_{t=o}^{r-L+1} \theta(M)^t \, \theta_{j-t}(K_1) + \sum_{t=r+L}^{j-r} \theta(M)^L \, \theta_{j-t}(K_1) \, \theta_{t-L}(K_1) +$$

$$+ \sum_{t=j-r+1}^{j+L} \theta(M)^{j+t-L} \, \theta_{t-L}(K_1)$$

$$\leq \theta(M)^{r+L} \, [2(r+L) \, \theta_j(K_1) + \sum_{t=r}^{j-r} \theta_{j-t}(K_1) \, \theta_t(K_1)].$$

A similar summand comes from the U with $l(U) = L + 1$. Taking into account that for $t \geq r$, $j - t \geq r$:

$$\theta_{j-t}(K_1) \, \theta_t(K_1) < \exp[j(h(K_1) + \frac{\varepsilon}{2})], \quad \text{we get}$$

$$\theta_j(M_1) \leq 4(L+j) \, \theta(M)^{r+L+1} \exp[j(h(K_1) + \frac{\varepsilon}{2})],$$

and because of (*) : $h(M_1) < h(K_1) + \varepsilon$

b) For the first property, we simply need r so large that

$$P \in C,\ Q \in Bl_r(K_1) \Rightarrow P \prec Q \quad ,$$

for the second property $\varepsilon' < \varepsilon$ and r such that for $P \in C'$,

$$Q_1, Q_2 \in Bl(K_1),\ l(Q_i) > r \Rightarrow |\mu_{Q_1}(P) - \mu_{Q_2}(P)| < \varepsilon.$$

The reader may check that for sufficiently large r and j (r also much bigger than L):

$$P \in C',\ Q_1, Q_2 \in Bl_j(M_1) \Rightarrow |\mu_{Q_1}(P) - \mu_{Q_2}(P)| < \varepsilon. \quad \square$$

<u>(26.17) Lemma:</u> Let M be an m.f.t.-subshift, $K \subsetneqq M$ a subshift, $0 < h < h(M)$ and $\varepsilon > 0$. Then there exists an m.f.t.-subshift $\bar{M} \subset M$ **satisfying**

$$\bar{M} \cap K = \emptyset$$

$$|h(\bar{M}) - h| < \varepsilon.$$

<u>Proof:</u> We choose a block $P \prec M$ with $P \nprec K$ and an $L \in \mathbb{N}$ so large that for $Q, Q' \in Bl(M)$ there exists a transition block $U \in Bl_L(M)$ with

(**) $\quad QUQ' \prec M;\ \ P \prec U.$

We take t so large that

$$(t+L)^{-1} \log \theta_t(M) > h(M) - \varepsilon$$

$$\frac{L}{t} \log \theta(M) < \frac{\varepsilon}{2}.$$

Now we choose a subset $C \subset Bl_t(M)$ such that with θ = card C

$$h - \varepsilon < (t+L)^{-1} \log \theta \leq h$$

and n so large that

$$\frac{\log(t+L)}{n(t+L)} + \left(\frac{1}{n} + \frac{L}{t}\right) \log \theta(M) < \varepsilon.$$

We let the f.t-subshift $\hat{M}$ (not necessarily mixing) have as defining block system $Bl_{n(t+L)}(\hat{M})$, which consists of all $n(t+L)$-subblocks of blocks of the general form

$$Q_1 \cdot U \cdot Q_2 \cdot U \cdot \ldots \cdot Q_n \cdot U \cdot Q_{n+1} \cdot U \cdot Q_{n+2}$$

where the Q_i run through C and $U \in Bl(M)$ is always a transition block with condition (**).

Obviously $\hat{M} \cap K = \emptyset$, and the number of $\hat{M}$-blocks can easily be estimated: $\theta_{k(t+L)}(\hat{M}) \geq \theta^k$, hence $h(\hat{M}) > h - \varepsilon$; $\theta_{n(t+L)}(\hat{M}) \leq (t + L)\, \theta(M)^{t+L}\, \theta(M)^{L(n-1)}\, \theta^{n-1}$, hence

$$h(\hat{M}) \leq [n(t+L)]^{-1} [\log(t+L) + (t+nL) \log \theta(M) + (n-1) \log \theta] <$$

$$< \frac{\log(t+L)}{n(t+L)} + \left(\frac{1}{n} + \frac{L}{t}\right) \log \theta(M) + \frac{\log \theta}{t+L} < h + \varepsilon.$$

Now, using lemma (26.16), we also find an m.f.t. subshift $\overline{M} \supset \hat{M}$ such that $\overline{M} \cap K = \emptyset$ and $h(\overline{M}) < h + \varepsilon$. □

<u>(26.18) Lemma:</u> Let $M \supsetneqq \overline{M}$ be m.f.t.-subshifts and $r, n \in \mathbb{N}$. Then there exist n M-blocks $a_1, \ldots, a_n$ of the same length t and m.f.t.-subshifts $N_{a_1}, \ldots, N_{a_n}$ with $M \supset N_{a_i} \supset \overline{M}$ such that

1. $\forall\ i \leq n \quad \forall\ P \in Bl_r(M) : P \prec a_i$

2. $a_i \prec N_{a_i}$, $a_i \not\prec \bigcup_{j \neq i} N_{a_j}$.

<u>Proof:</u> First we choose disjoint m.f.t. subshifts $N_i \subset M \setminus \overline{M}$ such that for $P \in Bl_r(M) : P \underset{d}{\prec} N_i$. (The existence of these

systems can be shown along the lines of the proof of lemma (26.16)). The m.f.t.-subshifts N_{a_i} can, by lemma (26.16.b), be obtained such that $N_{a_i} \supset N_i \cup \overline{M}$, $N_{a_i} \cap \bigcup_{j \neq i} N_j = \emptyset$.

Finally, we take t so large that

$$P \in Bl_r(M),\ Q \in B_t(N_i) \Rightarrow P \prec Q \quad (1 \leq i \leq n)$$

and

$$Bl_t(N_i) \cap Bl_t(N_{a_j}) = \emptyset \quad (i \neq j),$$

and choose some (no matter which) $a_i \in Bl_t(N_i)$. □

(26.19) Lemma: Suppose we have an m.f.t.-subshift M; $\delta > 0$, $\epsilon > 0$, $r \in \mathbb{N}$; a non-empty subshift $K \subsetneqq M$ with

$$P \in Bl_r(M) \Rightarrow P \underset{\frac{\epsilon}{2}\text{-reg.}}{\prec} K;$$

a finite set $C \subset Bl(M)$ with

$$P \in C \Rightarrow P \underset{d}{\prec} K;$$

and h, $h(K) < h < h(M)$ such that

$$[h(M) - h(K)]^{-1} (h - h(K)) < \frac{\epsilon}{4}.$$

Then there exists an m.f.t.-subshift $\overline{M}$ such that

1) $K \subset \overline{M} \subset M$
2) $h < h(\overline{M}) < h + \delta$
3) $P \in Bl_r(M) \Rightarrow P \underset{\epsilon\text{-reg}}{\prec} \overline{M}$
4) $P \in C \Rightarrow P \underset{d}{\prec} \overline{M}$.

Proof: Instead of an m.f.t.-subshift M with 1) - 4) it is sufficient to construct a subshift M with 1) 3) 4) and

$h(\overline{M}) > h$. For then we get successively:

An m.f.t.-subshift $M' \supset \overline{M}$ with 3) and 4) (by lemma (26.16.b)); an m.f.t.- subshift $\hat{M} \subset M'$ with $h < h(\hat{M}) < h + \delta$ (by lemma (26.17)) - $\hat{M} \vee K$ satisfies 1) - 4) but it is not yet an m.f.t.- subshift -; and an m.f.t.- subshift $\tilde{M} \supset \hat{M} \cup K$ with 1) - 4) by lemma (26.16.b). $\tilde{M}$ is the m.f.t.-subshift we wanted to have.

Let L be an M-transition length and $s,t \in \mathbb{N}$ such that

$$\alpha)\ [h(M) - h(K)]^{-1}\ [h\cdot(1 - \frac{2L}{s+t+2L})^{-1} - h(K)] < \frac{\varepsilon}{4} < \frac{s}{t+s},$$

$$\beta)\ (s + t + 2L)^{-1}\ (s + 2(L+r)) < \frac{\varepsilon}{3},$$

$$\gamma)\ P \in Bl_r(M),\ Q_1, Q_2 \in Bl_t(K) \Rightarrow |\mu_{Q_1}(P) - \mu_{Q_2}(P)| < \frac{\varepsilon}{2},$$

$$\delta)\ P \in C,\quad Q \in Bl_t(K) \Rightarrow P \prec Q.$$

For some $n \in \mathbb{N}$ let $\overline{M}$ be the f.t.- subshift having as defining block system $Bl_{n(t+s+2L)}(\overline{M})$, the set of all $n(t+s+2L)$-subblocks of blocks of the general form

$$Q_1U_1S_1U'_1Q_2U_2S_2U'_2\ldots Q_nU_nS_nU'_nQ_{n+1}U_{n+1}S_{n+1}U'_{n+1}Q_{n+2}$$

with $Q_i \in Bl_t(K)$, $S_i \in Bl_s(M)$, $U_i, U'_i \in Bl_L(M)$ transition blocks. Then for each $k \in \mathbb{N}$ $\theta_{(k+1)(t+s+2L)}(\overline{M}) \geq \theta_t(K)^k \theta_s(M)^k$, hence for $k \to \infty$

$$h(\overline{M}) \geq (t+s+2L)^{-1}(\log \theta_t(K) + \log \theta_s(M)) \geq$$

$$\geq (t+s+2L)^{-1}\ (t\cdot h(K) + s\cdot h(M)) =$$

$$= (t + s + 2L)^{-1}\ [(t+s)\cdot h(K) + s(h(M) - h(K))] =$$

$$= (1 - \frac{2L}{s+t+2L})\ [h(K) + \frac{s}{t+s}(h(M) - h(K))] > h$$

(using the two extremal terms of α).

Because of β), in a block $Q_i U_i S_i U_i'$, the relative frequencies of a block $P \in Bl_r(M)$ do not differ much from those in Q_i, so that, if n was chosen large enough, we have property 3) for $\overline{M}$. □

<u>(26.20) Lemma:</u> Let M be an m.f.t.-subshift, $\nu \in \mathfrak{M}_\sigma(M)$ ergodic, $r \in \mathbb{N}$ and $\varepsilon > 0$. Then there exists (constructively) an m.f.t.-subshift $\overline{M} \subset M$ such that for $P \in Bl_r(M)$ and $\overline{\nu} \in \mathfrak{M}_\sigma(\overline{M})$:

$$P \underset{d}{\prec} \overline{M}$$

$$|\overline{\nu}(P) - \nu(P)| < \varepsilon.$$

<u>Proof:</u> It is sufficient to construct a subshift $\overline{M}$ (instead of an m.f.t.-subshift) with the required properties and with

$$P \in Bl_r(M) \Rightarrow P \underset{\varepsilon\text{-reg}}{\prec} \overline{M};$$

for then we can apply lemma (26.16.b) and obtain an m.f.t.-subshift $\widetilde{M} \supset \overline{M}$ such that for all $P \in Bl_r(M)$:

$$P \underset{d}{\prec} \widetilde{M},\ P \underset{\varepsilon\text{-reg}}{\prec} \widetilde{M},$$

and since $\overline{M} \subset \widetilde{M}$, this implies also

$$\forall\ \widetilde{\nu} \in \mathfrak{M}_\sigma(\widetilde{M}) : |\widetilde{\nu}(P) - \nu(P)| < 2\varepsilon.$$

Let the M-transition length L be so large that for $Q, Q' \in Bl(M)$ there is $U \in Bl_L(M)$ with $Q \cdot U \cdot Q' \prec M$ and

$$P \in Bl_r(M) \Rightarrow P \prec U.$$

Let k be so large that there exists a block $R \in Bl_k(M)$ with

$$|\mu_R(P) - \nu(P)| < \frac{\varepsilon}{2} \quad (P \in Bl_r(M)),$$

and $\omega \in M$ a sequence of the form $\ldots R \cdot U \cdot R \cdot U \cdot R \cdot U \ldots$ where R is fixed and U always the same transition block of the type above. ω is periodic, so $\overline{M} := \overline{O(\omega)}$ is strictly ergodic. It is clear that, if k was chosen large enough, the unique element $\overline{\nu} \in \mathfrak{M}_\sigma(\overline{M})$ satisfies $|\overline{\nu}(P) - \nu(P)| < \epsilon$ $(P \in Bl_r(M))$; $P \underset{\epsilon\text{-reg}}{\prec} \overline{M}$ and $P \underset{d}{\prec} \overline{M}$ are clear. □

27. Combinatorial Construction of Minimal Sets

In this section we isolate the combinatorial part of the ideas used in the proofs of theorems (29.2), (31.2), (31.3) in order to get generators γ for which Λ_γ is strictly ergodic resp. minimal. All steps here are purely constructive. - The first construction is rather trivial.

(27.1) A strictly ergodic set.

Let M be an m.f.t.-subshift. We choose successively

$M_i, \nu_i, r_i, \epsilon_i$ $(i \geq 0)$ where

$M_o = M$; $M_i \subset M_{i-1}$ an m.f.t.-subshift

$\nu_i \in \mathfrak{M}_\sigma(M_i)$ ergodic

$r_i \in \mathbb{N}$; $r_i > 2r_{i-1}$ if $i \geq 1$

$\epsilon_i < 2^{-i} \theta_{r_i}(M_i)^{-1}$

such that for $P \in Bl_{r_i}(M_i)$, $Q \in Bl_{r_{i+1}}(M_{i+1})$:

1) $P \prec Q$

2) $|\mu_a(P) - \nu_i(P)| < \epsilon_i$.

This choice is possible by lemma (26.20).

Let now $\overline{M} = \bigcap_i M_i$. $\overline{M}$ is also characterized by the property $Bl_{r_i}(\overline{M}) = Bl_{r_i}(M_i)$ $(i \in \mathbb{N})$. Because 1) holds in each step, every r_i-block occurs densely in M_{i+1}, hence in $\overline{M}$, and $\overline{M}$ is minimal; because 2) holds for each i, $\overline{M}$ is uniquely ergodic by (26.7).

Let us observe two points: a) If we omit condition 1),

then $\bigcap_i M_i$ remains uniquely ergodic and the support of the unique measure is strictly ergodic.

b) If we omit 2), then $\bigcap M_i$ is minimal, but not necessarily strictly ergodic. Our next construction is of this type; we shall replace 2) by other conditions which guarantee that $\bigcap_i M_i$ is not strictly ergodic, but enable us to control the ergodic measures on $\bigcap M_i$

(27.2) A minimal set with two ergodic measures

We assume that we have an m.f.t.-subshift M and obtain a subshift $\overline{M} \subset M$ with the following properties:

a) $\overline{M}$ is minimal
b) $\mathfrak{M}_\sigma(\overline{M})$ contains exactly two ergodic measures, $\overline{\mu}^1$ and $\overline{\mu}^2$
c) All quasiregular points of $\overline{M}$ are generic for one of the two measures $\overline{\mu}^1$, $\overline{\mu}^2$.

Remark: The following construction can be changed so that the set of ergodic measures on $\overline{M}$ consists of two measures with equal entropy, or that it is of a given finite cardinality, or a convergent sequence with its limit point, or a perfect totally disconnected set in $\mathfrak{M}_\sigma(\overline{M})$ (all with the same entropy, if desired). Using the $\overline{d}$-process metric of Ornstein (see [141]) as in the construction in [74], one can also achieve the property that all the ergodic measures on $\overline{M}$ are isomorphic to Bernoulli systems; but with this condition we are not able to maintain a constant entropy.

We take a metric d on $\mathfrak{M}_\sigma(M)$ (recall (2.8)). We put $M_0 = M$ and construct disjoint m.f.t.-subshifts $M_0^1, M_0^2 \subset M_0$. Then

$$\varphi := d(\mathfrak{M}_\sigma(M_0^1), \mathfrak{M}_\sigma(M_0^2)) > 0.$$

In the (i-1)th step of the construction we have obtained an m.f.t.-subshift M_{i-1} and disjoint m.f.t.-subshifts $M_{i-1}^1, M_{i-1}^2 \subset M_{i-1}$. We find now

1. $t_{i-1} > 2^{i-1}$ so large and $2^{-i+1} > \varepsilon_{i-1} > 0$ so small that

$$Bl_{t_{i-1}}(M_{i-1}^1) \cap Bl_{t_{i-1}}(M_{i-1}^2) = \emptyset$$

$$\nu, \nu' \in \mathfrak{M}_\sigma(M_{i-1}),\ |\nu(P) - \nu'(P)| < \varepsilon_{i-1}\ \forall\ P \in Bl_{t_{i-1}}(M_{i-1})$$
$$\Longrightarrow \quad d(\nu,\nu') < 3^{-i}\varphi.$$

2. Ergodic measures $\nu_{i-1}^j \in \mathfrak{M}_\sigma(M_{i-1}^j)$ $(j = 1,2)$.

3. m.f.t.-subshifts $M_i^j \subset M_{i-1}$ $(j = 1,2)$ and $\overline{r}_i > t_{i-1}$ such that

   3 a) $Bl_{\overline{r}_i}(M_i^1 \cup M_i^2) \neq Bl_{\overline{r}_i}(M_{i-1})$

   3 b) $P \in Bl_{t_{i-1}}(M_{i-1}),\ Q \in Bl_{\overline{r}_i}(M_i^j) \Rightarrow P < Q$ $(j = 1,2)$.

   3 c) $P \in \bigcup_{t \leq t_{i-1}} Bl_t(M_{i-1}),\ Q \in \bigcup_{t \geq \overline{r}_i} Bl_t(M_i^j)$

$$\Longrightarrow \quad |\mu_Q(P) - \nu_{i-1}^j(P)| < \varepsilon_{i-1} (j=1,2)$$

   (This is possible by lemma (26.20). 3c) and 1. make M_i^1 and M_i^2 disjoint.)

4. A block $C_i \in Bl_{\overline{r}_i}(M_{i-1}) \setminus Bl_{\overline{r}_i}(M_i^1 \cup M_i^2)$

5. An M_{i-1}-transition length L_i which is so large that for $P^j \in Bl(M_i^j)$ $(j = 1,2)$ there are blocks $U_i^1, U_i^2 \in Bl_{L_i}(M_{i-1})$ with

a) $P^1 \cdot U_i^1 \cdot P^2 < M_{i-1}$, $P^2 \cdot U_i^2 \cdot P^1 < M_{i-1}$

b) $C_i < U_i^j$ $(j = 1,2)$, but only once, namely at the place $[\frac{L_i}{2}]$.

6. $r_i > 3(\overline{r}_i + L_i) \cdot \varepsilon_{i-1}^{-4}$ and bigger than the order of M_{i-1}, and two fixed blocks $S_i^j \in Bl_{r_i}(M_i^j)$ $(j = 1,2)$, and for these blocks two fixed transition blocks U_i^j $(j = 1,2)$ as in 5. ($S_i^1 \cdot U_i^1 \cdot S_i^2 \prec M_{i-1}$, $S_i^2 \cdot U_i^2 \cdot S_i^1 \prec M_{i-1}$).

7. M_i is the m.f.t.-subshift having as defining block system $Bl_{r_i}(M_i) =$

$$= Bl_{r_i}(M_i^1 \cup M_i^2) \cup \{Q \in Bl_{r_i}(M_{i-1}) \mid Q \prec S_i^1 U_i^1 S_i^2 \text{ or } Q \prec S_i^2 U_i^2 S_i^1\}.$$

The intuitive sense of this construction is: M_{i-1}^1, M_{i-1}^2 are disjoint. If we would construct $M_i^j \subset M_{i-1}^j$, $\bigcap_i M_i$ could not be minimal, since it contains $\bigcap_i M_i^j$ $(j = 1,2)$, which are invariant. Therefore we "inject" into the sequences in M_{i-1}^1 blocks of M_{i-1}^2 in bounded distances, but so rarely that the good frequencies are not considerably disturbed. So, a typical element of M_i^1 remains for a long time near M_{i-1}^1, then moves to M_{i-1}^2 to gather up as quickly as possible all t_{i-1}-blocks of M_{i-1}^2, and returns to M_{i-1}^1. These transitions are possible only if we first envelop $M_{i-1}^1 \cup M_{i-1}^2$ in an m.f.t.-subshift M_{i-1}.

$(M_i)_{i \in \mathbb{N}}$ is a decreasing sequence of m.f.t.-subshifts. Its intersection $\overline{M} = \bigcap_i M_i$ is characterized by $Bl_{t_i}(\overline{M}) = Bl_{t_i}(M_i)$. $\overline{M}$ has the properties a) - c):

a) Minimality follows from 3b,6,7 and theorem (26.7).

b) By 1, 2 and 3c, $(\nu_i^j)_{i \in \mathbb{N}}$ is a Cauchy sequence and if $\overline{\mu}^j$ denotes its limit, then $d(\overline{\mu}^1, \overline{\mu}^2) \geq \frac{1}{2}\varphi$. The rest of the assertion follows when we have proven c).

c) Let $\nu \in \mathfrak{M}_\sigma(\overline{M})$, $\omega \in \overline{M}$ quasiregular and generic for $\nu \in \mathfrak{M}_\sigma(\overline{M})$. Let $\delta > 0$. There exist $r \in \mathbb{N}$ and $\eta > 0$ such that for $\rho, \rho' \in \mathfrak{M}_\sigma(\overline{M})$:

$$\forall\, P \in Bl_r(M) : |\rho(P) - \rho'(P)| < 2\eta \Rightarrow d(\rho, \rho') < \delta.$$

If i_o is so large that $2^{-i_o+1} < \eta$ and $2^{i_o-1} \geq r$, then for all $i \geq i_o$, $t \geq \overline{r}_i$, $Q \in Bl_t(M_i^j)$, $P \in Bl_r(M)$: $|\mu_Q(P) - \overline{\mu}^j(P)| < \eta$, by 1 and 3c) which imply that $|\nu_i^j(P) - \nu_{i-1}^j(P)| < \varepsilon_{i-1}$ if $P \in Bl_r(M)$, $i \geq i_o$.

There is $n_o \in \mathbb{N}$ such that for all $n \geq n_o$ and all $P \in Bl_r(M)$

$$2\,|\,\mu_{\omega\langle 0,n\rangle}(P) - \nu(P)| < \eta$$

and, because of 6, an $n \geq n_o$ and $i \geq i_o$ such that

$$\varepsilon_{i-1}^2 \cdot r_i > n, \quad \varepsilon_{i-1}^2 \cdot n > r_i.$$

Since all the transition blocks U_i^j are recognizable by the block C_i (see 4 and 5), the sequence ω can be split up uniquely into the transition blocks and into blocks belonging to the subshifts M_i^j, the latter with length at least $\frac{3}{4} r_i$ by 6, so we have the following schema for ω:

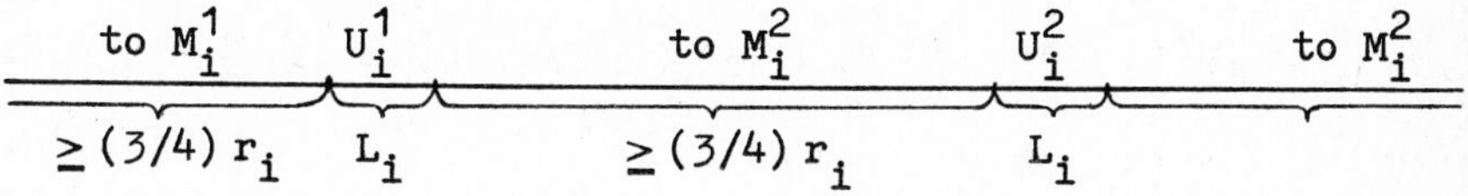

For the position of $\omega_{\langle 0,n\rangle}$ in this schema there exist the following two cases:

α) $\omega_{\langle 0,n\rangle}$ falls completely into an M_i^j-piece. Then $\omega_{\langle 0,n\rangle} \in Bl_{n+1}(M_i^j)$, and, since $n > \bar{r}_i$, and by the choice of i and n, $|\nu(P) - \bar{\mu}^j(P)| < 2\eta$ for $P \in Bl_r(M)$, so that $d(\nu,\bar{\mu}^j) < \delta$.

β) $\omega_{\langle 0,n\rangle}$ overlaps one of the transition blocks U. Then at the right of U begins an M_i^j-piece (j = 1 or 2). For the relative frequency

$$\mu_{\omega_{\langle 0,n+L_i+[\frac{1}{2} r_i]\rangle}}(P)$$

only the part $\omega_{\langle n+L_i,n+L_i+[\frac{1}{2}r_i]\rangle}$ is essential, since $(n + L_i) < 2\epsilon_{i-1}^2 r_i$, and this part is an M_i^j-block. So we get as in case α): $|\nu(P) - \bar{\mu}^j(P)| < 2\eta$, and $d(\nu,\bar{\mu}^j) < \delta$. Therefore ν must be one of the measures $\bar{\mu}^j$. □

28. Finite Generators for Ergodic Transformations (Theorem of Krieger)

In section 9 generators were defined and the existence of generators was proven for aperiodic transformations. Rohlin has shown that for a dynamical system $(\mathfrak{X},\Sigma,m,\Psi)$ with $h(\Psi) < \infty$ there is a countable generator α with $H(\alpha) < \infty$ (see [159], [149]). W.Krieger has used this result to construct a finite generator in the case of an ergodic transformation. He also added the following extra conditions (separately) for the generator:

a) If $\Sigma_o \subset \Sigma$ is an exhaustive σ-algebra, i.e. $\Psi^{-1}\Sigma_o \subset \Sigma_o$ and $\bigvee_{t\in\mathbb{Z}} \Psi^t\Sigma_o = \Sigma$; then the generator can be found to be Σ_o-measurable (see [117]).

b) The (α,Ψ)-name can be prescribed to be in any given mixing subshift of finite type M with $h(M) > h(\Psi)$. (see [119])

c) If an invariant measure ν on the shift space satisfies $h(\nu) > h(\Psi)$, then the generator α can be constructed such that μ_α is arbitrarily close to ν in the weak topology (see [116])

Condition b) is stronger than c) since it is possible to find an m.f.t.-subshift M with $h(M) > h(\Psi)$ and such that any invariant measure on M is weakly close to ν.

If the measure ν in c) is weakly mixing and $n \in \mathbb{N}$, one can have α such that the n-dimensional marginal distributions of ν and μ_α are equal. This will be shown in [199].

M.Denker has given a more direct proof for Krieger's theorem, which we follow here. We prove it with condition b) above.

The different elements of the proof are separated because this will make it easier to refer to single steps in the proofs of later theorems.

<u>(28.1) Theorem:</u> Let $(\mathfrak{X},m,\Psi)$ be an ergodic, aperiodic dynamical system and M a mixing subshift of finite type with $h(\Psi) < h(M)$. Then there exists an M-generator γ for Ψ.

<u>Proof:</u> First we take a sequence (ϵ_i) of positive numbers with $\frac{1}{3} > \epsilon_i \searrow 0$, an increasing sequence γ_i of finite partitions with $\bigvee_{i\in\mathbb{N}} \Sigma_\Psi(\gamma_i) = \Sigma$, and an m.f.t.-subshift $\overline{M} \subsetneqq M$ with $h(\overline{M}) > h(\Psi)$ ($\overline{M}$ exists by lemma (26.17).)

We shall recursively construct disjoint sets G_i, Z_i, H_i with

$$G_i \cup Z_i \cup H_i = \mathfrak{X}$$

$$G_i \subset G_{i+1}, \quad m(G_i) \nearrow 1$$

$$H_{i+1} \subset H_i.$$

In the i-th step we shall fix the restriction to G_i of the generator γ which is being constructed, i.e. γ_{G_i} or $\gamma|G_i$. Thereby we always insure that the following four properties hold:

a) $\quad h(\gamma \cap G_i,\Psi) > h(\Psi) - m(H_i)\cdot h(\overline{M})$

b) $\quad \gamma_i \overset{\epsilon_i}{\subset} \Sigma_\Psi(\gamma \cap G_i)$

c) $\quad \gamma|G_i$ is such that every extension to a partition $\hat{\gamma}$ of $\mathfrak{X}$ (i.e. every $\hat{\gamma}$ with $\hat{\gamma}|G_i = \gamma|G_i$) satisfies

$$(G_i,Z_i,H_i) \subset \Sigma_\Psi(\hat{\gamma}).$$

d) m_{H_i} - almost all $(\gamma_{H_i}, \Psi_{H_i})$-names are in M.

Then, from b) and c) we can conclude that for any $i \in \mathbb{N}$,

$$\Sigma_\Psi(\gamma) \overset{o}{\supset} \Sigma_\Psi(\gamma \cap G_i) \overset{\epsilon_i}{\supset} \gamma_i,$$

and by the choice of the sequence (γ_i), γ is a generator. From d) we know that it is an M-generator; d) is even stronger than necessary for this purpose. Condition a) says, roughly, that the part $\mathfrak{X} \setminus G_i$ which we have spared in step i is large enough, as far as entropy is concerned, to extend $\gamma | G_i$ to $\mathfrak{X} \setminus G_i$ such that γ is a generator. A more detailed motivation will be given later.

When in the step i $\gamma | G_i$ and in later steps $\gamma | H_i$ is determined, then for $x \in \mathfrak{X}$ we have fixed unconnected pieces of what is to become the (γ, Ψ)-name of x. The portion Z_i of $\mathfrak{X}$ will allow us to define $\gamma | Z_i$ such that the complementary parts of the (γ, Ψ)-name are M-transition blocks, so that we always obtain names in M. If M is the full shift space, we can put $Z_i = \emptyset$.

The properties a), b), and c), d) are of different nature and we show separately how they can be obtained.

Part I. Using a coding method we first get c) and d). Doing this, we leave considerable freedom in the construction, so that a) and b) can be achieved afterwards.

Let $a < M$, $a \nless \overline{M}$ a block and the M-transition length L so large that for $P < \overline{M}$ there are two blocks $U_P^j \in Bl_L(M)$ $(j = 1,2)$ such that

$P \cdot U_P^1 \cdot a < M$, $a \cdot U_P^2 \cdot P < M$;

a occurs in $P \cdot U_P^1 \cdot a$ only as the final piece;

a occurs in $a \cdot U_P^2 \cdot P$ only as the initial piece.

(The two last points need some reflection; we leave it to the reader.) Let $c = 2[2L + l(a)]$.

a, L, c and U_P^j will be used in all steps of the proof.

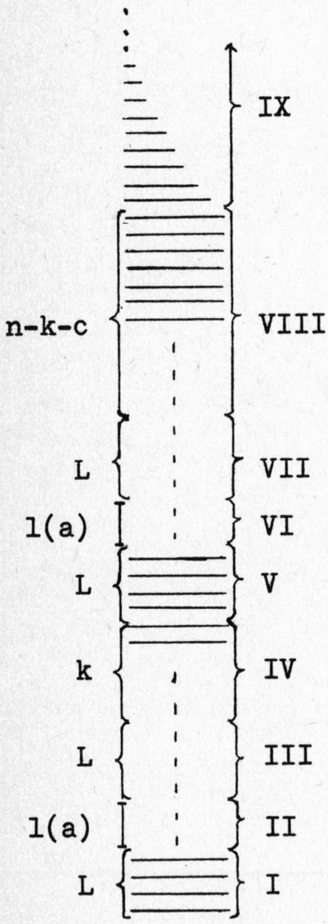

Now let $k_1 \in \mathbb{N}$ and $n_1 > 2(c+k_1)$, and F_1 a (Ψ, n_1)-Rohlin set. Using **the** (F_1, Ψ)-tower we divide $\mathfrak{X}$ into 9 parts, which we call $I_1 - IX_1$ as indicated in the picture (the index 1 in $I_1 - IX_1$, k_1, n_1 is omitted since the same figure will be used in later steps.) The numbers at the left hand side indicate how many levels are taken for the corresponding set. Then we put

$$G_1 = I_1 \cup II_1 \cup (VI_1 \text{ to } IX_1)$$

$$Z_1 = III_1 \cup V_1$$

$$H_1 = IV_1.$$

In the part $VIII_1 \cup IX_1$ γ is defined in such a way that the corresponding pieces of the later (γ, Ψ)-names are $\overline{M}$-blocks: The lowest level of VIII_1 is $\Psi^{c+k_1}F_1$. Let $x \in F_1$, $\gamma = (g_1, \ldots, g_r)$ the desired generator and $n(x) = \min\{j \geq 1 \mid \Psi^j x \in F_1\} \geq n_1$.

For almost all $x \in F_1$, we determine in a measurable way a block

$$(i_{c+k_1}(x), i_{c+k_1+1}(x), \ldots, i_{n(x)-1}(x)) < \overline{M}$$

and then put

$$\Psi^j(x) \in g_{i_j}(x) \quad (c + k_1 \leq j < n(x))$$

In the same sense it is understood that within II_1 and IV_1 the $(\gamma,\Psi,l(a))$-name is a: For $x \in F_1$, $0 \leq j < l(a)$; $a = (a_o,\ldots,a_{l(a)-1})$

$$\Psi^{L+j} x \in g_{a_j} \quad ; \quad \Psi^{3L+l(a)+k_1+j} x \in g_{a_j} .$$

In I_1 and VII_1 blocks of the type U_p^1 resp. U_P^2 are filled in as (γ,Ψ,L)-names. **We can do this** as soon as $\gamma|VIII_1 \cup IX_1$ is fixed. In III_1 and V_1, too, such blocks will be put in; but this is possible only in later steps, when the $(\gamma_{H_1}, \Psi_{H_1})$-name has been fixed over a complete "upcrossing through H_1."

Let us make sure now that property c) holds for $i = 1$. Let $\hat{\gamma}$ be an extension of $\gamma|G_1$. The bases of the tower sections II_1 and VI_1 are at a distance $2L + l(a) + k_1$, going from II_1 to VI_1, and at a distance greater than $n_1 - c \geq 2(c + k_1) - c > 2(2L+l(a) + k_1)$ going from VI_1 to II_1. Since in the (γ,Ψ)-names along $VII_1 \cup VIII_1 \cup IX_1 \cup I_1$ a does not appear, the entrance of x into the set $\Psi^L(F_1)$, i.e. the basis of II_1, is uniquely determined by the event "two $l(a)$-blocks in the $(\hat{\gamma},\Psi)$-name of x are a and appear at a distance $2L + l(a) + k_1$ from each other". But this allows us to identify the set F_1 and hence all levels of the (F_1,Ψ)-tower.

Property d), of course, depends on the later steps of

the construction. Suppose step (i-1) is finished. Now we consider only the part H_{i-1} with the induced transformation $\Psi_{H_{i-1}}$. We choose $k_i \in \mathbb{N}$, $n_i > 2(c+k_i)$ and a $(\Psi_{H_{i-1}}, n_i)$-Rohlin set F_i. As before, the $(F_i, \Psi_{H_{i-1}})$-tower is divided into 9 parts I_i to IX_i, and again

$$G_i \cap H_{i-1} = I_i \cup II_i \cup (VI_i \text{ to } IX_i)$$

$$Z_i \cap H_{i-1} = III_i \cup V_i$$

$$H_i = IV_i .$$

Using $\overline{M}$-blocks in $VIII_i \cup IX_i$, the block a in II_i and VI_i and transition blocks in I_i and VII_i, we fix $\gamma | (G_i \cap H_{i-1})$. But then it will happen at some places that for an entire "upcrossing through H_{i-1}" we know into which atom of γ a point enters, and there we can determine transition blocks which can be filled in at the upcrossings (under the transformation $\Psi_{H_{i-2}}$) through the sets III_{i-1} and V_{i-1}, both contained in Z_{i-1}. When this is accomplished, it may be possible to fill in transition blocks as names for upcrossings through the sets III_{i-2} and V_{i-2} under the transformation $\Psi_{H_{i-3}}$, where it was not possible before, and so on. So recursively we fill up as much of Z_{i-1} as possible. We get

$$G_i \cap (\mathfrak{X} \setminus H_{i-1}) \supset G_{i-1}$$

$$Z_i \cap (\mathfrak{X} \setminus H_{i-1}) \subset Z_{i-1}$$

Since obviously $m(H_i) \longrightarrow 0$, $m(Z_i) \longrightarrow 0$ will be sufficient in order that $m(G_i) \nearrow 1$, i.e. that finally the partition γ is defined on all of $\mathfrak{X}$.

To prove this fact, take $j < i$ and suppose that, going backwards, in the i-th step, we have filled in all possible transitions in the sets IV_t $(j < t < i)$. We consider the $(F_j, \Psi_{H_{j-1}})$-tower and delete all those points of the basis of IV_j (i.e. of $\Psi_{H_{j-1}}^{2L+1(a)} F_j$) for which the $(\gamma, \Psi_{H_{j-1}}, k_j)$-name is not yet fixed. Thus we obtain $R_j \subset \Psi_{H_{j-1}}^{2L+1(a)} F_j$. The part $\bar{R}_j = \bigcup_{s=o}^{k_j-1} \Psi_{H_{j-1}}^{s}(R_j)$ of IV_j can, within H_{j-1}, be linked with VI_j above and with II_j below. Let

$$m(\bar{R}_j) = \alpha\, m(H_j) ,$$

$$m(R_j) = k_j^{-1}\, \alpha \cdot m(H_j) .$$

Together with the deleted parts $H_j \setminus \bar{R}_j$, the undetermined places of H_{j-1} (i.e. the points which have not yet been assigned to an atom of γ), always occur in strings of the length $2L + k_j$ under the transformation $\Psi_{H_{j-1}}$. Now we look at what can be achieved in the $(F_{j-1}, \Psi_{H_{j-2}})$-tower: We decompose the $\Psi_{H_{j-1}}$-strings into pieces of length k_{j-1}, which have to be filled in in IV_{j-1}. The empty (or unlabelled) parts of length $2L+k_j$ can intersect with at most $2 + k_{j-1}^{-1} \cdot (k_j + 2L)$ upcrossings through IV_{j-1}. Hence the part of H_{j-1}, for which γ is not fixed on a complete upcrossing, has at most the measure (under m)

$$m(R_j) \cdot k_{j-1}[2+k_{j-1}^{-1}(2L+k_j)] \le \alpha\, n_j^{-1}\, m(H_{j-1})(2k_{j-1}+2L+k_j),$$

because $m_{H_{j-1}}(H_j) \le n_j^{-1}\, k_j$. Therefore it is sufficient for us to remark that later k_j and n_j will be subject to conditions (5'), (7'b) giving

$$\prod_{j\geq 2} n_j^{-1}(2k_{j-1} + 2L + k_j) = 0.$$

<u>Part II.</u> It remains to determine the lengths k_i and n_i and the sets F_i, and also to distribute the names within the parts $VIII_i \cup IX_i$ of G_i in such a way that the properties a) and b) come out. The important part hereby is only VIII_i. When $\gamma|VIII_i$ is found, we shall simply fill in an $\overline{M}$-compatible name in IX_i and then link it with the part II_i (through I_i) as it was said above.

The first step is somewhat simpler than the others. To make the understanding easier we give a separate proof of it.

1st step

(1) Let $h(\overline{M}) - h(\Psi) = f_1 \quad \tau_1 > 0$,

where $f_1 \geq 8$ is so large that

(2) $\tau_1^{-1} \cdot h(\overline{M}) = [h(\overline{M}) - h(\Psi)]^{-1} \cdot h(\overline{M}) f_1 > 18$

(3) and assume that $h(\gamma_1, \Psi) \geq h(\Psi) - \tau_1$.

(Such an assumption can always be made since we may pass to finer partitions γ_i.)

Let $\xi_1 < \epsilon_1$ be so small that for any partition ρ

(4) $\rho \overset{\xi_1}{\supset} \gamma_1 \Rightarrow h(\rho, \Psi) \geq h(\gamma_1, \Psi) - \tau_1$.

By (13.5) we find $n_1 \in \mathbb{N}$ so large that

(5) $\tau_1 \cdot n_1 > c \cdot h(\overline{M})$,

and that the set

$$S_1 = \cup \,[\text{atoms } A \text{ of } (\gamma_1)_o^{n_1-1} \mid \exp\{-n_1(h(\gamma_1,\Psi)+\tau_1)\} \leq m(A)]$$

(6) has measure $m(S_1) \geq 1 - \frac{1}{2}\xi_1$.

We abbreviate $\alpha_1 = (\gamma_1)_o^{n_1-1}$ and call the atoms of α_1 belonging to S_1 "good atoms". The number of good atoms is at most $\exp\{n_1(h(\gamma_1,\Psi) + \tau_1)\}$.

Because of (5) we may choose our k_1 such that

(7a) $\quad \dfrac{3\tau_1}{h(\overline{M})} \leq \dfrac{k_1}{n_1}$; $\quad$ (7b) $\quad \dfrac{k_1+c}{n_1} \leq \dfrac{6\tau_1}{h(\overline{M})}$.

(7b) together with (2) gives $n_1 > 3(k_1+c)$ (conditions postulated in part I).

Next we construct a (Ψ,n_1,ξ_1)-Rohlin set $F_1 \subset S_1$, which is possible by (6) and (26.3.b). $\alpha_1|F_1$ consists only of traces of good atoms (here we do not count empty atoms), so we have a bound for its cardinality. We want to achieve that $\alpha_1|F_1$ can be reconstructed from $\gamma \cap G_1$, more precisely that

(8) $\qquad \alpha_1 \cap F_1 \subset \Sigma_\Psi(\gamma \cap G_1)$;

for in this case also $\gamma_1 \overset{\xi_1}{\subset} \gamma_1 \cap \bigcup_{i=o}^{n_1-1} \Psi^i F_i \subset \Sigma_\Psi(\gamma \cap G_1)$, and property b) will be satisfied. But because of (4) and (3) also

$$h(\gamma \cap G_1,\Psi) \geq h(\gamma_1,\Psi) \geq h(\gamma_1,\Psi) - \tau_1 \geq h(\Psi) - 2\tau_1.$$

Then property a) follows because $m(H_1) \geq n_1^{-1}k_1(1 - \xi_1)$ and, by (7a)

$$2\tau_1 \leq \tfrac{2}{3}\, n_1^{-1}\cdot k_1 h(\overline{M}) < n_1^{-1}k_1(1-\xi_1)h(\overline{M}) \leq m(H_1)\cdot h(\overline{M}).$$

So, we only have to take care of (8); an equivalent condition is

$$\Psi^{c+k_1}\alpha_1 \cap \Psi^{c+k_1}F_1 = \Psi^{c+k_1}(\alpha_1 \cap F_1) \subset \Sigma_\Psi(\gamma \cap G_1),$$

and **we get this** if to each atom of $\Psi^{c+k_1}(\alpha_1|F_1)$ we adjoin a different $\overline{M}$-block of length n_1-c-k_1 which serves as the (γ,Ψ,n_1-c-k_1)-name of every point of the atom. ($c+k_1$ is the lowest level of VIII_1, n_1-c-k_1 the "upcrossing time" through VIII_1). It follows from (7b), (1) that sufficiently many such $\overline{M}$-blocks are available:

$$(n_1-k_1-c)\,h(\overline{M}) = n_1(1-n_1^{-1}(k_1+c))h(\overline{M}) \geq n_1(1-h(\overline{M})^{-1}\cdot 6\tau_1)h(\overline{M})$$
$$= n_1(h(\overline{M})-6\tau_1) = n_1(h(\Psi)+(f_1-6)\tau_1) \geq n_1(h(\gamma_1,\Psi)+\tau_1),$$

i.e. there are more $\overline{M}$-blocks of length n_1-k_1-c than atoms of positive measure in $\alpha_1|F_1$.

Induction step:

Let now $i \geq 2$ and $G_{i-1}, Z_{i-1}, H_{i-1}, \gamma \cap G_{i-1}$ be constructed in such a way that a) b) c) hold and

(1') $h(\gamma \cap G_{i-1},\Psi) = h(\Psi) - m(H_{i-1})\cdot h(\overline{M}) + f_i\cdot\tau_i$

with $\tau_i > 0$, where $f_i \geq 8$ is so large that

(2') $\tau_i^{-1}\cdot h(\overline{M})\cdot m(H_{i-1}) = [h(\gamma \cap G_{i-1},\Psi) + m(H_{i-1})h(\overline{M}) - h(\Psi)]^{-1}\cdot$
$\cdot\, h(\overline{M})\cdot m(H_{i-1})f_i > 18.$

Assume that

$\gamma_i \supset \gamma \cap G_{i-1}$ and

(3') $h(\gamma_i,\Psi) \geq h(\Psi) - \tau_i.$

Let $\xi_i - \epsilon_i$ be so small that for any partition ρ

(4') $\rho \overset{\xi_i}{\supset} \gamma_i \Rightarrow h(\rho,\Psi) \geq h(\gamma_i,\Psi) - \tau_i,$

and $q_i \in \mathbb{N}$ so large that for

(5") $n_i = \lceil q_i\cdot m(H_{i-1})\rceil \geq 6,$ and for

$$S_i' = \cup \{\text{atoms } A' \text{ of } (\gamma_i)_o^{q_i-1} \mid \exp[-q_i(h(\gamma_i,\Psi)+\tau_i)] \leq m(A')\},$$

$$S_i'' = \cup \{\text{atoms } A'' \text{ of } (\gamma \cap G_{i-1})_o^{q_i-1} \mid m(A'') \leq \exp[-q_i(h(\gamma \cap G_{i-1},\Psi) - \tau_i)]\},$$

$S_i = S_i' \cap S_i''$ we have

(6') $m(S_i) \geq 1 - \frac{1}{2}\xi_i \cdot m(H_{i-1})$,

(5') $n_i\tau_i \geq (k_{i-1}+c) \cdot h(\overline{M}) \cdot m(H_{i-1})$.

Furthermore assume q_i to be so large that there exists (use (26.4)) a set $F_i \subset S_i \cap H_{i-1}$ which is at the same time a (Ψ, q_i, ξ_i)- Rohlin set and a $(\Psi_{H_{i-1}}, n_i)$-Rohlin set.

We abbreviate

$$\alpha_i = (\gamma_i)_0^{q_i-1}, \ \beta_i = (\gamma \cap G_{i-1})_o^{q_i-1},$$

and find the number $k_i \in \mathbb{N}$ $(q_i > n_i > k_i)$ with

(7'a) $\dfrac{3\tau_i}{m(H_{i-1})h(\overline{M})} < \dfrac{k_i}{n_i}$; (7'b) $\dfrac{k_i+c}{n_i} < \dfrac{5\tau_i}{m(H_{i-1})h(\overline{M})}$

This is possible by (5')

By the choice of n_i, k_i and F_i, the sets I_i to IX_i and G_i, H_i, Z_i are determined (part I of the proof). In IX_i names again are chosen arbitrarily. $\gamma|VIII_i$ will be found such that

(8') $\alpha_i \cap F_i \subset \Sigma_\Psi(\gamma \cap G_i)$

and hence $\gamma_i \cap \bigcup_{j=o}^{q_i-1} \Psi^j F_i \subset \Sigma_\Psi(\gamma \cap G_i)$. When this holds, we can show as in the first step that the properties a) and b) are satisfied for i.

One could try to make $\gamma_{H_{i-1}}$ even a generator for $\Psi_{H_{i-1}}$. This is possible in case $h(\Psi) = 0$, but impossible in all other cases. For by theorem (26.12) $h(\Psi_{H_{i-1}}) = m(H_{i-1})^{-1} h(\Psi)$. On the other hand γ may have at most $\theta(M)$ atoms. Hence we shall necessarily have $h(\gamma_{H_{i-1}}, \Psi_{H_{i-1}}) \leq \log \theta(M)$ or, since we even want property d), $h(\gamma_{H_{i-1}}, \Psi_{H_{i-1}}) \leq h(M)$. So we have to use the good (almost generating) properties of $\gamma \cap G_{i-1}$. Corollary (26.13) says that

$$h(\gamma,\Psi) \leq h(\gamma \cap (G_{i-1} \cup Z_{i-1}), \Psi) + m(H_{i-1})\, h(\gamma_{H_{i-1}}, \Psi_{H_{i-1}}) \leq$$

$$\leq h(\gamma \cap G_{i-1}, \Psi) +$$

$$+ m(Z_{i-1} \cup H_{i-1})\, h(\gamma \cap Z_{i-1} | Z_{i-1} \cup H_{i-1}, \Psi_{Z_{i-1} \cup H_{i-1}}) +$$

$$+ m(H_{i-1})\, h(M) .$$

In this estimate we must even omit the middle term of the right hand side, because within Z_{i-1} we do not take care that γ should have other properties than good names; and since in the essential part VIII_i we want to use only $\overline{M}$-names, a necessary condition for our proof is

$$h(\gamma \cap G_{i-1}, \Psi) \geq h(\Psi) - m(H_{i-1})\, h(\overline{M}),$$

which we have by property a). We even have the ">" relation because some part of H_{i-1} is needed for the coding techniques exhibited in part I of the proof.

The use of the properties of $\gamma \cap G_{i-1}$ is as follows: For any atom A' of α_i with $A' \subset S_i$, there is an atom $A'' \subset S_i''$ of β_i with $A' \subset A''$ (since α_i refines β_i), and

$$m(A') \geq m(A'') \exp\{-q_i[h(\gamma_i,\Psi)-h(\gamma \cap G_{i-1},\Psi) + 2\tau_i]\}.$$

Therefore in an atom $A'' \in \beta_i|S_i''$ the number of atoms $A' \in \alpha_i|S_i'$ is at most

$$\exp\{q_i[h(\gamma_i,\Psi) - h(\gamma \cap G_{i-1},\Psi) + 2\tau_i]\} =$$

$$= \exp\{q_i[h(\gamma_i,\Psi) - h(\Psi) + m(H_{i-1})\cdot h(\overline{M}) - (f_i-2)\tau_i]\} \leq$$

$$\leq \exp\{q_i[m(H_{i-1})\cdot h(\overline{M}) - 6\tau_i]\} \geq$$

$$\leq \exp\{n_i[h(\overline{M}) - m(H_{i-1})^{-1} 5\tau_i]\} \leq \exp\{(n_i-k_i-c)h(\overline{M})\}$$

by a calculation similar to the one in step 1.

Since $\beta_i \subset \Sigma_\Psi(\gamma \cap G_{i-1}) \subset \Sigma_\Psi(\gamma)$ and $F_i \subset \Sigma_\Psi(\gamma)$ (by part I), $\beta_i \cap F_i \subset \Sigma_\Psi(\gamma)$, no matter how we fix $\gamma|VIII_i$. This means that we are through if we are able to recognize the different atoms of $\alpha_i|F_i$ which belong to the same atom of $\beta_i|F_i$. But by the estimate above, for fixed $A'' \cap F_i$ we can adjoin to every $A' \cap F_i \subset A'' \cap F_i$ a different $\overline{M}$-block of length $n_i - k_i - c$. If this block is $P = (p_0,\ldots,p_{n_i-k_i-c-1})$, we put

$$\Psi_{H_{i-1}}^{k_i-c+j}(A' \cap F_i) \subset \gamma_{p_i} \quad (0 \leq j < n_i - k_i - c),$$

as we did in step 1. Here it does not matter that we use the induced transformation, since $H_{i-1} \in \Sigma_\Psi(\gamma)$. □

A consequence of the proof of theorem (28.1) is the following proposition which we leave as an exercise to the reader:

<u>(28.2) Proposition:</u> Let $(\mathfrak{X},m,\Psi)$ be ergodic and $\alpha_o = (A_o,\ldots,A_s)$ a partition such that $m(A_o) > 0$. Let $\eta = m(A_o)^{-1}[h(\Psi) - h(\alpha_o,\Psi)]$.

a) If $n \in \mathbb{N}$ and $\log n > \eta$, then there is a partition $(B_1,\dots,B_n)$ of A_o such that $\alpha = (B_1,\dots,B_n,A_1,\dots,A_s)$ is a generator for Ψ.

b) If $\log s > \eta$ and α_o has the property: For any partition $\beta = (B_1,\dots,B_s)$ of A_o and $\alpha_\beta = (A_1 \cup B_1,\dots,A_s \cup B_s)$ one has $\alpha_o \subset \Sigma_\Psi(\alpha_\beta)$, then there exists a partition β of A_o with s elements such that α_β is a generator for Ψ.

We show now that the generator theorem can be improved in the case of an ergodic measure ν on a top.dynamical system (X,T). Recall that $\alpha = (A_i)_{i\in I}$ is a ν-continuity partition if for all i $A_i \in B^o_\nu$, i.e. $\nu(\mathrm{bd}\, A_i) = 0$. We write $\mathrm{bd}\,\alpha = X \setminus \bigcup_i \mathrm{int}\, A_i$ (this is not $\bigcup \mathrm{bd}\, A_i$ in general!). Thus α is a ν-continuity partition if and only if $\nu(\mathrm{bd}\,\alpha) = 0$. We call α open if every $A_i \in \alpha$ is open (an open partition is always a ν-continuity partition). Let $X_\alpha = X \setminus \bigcup_{i\in\mathbb{Z}} T^i\, \mathrm{bd}\,\alpha = \bigcap_{i\in\mathbb{Z}} T^i(X \setminus \mathrm{bd}\,\alpha)$. If α is open, this is just the set where $\Phi_\alpha x$ is defined. It is clear that X_α is a G_δ, $\nu(X_\alpha) = 1$ if $\nu(\mathrm{bd}\,\alpha) = 0$, X_α is dense if ν is positive on all nonempty open sets and $\nu(\mathrm{bd}\,\alpha) = 0$, and that Φ_α is continuous on X_α (see also (15.2)). On the other hand we are interested in the continuity of the inverse mapping Φ_α^{-1} on a subset of Λ_α. Let us call for the moment $g(x,s)$ the atom of $(\alpha)^s_{-s}$ to which x belongs. For a point $x \in X_\alpha$ it is clear that $\Phi_\alpha^{-1}|\Phi_\alpha(X_\alpha)$ is uniquely defined and continuous at $\Phi_\alpha x$ if and only if

$$\lim_{s\to\infty} \mathrm{diam}\; g(x,s) = 0.$$

Therefore we consider

$$X'_\alpha = \{x \in X \mid \lim_{s\to\infty} \operatorname{diam} g(x,s) = 0\}$$

X'_α is invariant (so its measure is 0 or 1), and $\Phi_\alpha : X'_\alpha \longrightarrow X'_\alpha$ is a homeomorphism. We see that X'_α is a G_δ- set if α is open; for $x \longmapsto \operatorname{diam} g(x,s)$ is defined on $X \setminus \operatorname{bd}(\alpha)^s_{-s}$, and $\operatorname{diam} g(x,s) \geq \operatorname{diam} g(x,s+1)$ if $x \in X \setminus \operatorname{bd}(\alpha)^{s+1}_{-(s+1)}$. Hence for any $\varepsilon > 0$ $\{x \in X \mid \exists\, s \in \mathbb{N} : \operatorname{diam} g(x,s) < \varepsilon\}$ is open, and

$$X'_\alpha = \bigcap_{\varepsilon>0} \{x \in X \mid \exists\, s \in \mathbb{N} : \operatorname{diam} g(x,s) < \varepsilon\} \subset X_\alpha$$

is a G_δ-set.

(28.3) Proposition ([51]): If ν is an ergodic, aperiodic measure on a top.dynamical system (X,T), and M an m.f.t.-subshift such that $h_\nu(T) < h(M)$, then there exists an open ν-continuity partition γ of X which is an M-generator and with $\nu(X'_\gamma) = 1$; hence Φ_γ is a homeomorphism on a G_δ-set of measure 1, and on a residual set in the case that ν is positive on all nonempty open sets.

Proof: a) To get a generating ν-continuity partition γ, we have to repeat the proof of (28.1), always watching that all occuring sets are ν-continuity sets. At the end we replace all atoms of γ by their interiors to have an open partition.

The partitions γ_i, with which we begin, can be chosen as ν-continuity partitions, since B^o_ν is dense in B_ν in the metric $\nu(A \,\Delta\, B)$. For all other sets ν-continuity is obtained from the following considerations:

1. B^o_ν is a T-invariant algebra. If $F \in B^o_\nu$, the (F,T)-

tower is a ν-continuity partition.

2. If $A \in \overset{\circ}{B}_\nu$, $A_i \in \overset{\circ}{B}_\nu$ $(i \in \mathbb{N})$ is a sequence of disjoint subsets of A, and $\sum \nu(A_i) = \nu(A)$, then for any subset $K \subset \mathbb{N}$, $\bigcup_{i \in K} A_i \in \overset{\circ}{B}_\nu$.

3. If $A, B \in \overset{\circ}{B}_\nu$ with $B \subset A$ then $T_A B \in \overset{\circ}{B}_\nu$; if $A \in \overset{\circ}{B}_\nu$ and γ is a ν-continuity partition, then $(\gamma_A)_s^t$ is a ν-continuity partition of A.

This comes immediately from 1, 2 and the fact that

$$T_A B = \bigcup_{i=1}^{\infty} T^i\{B \cap [T^{-i}A \setminus \bigcup_{j=1}^{i-1} T^{-j}A]\},$$

$$T_A(A \setminus B) = \bigcup_{i=1}^{\infty} T^i\{(A \setminus B) \cap [T^{-i}A \setminus \bigcup_{j=1}^{i-1} T^{-j}A]\}, \text{ and}$$

$$\nu(T_A B) + \nu(T_A(A \setminus B)) = \nu(A).$$

4. If in lemma (26.4) $A \in \overset{\circ}{B}_\nu$ is open and $Q \in \overset{\circ}{B}_\nu$, then the Rohlin set $F \subset A \cap Q$ can be chosen such that $F \in \overset{\circ}{B}_\nu$ is open as well as $T_A^j F$ $(1 \le j < n)$.

To prove this, we go back to the proof of (26.4). Obviously $B_q \in \overset{\circ}{B}_\nu$. Let $0 < 3\tau < 4^{-1}\epsilon$. We take $\hat{F} \subset B_q \cap Q$ as a $(T_A, n, \delta + 4^{-1}\epsilon + \tau)$-Rohlin set, and t so large that for

$$B_t' = \{x \in \text{int}(B_q \cap Q) \mid \sum_{j=1}^{t-1} 1_A(T^j x) \ge n - 1\}$$

we have $m(\hat{F} \cap B_t') \ge n^{-1} m(A)(1 - \delta - 4^{-1}\epsilon - 2\tau)$. B_t' is open, and it is easy to see that on B_t' all the mappings $T_A, T_A^2, \ldots, T_A^{n-1}$ are continuous. Now we choose $K \subset \hat{F} \cap B_t'$ compact with $m(K) > n^{-1} \cdot m(A) \cdot (1 - \delta - 2^{-1}\epsilon)$, and get some $\delta > 0$ such that $B_\delta(K) \in \overset{\circ}{B}_\nu$, $\overline{B_\delta(K)} \subset B_t'$ and $\overline{B_\delta(K)}$ is a $(T_A, n, 2^{-1}\epsilon + \delta)$-Rohlin set. By the continuity of $T_A, \ldots, T_A^{n-1}$

all the sets $T_A^j B_\delta(K)$ $(1 \leq j < n)$ are open ν-continuity sets. $B_\delta(K)$ is the desired F.

5. As a consequence of 4., we can obtain the sets

$$H_i = \bigcup_{j=l(a)+2L}^{l(a)+2L+k_i-1} T_{H_{i-1}}^j F_i \qquad (i \in \mathbb{N})$$

as open ν-continuity sets.

6. Finally, since $G_1 \subset G_2 \subset G_3 \ldots$, $\nu(G_i) \nearrow 1$ and all the atoms of $\gamma | G_{i+1} \setminus G_i$ will be ν-continuity sets, we conclude from 2. that $\nu(\mathrm{bd}\ \gamma) = 0$.

b) We show that, in order to get also $\nu(X_\gamma') = 1$, it is sufficient to take the γ_i such that $\lim \operatorname{diam} \gamma_i = 0$.

Let us call $\Phi_\gamma(x,t)$ the centred subblock of length $(2t+1)$ of $\Phi_\gamma(x)$ (i.e. the $(\gamma,T,2t+1)$-name of $T^{-t}x$). From the construction of $\gamma \cap G_1$ in the proof of (28.1) it results that, knowing $\Phi_\gamma(x,n_1)$, we can decide

- if $x \in \bigcup_{j=o}^{n_1-1} T^j F_1$ and, if the answer is positive, in which level of the (F,T)-tower x lies. We call this level $j_1(x)$.
- in the case that $x \in \bigcup_{j=o}^{n_1-1} T^j F_1$, in which atom of $(\gamma_1)_o^{n_1-1} | F_1$ $T^{-j_1(x)}(x)$ lies (since $(\gamma_1)_o^{n_1-1} | F_1$ is coarser than $(\gamma)_o^{n_1-1} | F_1)$, and hence to which atom of γ_1 x belongs.

Consequently, $\gamma_1 \cap (X_\gamma \cap \bigcup_{j=o}^{n_1-1} T^j F_1) \subset$

$$\subset (\gamma)_{-n_1}^{n_1} \cap (X_\gamma \cap \bigcup_{j=o}^{n_1-1} T^j F) \subset (\gamma)_{-n_1}^{n_1} \cap X_\gamma .$$

For the higher steps we remark first that from the construction of the Rohlin sets F_i in part a 4) of the proof it follows that there are numbers $t_i \in \mathbb{N}$ with the property

$$\bigcup_{j=o}^{t_i} T^j F_i \supset \bigcup_{j=o}^{n_i-1} T^j_{H_{i-1}} F_i \quad (i \geq 2),$$

that means that in the time interval $\langle 1, t_i \rangle$ a point $x \in F_i$ returns at least n_i-1 times to H_{i-1}.

Now we put

$$\bar{t}_1 = n_1$$

$$\bar{t}_i = \bar{t}_{i-1} + t_i + q_i \quad (i \geq 2)$$

and assume that by induction it is known that $G_{i-1} \cap X_\gamma$ and $H_{i-1} \cap X_\gamma$ are $(\gamma)_{-\bar{t}_{i-1}}^{\bar{t}_{i-1}}$ - measurable.

If we know $\Phi_\gamma(x,\bar{t}_i)$, then for any $y \in \{T^{-(t_i+q_i)}x,\ldots,T^{(t_i+q_i)}x\}$ we know $\Phi_\gamma(y,\bar{t}_{i-1})$, and hence we know if $y \in H_{i-1}$ or not.

If $x \in \bigcup_{j=o}^{q_i-1} T^j F_1$ (say $x \in T^{j_i(x)} F_i$), we can follow a string of at least n_i entrances of x into H_{i-1} during the time $\langle -q_i, + t_i \rangle$, and from the coding (by a possible double occurence of the block a along this string) we also get the number $j_i(x)$. But then the $(\gamma_{H_{i-1}}, T_{H_{i-1}}, n_i)$-name and the $(\gamma | G_{i-1}, T, q_i)$-name of $T^{-j_i(x)}(x)$ together indicate to which atom of $(\gamma)_o^{q_i-1} | F_i \quad T^{-j_i(x)}(x)$ belongs. This atom is contained in an atom of

$(\gamma_i)_o^{q_i-1} | F_i$, so we know to which atom of γ_i x belongs. Since we see from the discussion that we can reconstruct $F_i, T_{H_{i-1}} F_i, \ldots, T_{H_{i-1}}^{n_i-1} F_i$ from $(\gamma)_{-\bar{t}_i}^{\bar{t}_i}$, we get the $(\gamma)_{-\bar{t}_i}^{\bar{t}_i} \cap X_\gamma$ - measurability of $G_i \cap X_\gamma$ and $H_i \cap X_\gamma$, which is necessary for the induction, and

$$\gamma_i \cap (X_{\gamma_i} \cap \bigcup_{j=o}^{q_i-1} F_i) \subset (\gamma)_{-\bar{t}_i}^{\bar{t}_i} \cap (X_\gamma \cap \bigcup_{j=o}^{q_i-1} F_i) \subset (\gamma_i)_{-\bar{t}_i}^{\bar{t}_i} \cap X_\gamma .$$

Finally, $\nu(\bigcup_{j=o}^{q_i-1} F_i) > 1 - \xi_i$ and $\xi_i \to 0$; so that for a.e. $x \in X_\gamma$ there are infinitely many i with $x \in \bigcup_{j=o}^{q_i-1} F_i$; but for all these i diam $g(x,\bar{t}_i) \leq$ diam $\gamma_i \to 0$. □

29. Strictly Ergodic Embedding (Theorem of Jewett and Krieger)

(29.1) Definition: Let $\Phi : (\mathfrak{X},m,\Psi) \to (X,\mu,T)$ be an m.t. conjugacy, where $(\mathfrak{X},m,\Psi)$ is an ergodic m.t.dynamical system and (X,T) a strictly ergodic top.dynamical system with the unique invariant measure ν. Then we call Φ a strictly ergodic embedding of $(\mathfrak{X},m,\Psi)$.

It may be thought that such an embedding exists only for a very restricted class of m.t.dynamical systems, and indeed this was the general opinion for a long time. In all examples of strictly ergodic systems one had found, the structure of the transformation was not too complicated. Thus, they all had topological entropy 0, and non-empty point spectrum. This is the case for the torus rotation by an irrational angle, the Sturmian sequences of Hedlund ([85] and Morse, Hedlund [87]) which are isomorphic to such rotations; the Morse sequence (see Kakutani [101]) which has a continuous part in the spectrum and a discrete part consisting of all dyadic rationals, and its generalizations by Kakutani [101] and Keane [105]; the Toeplitz sequences of Jacobs and Keane [98] with pure point spectrum which can be any group of rational numbers; the Sturm-Toeplitz sequences of Grillenberger [71] which can have any countable group of irrational numbers in the pure point spectrum.

The first examples of weakly mixing strictly ergodic systems were given by Jacobs [97], Kakutani (in[201]) and Petersen [153]). The question of Parry, whether a strictly ergodic system had necessarily entropy 0, was answered by Hahn and Katznelson [78] who constructed systems with arbitrarily large finite entropy. A simpler construction which gives strictly ergodic subshifts of the n-shift with entropy arbitrarily close to log n, and also with infinite entropy, is given in Grillenberger [72]. In [73] a K-system is constructed, but this is not the best one can do: with the same method and using Ornstein's theory of the $\bar{d}$-metric of processes, it is possible to obtain a system which is conjugate to a Bernoulli shift (see Grillenberger and Shields [74]).

Most of these examples are constructive in the sense of Turing machines. The machine can write up either a sequence whose orbit closure has the mentioned properties or equivalently the set of all blocks which occur in the system (if the system is a subshift). Though the theorem of Jewett makes the general situation completely clear, the former work still is of interest by this constructive character.

In 1969 came the completely unexpected theorem of Jewett ([100]) that any weakly mixing m.t.dynamical system has a strictly ergodic embedding. This result was then extended by Krieger [118] to the ergodic case, and also with another method by Hansel and Raoult [84]. The corresponding theory for continuous time flows was done by Jacobs [96] and Denker and Eberlein [50].

We give here the proof of Denker [48] for the theorem of Jewett-Krieger, with a systematic treatment of the combinatorial part which is helpful for later generalizations. We restrict ourselves to the case of finite entropy in which all essential difficulties are contained.

<u>(29.2) Theorem:</u> Let $(\mathfrak{X},m,\Psi)$ be an aperiodic ergodic measure theoretic dynamical system, $\overline{M} \subsetneqq M$ two m.f.t.-subshifts and $\overline{\gamma}$ an $\overline{M}$-generator for Ψ. Then, for any $\epsilon > 0$, there exists an M-generator γ for Ψ for which Λ_γ is strictly ergodic and $\|\gamma,\overline{\gamma}\| < \epsilon$.

To prove this theorem, we improve the generator $\overline{\gamma}$ step by step. The main work is contained in the following lemma:

(29.3) Lemma: Let $(\mathfrak{X},m,\Psi)$, M, $\overline{M}$ and $\overline{\gamma}$ be as in the theorem, $\varepsilon > 0$, $r \in \mathbb{N}$. Then there exists an M-generator γ' such that

a) $\|\gamma,\gamma'\| < \varepsilon$

b) $P \in Bl_r(M) \Longrightarrow P \underset{\varepsilon\text{-reg}}{<} \Lambda_{\gamma'}$

c) $P \in Bl_r(M) \Longrightarrow P \underset{d}{<} \Lambda_{\gamma'}$

d) $h(\Lambda_{\gamma'}) < h(\Psi) + \varepsilon$

Also there exist m.f.t.-subshifts $\overline{M}' \subsetneqq M' \subset M$ with $\Lambda_{\gamma'} \subset \overline{M}'$ and such that the properties b), c), d) hold with M' instead of $\Lambda_{\gamma'}$.

Remark: Properties c), d) are not important for the proof of theorem (29.2); we prove them here because from this lemma we shall later derive a version for aperiodic transformations where the corresponding properties are needed essentially. In the proof we put c), d) at the end, so that the reader can omit them.

Proof of (29.3): We prove only properties a) - d) for γ'; when this is done, we can find $\overline{M}'$ and M' by the method used in lemma (26.16).

The construction of γ' follows the lines of the proof of theorem (28.1). The part of γ' which is fixed in the first step will already give the properties a) - d). All later steps serve only to make γ' a generator and remain unmodified as in the proof of (28.1). We can replace all the γ_i used there by $\overline{\gamma}$, since now we have already a generator available.

Enlarging $\overline{M}$ a little bit within M (see (26.17) and (26.16)) we may assume that $Bl_r(\overline{M}) = Bl_r(M)$ and that $h(\overline{M}) > h(\Psi)$.

Let $a < M$, $a \nless \overline{M}$ be a block and the M-transition length $L \in \mathbb{N}$ so large that

(1) for $P < \overline{M}$ there exist $U_P^1, U_P^2 \in Bl_L(M)$ such that

$P\cdot U_P^1\cdot a < M, \quad a\cdot U_P^2\cdot P < M$ and

$a < P\cdot U_P^1\cdot a$ only as final part;

$a < a\cdot U_P^2\cdot P$ only as initial part.

For the connections with the block a we use only transition blocks of these both types. Since we restrict ourselves to this first step, we do not explicit the subscript 1 at the letters $\tau, f, \delta, \varepsilon, \eta, \xi, k, n, F, S$, I-IX, most of which occured already in the proof of (28.1) and have here almost the same meaning.

(2) Let $h(\bar{M}) = h(\Psi) + f\cdot\tau$, where $\tau > 0$ and $f \geq 8$ is so large that

(3) $$\tau^{-1}h(\bar{M}) = [h(\bar{M}) - h(\Psi)]^{-1}\cdot h(\bar{M})\cdot f > 18,$$

and $\delta, \eta > 0$ so small that

(4) $$\frac{\delta}{\tau} < \eta < \frac{\varepsilon}{10} \wedge \frac{\tau}{h(\bar{M})}$$

(5) Let further ξ be so small that

(5a) $$\xi < \frac{\tau}{2h(\bar{M})} \wedge \frac{\varepsilon}{5}$$

(5b) $$\bar{\gamma} \overset{\xi}{\subset} \rho \Longrightarrow h(\rho,\Psi) > h(\Psi) - \delta$$

(6) and $s \in \mathbb{N}$ with (6a) $2s^{-1}(L+r) < 10^{-1}\varepsilon$ so large that for any $t \geq s$ there exists a block $Q_t \in Bl_t(M)$ with

(6b) $$P \in Bl_r(M) \Rightarrow \begin{cases} P < Q_t \\ |\mu_{Q_t}(P) - \mu_{\bar{\gamma}}(P)| < \frac{\varepsilon}{10}. \end{cases}$$

For every $t \geq s$ we fix such a block Q_t.

(7) Let $c = 2(2L + l(a))$; choose the numbers $n, k \in \mathbb{N}$, $k < n$ such that

(7a) $\frac{c}{k} < \frac{\tau}{h(\overline{M})}$, (7b) $\frac{\delta}{\tau} < \frac{k+c}{n} < \eta$, (7c) $\frac{L+s}{n} < \frac{\xi}{8}$,

and, if we put

$$S = \bigcup [\text{atoms } A' \text{ of } (\overline{\gamma})_0^{n-1} \mid m(A') \geq \exp\{-n(h(\Psi)+10^{-1}\epsilon)\},$$
$$\text{and for } x \in A',\ Q \text{ the } (\overline{\gamma},\Psi,n)\text{-name of } x,$$
$$P \in Bl_r(M) : |\mu_Q(P) - \mu_{\overline{\gamma}}(P)| < 5^{-1}\epsilon]$$

$$\cap \bigcup [\text{atoms } A'' \text{ of } (\overline{\gamma})_0^{k+c-1} \mid m(A'') \geq \exp\{-(k+c)(h(\Psi)+\tau)\}]$$

(7d) then $m(S) > 1 - 8^{-1}\xi$.

Further we require that $L+s$ divides n and that (this serves only to get property d))

(7e) $n^{-1}(2L \log \theta(M) + \theta(M) \log(k+c) + \log 2(n+L+s)) < 4^{-1}\epsilon$.

We call the $(\overline{\gamma},\Psi,n)$-names of points $x \in S$ good blocks, and the atoms of $(\overline{\gamma})_0^{n-1}$ which belong to such blocks, good atoms.

Let $F \subset S$ be a $(\Psi, n+s+L, 8^{-1})$-Rohlin set. Because of (7c),

(8) $m(\bigcup_{j=0}^{n-1} \Psi^j F) > 1 - 4^{-1}\xi$.

Note that

(9) $(\overline{\gamma})_0^{k+c-1} | F$ has at most $\exp(k+c)(h(\Psi)+\tau)$ atoms.

The partition (I,...,IX) of $\mathfrak{X}$ is defined in the same way as in (28.1) with the (F,Ψ)-tower. Part IX is of great importance this time. The new blocks we put in there must not be too many, because the topological entropy of $\Lambda_{\gamma'}$ is not allowed to be large by d); also, in these new blocks all r-blocks must have good frequencies, in order that we have b), c). We do this as follows:

Since F is a $(\Psi, n+s+L)$-Rohlin set, every upcrossing through IX has a length $w \geq L + s$, where w is not constant and perhaps unbounded. The interval $n, n+1, \ldots, n+w-1$ of $\mathbb{Z}$ is divided into intervals of length between $L + s$ and $2(L+s) - 1$, but so that only the last interval may have a length $> L + s$

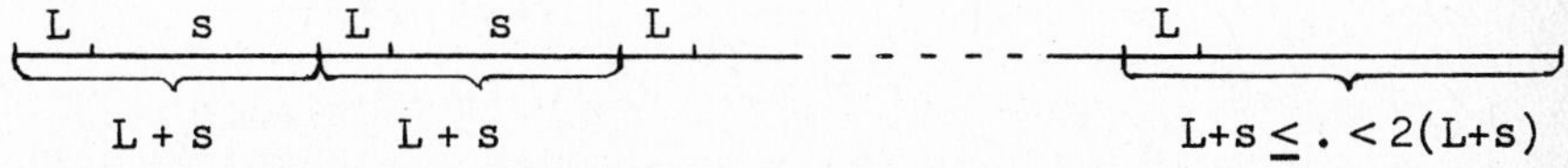

The initial parts of these intervals serve only for $\overline{M}$-compatible transitions. Into all s-intervals we put Q_s, and into the last remaining interval some block Q_t, according to its length. Now we are able to fix all transition blocks except for the first interval, and in doing this we always use the same transition between the blocks Q_s. We note that at this moment we have already property c) for γ' since, independently of what we shall do later, in any block Q of length $n + 4(s+L)$ with $\mu_{\gamma'}(Q) > 0$, all r-blocks of M will appear, because Q_s or one of the blocks Q_t appears, and because of (6b).

In part VIII $\overline{\gamma}$ is not changed, i.e. $\gamma'_{VIII} = \overline{\gamma}_{VIII}$. Then property a) is valid for γ', since

$$m(VIII) = (n-k-c)m(F) \geq (1-4^{-1}\xi)n^{-1}(n-k-c) \geq (1-4^{-1}\xi)(1-\eta)$$

$$\geq 1 - 4^{-1}\xi - \eta > 1 - 5^{-1}\epsilon$$

For $Q_1, Q_2 \in Bl(\Lambda_{\gamma'})$ which are sufficiently long, it is possible to see that

$$|\mu_{Q_1}(P) - \mu_{Q_2}(P)| < \epsilon \quad (P \in Bl_r(M)),$$

which shows that γ' will have property b). For, by the conditions imposed on S and since $F \subset S$, for any $x \in F$ the $(\bar{\gamma},\Psi,n)$-name of x has good frequencies for all r-blocks. These names are changed only over an interval of length $k + c$, and $n^{-1}(k+c) < \eta < 10^{-1}\varepsilon$.

In part IX only blocks UQ_t are put in, where Q_t has good frequencies, $l(U) = L$, $t \geq s$ and $s^{-1}(L+2r) < 10^{-1}\varepsilon$. We leave the exact estimates to the reader.

What we need in order that $\gamma'|G_1$ (as before $G_1 = \mathrm{I} \cup \mathrm{II} \cup (\mathrm{VI}$ to $\mathrm{IX})$, $H_1 = \mathrm{IV}$) can be extended to a generator γ' is the entropy inequality

$$h(\gamma' \cap G_1,\Psi) > h(\Psi) - m(H_1)\cdot h(\bar{M}).$$

To check this we put $R = \bigcup_{j=0}^{k+c-1} \Psi^j F$ (= I to VII).

Then
$$h(\bar{\gamma}_R,\Psi_R) \leq h((\bar{\gamma})_0^{k+c-1} \cap F|R,\Psi_R)$$
$$\leq m_R(F)\, h((\bar{\gamma})_0^{k+c-1}|F,\Psi_F) + h((F,R\setminus F),\Psi_R)$$
$$\leq m_R(F)\, H((\bar{\gamma})_0^{k+c-1}|F) \leq h(\Psi) + \tau \quad \text{by (26.13) and (9).}$$

Using $\delta < \tau \cdot n^{-1}(k+c) < 2m(R)\tau$, (5b), (7b), (7a), (2), (26.13) and the fact that

$$[\bar{\gamma} \cap (\mathfrak{X}\setminus \mathrm{IX})] \cap (\mathfrak{X}\setminus R) = (\bar{\gamma} \cap \mathrm{VIII}) \vee (R,\mathrm{VIII},\mathrm{IX}) \subset$$
$$\subset \Sigma_\Psi(\gamma' \cap G_1),$$

we obtain

$$h(\Psi) < h(\bar{\gamma}\cap(\mathfrak{X}\setminus \mathrm{IX}),\Psi)+\delta \leq$$
$$\leq h([\bar{\gamma}\cap(\mathfrak{X}\setminus \mathrm{IX})] \cap (\mathfrak{X}\setminus R),\Psi) + m(R)\, h(\bar{\gamma}_R,\Psi_R) + \delta \leq$$
$$\leq h(\gamma'\cap G_1,\Psi) + m(R)\cdot(h(\Psi) + 3\tau) \leq h(\gamma'\cap G_1,\Psi) + m(H_1)\cdot h(\bar{M}).$$

Finally, property d) has to be checked. The number of possible (γ',Ψ,n)-names of points $x \in F$ is at most

$$\exp\{n[h(\Psi) + 10^{-1}\varepsilon]\} \cdot \theta(M)^{k+c}$$

by the condition for S, since these names coincide with $(\overline{\gamma},\Psi,n)$-names except on the initial piece of length $k + c$.

For the moment, we call the pieces of (γ',Ψ)-names corresponding to upcrossings through IX "IX-names". From these IX-names we cut successively blocks of length n, as long as it is possible without overlapping one of the final parts $U \cdot Q_t$. All the blocks of length n we get in this way are of the form $\tilde{Q} = U' \cdot Q_s \cdot U \cdot Q_s \cdot \ldots \cdot U \cdot Q_s$, since $L + s$ divides n. By the conformity of the blocks (only the initial U' may vary), the number of these $\tilde{Q}$ is at most $\theta(M)^L$. The ends of the IX-names are of length at most $n + 2(s + L)$ and for each length the corresponding block is unique (except perhaps the initial transition if we were not able to cut off anything). So we have at most $(n + 2(L+s)) \cdot \theta(M)^L$ blocks of this type, which we call "irregular". Splitting up a (γ',Ψ)-name into n-blocks and irregular blocks as above, we see that an irregular block always lies between two n-blocks. So, for very large N, the number of (γ',Ψ,N)-names can be estimated by

$$[\exp\{n[h(\Psi) + 10^{-1}\varepsilon]\} \cdot \theta(M)^{k+c} + \theta(M)^L]^{n^{-1}N+1} \cdot$$

$$\cdot [(n+2(L+s))\, \theta(M)^L]^{n^{-1}N}\, 2(n+L+s).$$

(The last factor comes in because the beginning of the (γ',Ψ,N)-name does not always coincide with the beginning of one of the n-blocks or the irregular blocks). Using (7e), we get easily property d). □

<u>Proof of theorem (29.2)</u>: By recursion we construct a sequence $\bar{\gamma} = \gamma_o, \gamma_1, \gamma_2, \ldots$ of generators for Ψ with $\|\gamma_{i-1}, \gamma_i\| < \varepsilon_i$, where ε_i tends to zero so quickly that $\gamma := \lim \gamma_i$ is again a generator (see lemma (26.10)). To obtain the strict ergodicity of Λ_γ, we also construct m.f.t.-subshifts $M_i, \overline{M}_i$ $(M_o = M, \overline{M}_o = \overline{M})$ such that

$$\Lambda_{\gamma_i} \subset \overline{M}_i \subsetneqq M_i \subset M_{i-1}$$

and

$$P \in \bigcup_{j \leq i} Bl_j(M) \Rightarrow P \underset{2^{-i}\text{reg}}{<} M_i .$$

By (26.7), this makes $\bigcap_i M_i$ uniquely ergodic, and since $\Lambda_\gamma \subset \bigcap M_i$, μ_γ is the unique invariant measure on $\bigcap M_i$, so its support Λ_γ is strictly ergodic. □

30. Finite Generators for Aperiodic Transformations

It is clear that an analogue of Krieger's generator theorem must be valid in the case of an aperiodic transformation, although the entropy condition can not simply be $h(\Psi) < h(M)$, but must bound the entropy $h_x(\Psi)$ on a.e. ergodic fibre of $\mathfrak{X}$. This theorem has not been written up so far. The proof is the same as for theorem (28.1), but it offers some additional technical difficulties which must be treated carefully. Roughly speaking, they come from the applications of convergence theorems (ergodic theorem and theorem of Shannon-McMillan-Breiman) which do not give uniform speed of convergence on all fibres.

(30.1) Theorem: Let $(\mathfrak{X},\Sigma,m,\Psi)$ be an aperiodic measure theoretic dynamical system and M an m.f.t.-subshift such that for almost all $x \in \mathfrak{X}$:

$$h_x(\Psi) < h(M)$$

or, equivalently for any Ψ-invariant set $D \in \Sigma$ with $m(D) > 0$:

$$h(\Psi_D) < h(M).$$

Then there exists an M-generator γ for Ψ.

Proof: First (26.17) allows us to find a sequence (M_i) of disjoint m.f.t.-subshifts $M_i \subset M$ with $h(M_i) < h(M_{i+1}) \nearrow h(M)$. Let $Y_i = \{x \in \mathfrak{X} \mid h_x(\Psi) \leq h(M_{i-1})\}$ $(i \geq 2)$. It is sufficient to show that there exist invariant sets $\overline{Y}_i \subset Y_i$ with $m(\overline{Y}_i) > m(Y_i)\,(1 - 2^{-i})$ and M_i-generators ρ_i for $\Psi_{\overline{Y}_i}$. For in this case $m(\bigcup_{i\geq 2} \overline{Y}_i) = 1$, and if we put

$$\gamma|\overline{Y}_i \setminus \bigcup_{1<j<i} \overline{Y}_j = \rho_i|\overline{Y}_i \setminus \bigcup_{1<j<i} \overline{Y}_j,$$

then γ is an M-generator for Ψ. Therefore we may restrict ourselves to the following

(30.2) Proposition: If $h(M) > h_{sup}(\Psi)$ and $0 < \varepsilon < 1$, then

there are an invariant set $Y \subset \mathfrak{X}$ with $m(Y) > 1 - \varepsilon$ and an M-generator γ for $(Y, \Sigma_Y, m_Y, \Psi_Y)$.

Proof of (30.2): We take a sequence (ε_i) of positive numbers with $\sum \varepsilon_i < 3^{-1}\varepsilon$, an increasing sequence (γ_i) of finite partitions with $\bigvee_{i\in\mathbb{N}} \Sigma_\Psi(\gamma_i) = \Sigma$, and an m.f.t.-system $\overline{M} \subsetneqq M$ with $h(\overline{M}) > h_{sup}(\Psi)$.

We construct sets X_i, G_i, Z_i, H_i $(i \geq 1)$ such that X_i is Ψ-invariant, $X_{i+1} \subset X_i$, $m(X_{i+1}) \geq m(X_i) - 3\varepsilon_i$ (where $X_o = \mathfrak{X}$);

G_i, Z_i, H_i are disjoint

$G_i \cup Z_i \cup H_i = X_i$

$G_i \cap X_{i+1} \subset G_{i+1}$

$H_{i+1} \subset H_i \cap X_{i+1}$

G_i, Z_i, H_i are m_{X_i}-independent of all Ψ_{X_i}-invariant sets, i.e. $m_x(G_i) = m_{X_i}(G_i)$ a.e. on X_i, and similarly for Z_i and H_i.

$m_{X_i}(G_i \cap X_i) \nearrow 1.$

When this is done, Y will be the set $\bigcap_i X_i$.

In the i-th step we determine $\gamma|G_i$ provisionally - provisionally in so far as in the next step $\gamma|G_i \setminus G_{i+1}$ has to be left away. In each step we take care of the following properties (we abbreviate

$$\eta_i = \operatorname*{ess\,inf}_{x\in X_i} [h_x(\gamma \cap G_i,\Psi) - h_x(\Psi) + m_x(H_i)\cdot h(\overline{M})]) :$$

a) $\eta_i > 0$

b) $\gamma_i|X_i \overset{\epsilon_i}{\subset} \Sigma_{\Psi_{X_i}}(\gamma \cap G_i)$

c) $\gamma|G_i$ is such that every extension to a partition $\hat{\gamma}$ of X_i satisfies $(G_i,Z_i,H_i) \subset \Sigma_{\Psi_{X_i}}(\hat{\gamma})$

d) $m_{H_i\cap Y}$-almost all $(\gamma_{H_i\cap Y},\Psi_{H_i\cap Y})$-names are in M.

Part I: Properties c) and d).
We use the same coding method as in the proof of (28.1). The $\Psi_{H_{i-1}}$-Rohlin set $F_i \subset H_{i-1}$ is always chosen such that $m_x(F_i) = \text{const}$ for all $x \in \bigcup_{j\in\mathbb{N}} \Psi^j F_i$. The numbers k_i, n_i will not depend on $x \in X_{i-1}$, therefore $m_x(G_i)$, $m_x(Z_i)$, $m_x(H_i)$ are independent of $x \in X_i$. L and c are the same as before.

Part II: Properties a) and b)
If in the proof of theorem (28.1) we put $G_o = Z_o = \emptyset$, $H_o = \mathfrak{X}$, the first step is a special case of the induction step; it served only to explain the procedure. Here we only show the general step.

So, let $i \geq 1$ and suppose the proof has come to step $(i - 1)$; i.e. we have determined $\gamma \cap G_{i-1}$ and H_{i-1} in such a way that a) b) c) hold. We put

(1) $\eta_{i-1} = f_i \tau_i$

with $\tau_i > 0$ and $f_i \geq 11$ so large that

(2) $\tau_i^{-1} h(\overline{M})\, m_{X_{i-1}}(H_{i-1}) = \eta_{i-1}^{-1} f_i\, h(\overline{M})\, m_{X_{i-1}}(H_{i-1}) > 18.$

To be able to use a), we must split up X_{i-1} into invariant parts on which $h_x(\Psi)$ oscillates very little in comparison with η_{i-1}. Let

$$\pi_i = (p_i^o, \dots, p_i^{R_i})$$

be the Ψ-invariant partition of $\mathfrak{X}$ into the sets

$$p_i^o = \mathfrak{X} \setminus X_{i-1}$$

$$p_i^r = \{x \in X_{i-1} \mid (r-1)\cdot\tau_i \leq h_x(\Psi) < r\cdot\tau_i\}$$

$$(0 < r \leq \lceil \tau_i^{-1}\, h_{sup}(\Psi) \rceil =: R_i)$$

(Some of the p_i^r may be empty; we omit them from the following consideration.) Since we may refine γ_i, we assume

$$(\gamma \cap G_{i-1}) \vee \pi_i \subset \gamma_i$$

$$h(\gamma_i, \Psi) > h(\Psi) - \epsilon_i\, \tau_i.$$

Let $\overline{X}_i = \{x \in X_{i-1} \mid h_x(\gamma_i, \Psi) > h_x(\Psi) - \tau_i\}$.
Then $m(\overline{X}_i) \geq m(X_{i-1}) - \epsilon_i$. On $\overline{X}_i$ we have the partition

$$\overline{\pi}_i = (\overline{p}_i^1, \dots, \overline{p}_i^{R_i}) \quad \text{with } \overline{p}_i^j = p_i^j \cap \overline{X}_i.$$

Now we choose

(3a) $\xi_i < \epsilon_i \wedge h(\overline{M})^{-1} \tau_i$ so small that

(3b) $\quad \rho|\overline{X}_i \overset{\xi_i}{\supset} \gamma_i|\overline{X}_i \Rightarrow h(\rho|\overline{X}_i,\Psi_{\overline{X}_i}) \geq h(\gamma_i|\overline{X}_i,\Psi_{\overline{X}_i}) - \epsilon_i\tau_i$

and numbers $q_i > n_i > k_i \in \mathbb{N}$ such that

(4a) $\quad R_i < \exp(q_i\tau_i)$, $\qquad$ (4b) $\quad n_i = \lceil m_{\overline{X}_i}(H_{i-1})\cdot q_i \rceil \geq 6$,

(4c) $\dfrac{c}{n_i} < \dfrac{\tau_i}{3m_{\overline{X}_i}(H_{i-1})h(\overline{M})}$, (4d) $\dfrac{4\tau_i}{m_{\overline{X}_i}(H_{i-1})h(\overline{M})} < \dfrac{k_i}{n_i}$,

(4e) $\quad \dfrac{k_i+c}{n_i} < \dfrac{5\tau_i}{m_{\overline{X}_i}(H_{i-1})h(\overline{M})}$,

and such that, if we define

$$S_i' = \bigcup_{r\geq 1} \bigcup \{\text{atoms } A' \text{ of } [(\gamma_i)_o^{q_i-1} | \overline{p}_i^r] \,|\, \exp[-q_i(r+1)\tau_i] \leq m(A')\},$$

$$S_i'' = \bigcup_{r\geq 1} \bigcup \{\text{atoms } A'' \text{ of } [(\gamma\cap G_{i-1})_o^{q_i-1} | \overline{p}_i^r] \,|$$

$$m(A'') \leq \exp[-q_i((f_i+r-2)\tau_i - m_{\overline{X}_i}(H_{i-1})h(\overline{M}))]\} ,$$

$S_i = S_i' \cap S_i''$, $\qquad$ we have from (13.5)

(5) $\quad m_{\overline{X}_i\cap H_{i-1}}(S_i) > 1 - 4^{-2}\,\xi_i^2$

(for the conditions on S_i'' notice that for $x \in \overline{p}_i^r$:

$$h_x(\gamma \cap G_{i-1},\Psi) \geq \eta_{i-1} + (r-1)\tau_i - m_{\overline{X}_i}(H_{i-1})h(\overline{M}));$$

furthermore q_i is supposed to be so large that we can apply (26.5) with $\overline{X}_i$ instead of $\mathfrak{X}$, $\overline{X}_i \cap H_{i-1}$ instead of A, S_i for Q, and $4^{-1}\xi_i$ instead of δ and ϵ. The invariant set $\hat{X}$ of (26.5) will be called $\hat{X}_i$, and the Rohlin set F_i. So we have

(6a) $m(\hat{X}_i) > (1 - \frac{1}{2}\xi_i)\, m(\bar{X}_i) \geq m(\bar{X}_i) - \frac{1}{2}\xi_i$,

(6b) $F_i \subset S_i \cap \hat{X}_i \cap H_{i-1}$;

(6c) F_i is a uniform $(\psi_{\hat{X}_i}, q_i, 1 - \frac{1}{2}\xi_i)$-Rohlin set;

(6d) F_i is a uniform $(\psi_{\hat{X}_i \cap H_{i-1}}, n_i, 1 - \frac{1}{2}\xi_i)$-Rohlin set.

By (6a), (6c), (3b) $h(\gamma_i \cap \bigcup_{j=o}^{q_i-1} \psi^j F_i | \hat{X}_i, \psi_{\hat{X}_i}) \geq$

$\geq h(\gamma_i | \bar{X}_i, \psi_{\bar{X}_i}) - \epsilon_i \tau_i$, and if we put

(7) $$X_i = \{x \in \hat{X}_i | h_x(\gamma_i \cap \bigcup_{j=o}^{q_i-1} \psi^j F_i, \psi) \geq h_x(\gamma_i, \psi) - \tau_i\},$$

then $m(X_i) \geq m(\hat{X}_i) - \epsilon_i > m(X_{i-1}) - 2\epsilon_i - \xi_i > m(X_{i-1}) - 3\epsilon_i$.

Noting that for good atoms

$$A' \in (\gamma_i)_o^{q_i-1} | \bar{P}_i^r; \quad A'' \in (\gamma \cap G_{i-1})_o^{q_i-1} | \bar{P}_i^r$$

(an atom is good if it belongs to S_i' resp. S_i''), we obtain

$$m(A'')^{-1} m(A') \geq \exp\{-q_i[m_{\bar{X}_i}(H_{i-1})h(\bar{M}) - (f_i-3)\tau_i]\},$$

we see that the number of good atoms A' contained in some atom of $(\gamma \cap G_{i-1})_o^{q_i-1}$ is bounded by

$R_i \exp\{q_i[m_{\bar{X}_i}(H_{i-1})h(\bar{M}) - (f_i-3)\tau_i]\} \leq \exp\{(n_i - k_i - c)h(\bar{M})\}$;

the last estimate comes from (4a,b,e) and $f_i \geq 11$:

$$q_i^{-1} n_i = q_i^{-1} \lceil m_{\bar{X}_i}(H_{i-1}) q_i \rceil \leq \frac{6}{5} m_{\bar{X}_i}(H_{i-1}), \quad \text{so}$$

$$q_i[m_{\bar{X}_i}(H_{i-1})h(\bar{M}) - (f_i - 4)\tau_i] \leq$$

$$\leq n_i \cdot h(\bar{M}) \cdot (1 - \frac{6\tau_i q_i}{h(\bar{M}) n_i}) \leq n_i h(\bar{M}) \cdot (1 - \frac{k_i + c}{n_i}).$$

Now we define $G_i, Z_i, H_i \subset X_i$ exactly as in the proof of theorem (28.1) and determine $\gamma | G_i$ in such a way that

$$(8) \quad \gamma_i \mid \bigcup_{j=o}^{q_i-1} \psi^j F_i \subset \Sigma_{\psi_{X_i}} (\gamma_i \cap G_i | X_i);$$

the calculation above shows that we have enough blocks in $Bl_{n_i-k_i-c}(\bar{M})$ to do this. Then we have properties b) and c). a) still must be checked: For $x \in X_i$,

$$h_x(\gamma \cap G_i, \psi) = h_x(\psi) + m_x(H_i) \cdot h(\bar{M}) \geq$$

$$\geq h_x(\gamma_i, \psi) - \tau_i - h_x(\psi) + \frac{k_i}{n_i} \frac{n_i}{q_i} h(\bar{M}) -$$

$$- \left(\frac{k_i}{q_i} - m_x(H_i)\right) h(\bar{M}) \geq$$

$$\geq - 2 \tau_i + \frac{k_i}{n_i} \cdot m_{\bar{X}_i}(H_{i-1}) \cdot h(\bar{M}) - \tau_i \geq \tau_i > 0$$

by (8), (7), (4b,d), (3a), and since from

$$k_i \cdot m_{X_i}(F_i) = m_{X_i}(H_i)$$

$$q_i \cdot m_{X_i}(F_i) \geq 1 - \frac{1}{2} \xi_i$$

it follows that $\frac{k_i}{q_i} \leq (1 - \frac{1}{2} \xi_i)^{-1} m_{X_i}(H_i) \leq$

$$\leq m_{X_i}(H_i) + \xi_i \leq m_{X_i}(H_i) + h(\bar{M})^{-1} \tau_i .$$

Thus the proof is finished. $\square$

31. Embedding Theorems for Aperiodic Transformations

To obtain a generalization of the theorem of Jewett-Krieger (29.2) to aperiodic systems, there are two possibilities. Considering the transformation as a bundle of ergodic transformations, it is natural to look for a top. dynamical system which is a bundle of strictly ergodic systems in the sense that every point has a strictly ergodic orbit closure. **We call this a pointwise strictly ergodic embedding. G. Hansel went this way (see [83]). On the other hand, one may try to preserve the minimality of the image system.**

In this section we construct both kinds of embeddings (not in full generality, bien entendu, but with the restriction that $h_{sup}(\Psi) < \infty$). We shall see that they are closely related, since the existence of both follows from a lemma concerning the improvement of generators, similar to the lemma which was the main step in the proof of the theorem of Jewett-Krieger. We also give restrictions on the richness of invariant measures on the compact image space in terms of continuity conditions which will be discussed later.

(31.1) Lemma: Let $(\mathfrak{X}, m, \Psi)$ be aperiodic, $\overline{M} \subsetneq M$ m.f.t.-subshifts, $\overline{\gamma}$ an $\overline{M}$-generator for Ψ, $\varepsilon > 0$, $r \in \mathbb{N}$;

$$(X_i \mid 1 \leq i \leq \lceil \varepsilon^{-1} h_{sup}(\Psi) \rceil + 1)$$

the Ψ-invariant partition of $\mathfrak{X}$ into the sets

$$X_i = \{x \in \mathfrak{X} \mid (i-1)\,\varepsilon \leq h_x(\Psi) < i\varepsilon\},$$

$(X^j \mid 1 \leq j \leq J)$ a Ψ-invariant partition of $\mathfrak{X}$ into sets X^j with

$$x,y \in X^j,\ P \in Bl_r(M) \Rightarrow |\mu_{x,\bar{\gamma}}(P) - \mu_{y,\bar{\gamma}}(P)| < 10^{-1}\varepsilon$$

$$(1 \leq j \leq J)$$

and $X_{ij} = X_i \cap X^j$ (Some of the X_{ij} will be empty; in the following we consider only those (ij) with $m(X_{ij}) > 0$).

Then there exist an M-generator γ' for Ψ and m.f.t.-subshifts M_{ij} such that

a) $\Lambda_{\gamma'|X_{ij}} \subset M_{ij} \subset M$

b) $M_{i_1,j_1} \cap M_{i_2,j_2} = \emptyset \quad ((i_1,j_1) \neq (i_2,j_2))$

c) $\|\bar{\gamma}|X_{ij}, \gamma'|X_{ij}\| < \varepsilon$

d) $P \in Bl_r(M) \Rightarrow P \underset{d}{<} M_{ij}$

e) $P \in Bl_r(M) \Rightarrow P \underset{\varepsilon\text{-reg}}{<} M_{ij}$

f) $h(M_{ij}) < (i+1)\,\varepsilon$

Proof: The relation between lemma (31.1) and theorem (30.1) is the same as that between lemma (29.3) and theorem (28.1). The proof is essentially a combination of those for lemma (29.3) and theorem (30.1). Therefore we go into details only at those points where new considerations are necessary.

1. This time we need a somewhat more complicated coding technique. We use (26.18) to find M-blocks a, b_{ij} with the same length t and m.f.t.-subshifts $N = N_a$ and $N_{b_{ij}}$ such that

$$a < N_a; \quad a \nless \bigcup_{i,j} N_{b_{ij}}; \quad \overline{M} \subset N_a,$$

and similarly for all blocks b_{ij}. L is chosen as a

common transition length for N_a and $N_{b_{ij}}$ and so large that for $Q < \overline{M}$ there are blocks $U_1, U_2 \in Bl_L(N)$ such that

> $QU_1a < N$, $aU_2Q < N$ and a occurs in these composed blocks only as the final resp. the initial block;

if $Q, Q' < \overline{M}$ then for each of the blocks b_{ij} there exists a block $U \in Bl_L(M)$ such that

> $Q \cdot U \cdot Q' < N_{b_{ij}}$,
>
> b_{ij} appears exactly once in U, namely at the place $[\frac{L}{2}]$, and the initial and final t-subblocks of U are $\overline{M}$-blocks.

These blocks serve us to do the following: The different parts of the (F_1, Ψ)-tower have the same task as in the proof of (29.3); but additionally we use the transition intervals within IX to separate the different subshifts $\Lambda_{\gamma'|X_{ij}}$. This is achieved quite simply if, as long as we are in the part X_{ij} of $\mathfrak{X}$, we always take transition blocks U in which b_{ij} appears. The distance between two appearances of b_{ij} in a $(\gamma'|X_{ij}, \Psi)$-name then will be at most $n+2(L+s)$, so that $b_{ij} \underset{d}{<} \Lambda_{\gamma'|X_{ij}}$. Since $\overline{M} \underset{\neq}{\subset} N$, in all later steps we may restrict ourselves to use only N-blocks instead of M-blocks within $H_1 \cup Z_1$. Consequently, in each $\omega \in \Lambda_{\gamma'|X_{ij}}$, $b_{i'j'}$ will not appear for $(i',j') \neq (i,j)$; therefore the different $\Lambda_{\gamma'|X_{ij}}$ will have empty intersections. (Here it is indeed necessary to have $b_{ij} \underset{d}{<} \Lambda_{\gamma'|X_{ij}}$;

for otherwise there would be an $\omega \in \Lambda_{\gamma'|X_{ij}}$ with $b_{ij} \nprec \omega$).

2. Now we consider a given X_{ij}. Since, as in the proof of the generator theorem (30.1) in each step of the construction of γ' we loose a part of the space, at first we shall only get $\gamma'|Y_{ij}$ for some invariant set Y_{ij} with $m(Y_{ij}) > (1 - \frac{\varepsilon}{2})m(X_{ij})$; but on Y_{ij} γ' will be so good that $h(\Lambda_{\gamma'|Y_{ij}}) < (i+1)\varepsilon$, $\|\gamma'|Y_{ij}, \bar{\gamma}|Y_{ij}\| < \frac{\varepsilon}{2}$ and $h_{\sup}(\Psi_{X_{ij}}) - h(\Lambda_{\gamma'|Y_{ij}}) < \frac{\varepsilon}{4}[h(M) - h(\Lambda_{\gamma'|Y_{ij}})]$.

When this is achieved, then by (26.19) we can find an m.f.t.-subshift $M_{ij} \supsetneqq \Lambda_{\gamma'|Y_{ij}}$ such that b) d) e) f) hold for M_{ij}, and $b_{ij} \underset{d}{\prec} M_{ij}$, $b_{i'j'} \nprec M_{ij}$ $((i',j') \neq (i,j))$. (This is possible by the discussion in 1 and makes the different M_{ij} disjoint), and

$$h_{\sup}(\Psi_{X_{ij}}) < h(M_{ij}) < (i+1)\varepsilon;$$

by (26.17) we also get an m.f.t.-subshift $M'_{ij} \subset M_{ij} \setminus \Lambda_{\gamma'|Y_{ij}}$ with

$$h(M'_{ij}) > h_{\sup}(\Psi_{X_{ij}}).$$

Then $\gamma'|X_{ij} \setminus Y_{ij}$ is determined (using (30.1)) as an M'_{ij}-generator for $\Psi_{X_{ij}\setminus Y_{ij}}$, and we have properties a) b) and c).

3. So, knowing how to separate the $\Lambda_{\gamma'|X_{ij}}$ with the blocks b_{ij} and what to do with $X_{ij} \setminus Y_{ij}$, we are able

to restrict ourselves to a single X_{ij} and assume $\mathfrak{X} = X_{ij}$.

Now we give our usual set of constants which essentially is the same as in the proof of (29.3). Again we write only the first step of the proof; the iteration step is exactly the same as in the generator theorem. Hence we may omit the index 1. Some of the conditions are slightly different from those before. Therefore, in order to avoid confusion, we repeat most of what is unchanged:

We have already L, a and $c = 2(2L+l(a))$.
Let $h(\overline{M}) - h_{sup}(\Psi) = f\cdot\tau$, where $\tau > 0$, $f \geq 8$ and

$$\tau^{-1}\cdot h(\overline{M}) = [h(\overline{M}) - h_{sup}(\Psi)]^{-1}\cdot h(\overline{M})\cdot f > 18$$

(We may assume that $\overline{M} \supsetneqq \Lambda_{\overline{\gamma}}$).

Let $\delta,\eta > 0$ be so small that $2\frac{\delta}{\tau} < \eta < \frac{\varepsilon}{10} \wedge \frac{\tau}{h(\overline{M})}$.
Let ξ be so small that

$$\xi < \frac{\tau}{2h(\overline{M})} \wedge \frac{\varepsilon}{5}$$

$$(*) \qquad \overline{\gamma} \overset{\xi}{\subset} \rho \Rightarrow m\{x\in\mathfrak{X} \mid h_x(\rho,\Psi) \geq h_x(\Psi) - \delta\} > 1 - 10^{-1}\varepsilon.$$

(To get this, take ξ so small that

$$\overline{\gamma} \overset{\xi}{\subset} \rho \Rightarrow h(\overline{\gamma},\Psi) - h(\rho,\Psi) = \int (h_x(\Psi) - h_x(\rho,\Psi))\,dm < 10^{-1}\varepsilon\delta.)$$

$s \in \mathbb{N}$ and the blocks $Q_t \in Bl_t(\overline{M})$ $(t \geq s)$ are as in (29.3), (6a,b). Let $k < n \in \mathbb{N}$ be such that

$$\frac{c}{k} < \frac{\tau}{h(\overline{M})},\quad 2\,\frac{\delta}{\tau} < \frac{k+c}{n}\,,\quad \frac{L+s}{n} < \frac{\xi}{8},$$

$$n^{-1}(2L\cdot\log\theta(M) + \theta(M)\cdot\log(k+c) + \log 2(n+L+s)) < 4^{-1}\varepsilon$$

and if we put

$$S = \bigcup\{\text{atoms } A' \text{ of } (\overline{\gamma})_0^{n-1} \mid m(A') \geq \exp[-n(h_{\sup}(\Psi)+10^{-1}\varepsilon)],$$

and for $x \in A'$, Q the $(\overline{\gamma},\Psi,n)$-name of x,

$$P \in Bl_r(M) : |\mu_Q(P) - \mu_{\overline{\gamma}}(P)| < 5^{-1}\varepsilon\}$$

$$\cap \bigcup\{\text{atoms } A'' \text{ of } (\overline{\gamma})_0^{k+c-1} \mid m(A'') \geq \exp[-(k+c)(h_{\sup}(\Psi)+\tau)]\}$$

then $m(S) > 1 - 2^{-8}\xi^2$. (We use (13.5) and the ergodic theorem).

Then for $\hat{X} = \{x \in \mathfrak{X} \mid m_x(S) > 1 - 2^{-4}\xi\}$: $m(\hat{X}) > 1-2^{-4}\xi$.

We choose $\hat{F} \subset S \cap \hat{X}$ as a uniform $(\Psi_{\hat{X}}, n+s+L, 2^{-3}\xi)$-Rohlin set.

If $\widehat{IX} = \hat{X} \setminus \bigcup_{j=0}^{n-1} \Psi^j \hat{F}$, then

$m(\widehat{IX} \cup (\mathfrak{X} \setminus \hat{X})) \leq m(\mathfrak{X} \setminus \hat{X}) + m_{\hat{X}}(\widehat{IX}) < 2^{-1}\xi$, and by (*) the set

$$Y_1 = \{x \in \hat{X} \mid h_x(\Psi) - \delta \leq h_x(\overline{\gamma} \cap (\hat{X} \setminus \widehat{IX}), \Psi)\}$$

has measure $m(Y_1) \geq 1 - 2^{-4}\xi - 10^{-1}\varepsilon > 1 - 5^{-1}\varepsilon$.

We put $F = \hat{F} \cap Y_1$ and, after dividing Y_1 into parts $I,\dots,IX$, G_1, Z_1, H_1,

$R = \bigcup_{j=0}^{k+c-1} \Psi^j F$ as in (29.3), $\gamma'_{VIII} = \overline{\gamma}_{VIII}$.

For $x \in Y_1$ we obtain by the same calculation as in (29.3):

$$h_x(\Psi) \leq h_x(\gamma' \cap G_1, \Psi) + m_x(R)\cdot(h_x(\Psi) + 3\tau) \leq$$

$$\leq h_x(\gamma' \cap G_1, \Psi) + m_x(H_1)\cdot h(\overline{M}) - \tau\cdot m_x(H_1)$$

Now we are in the position to enter into the iteration step in the proof of (30.2), and thus the proof is finished. □

Pointwise strictly ergodic embedding for aperiodic systems

(31.2) Theorem (Hansel): Let $(\mathfrak{X},m,\Psi)$ be aperiodic, $\overline{M} \subsetneqq M$ m.f.t.-subshifts, $\overline{\gamma}$ an $\overline{M}$-generator for Ψ and $\epsilon > 0$.
Then there exists an M-generator γ for Ψ such that

a) $\|\gamma,\overline{\gamma}\| < \epsilon$
b) Each point in Λ_γ is strictly ergodic.
c) $\omega \mapsto \mu_\omega$ is a continuous function on Λ_γ.
d) $\omega \mapsto h(\mu_\omega)$ is continuous on Λ_γ
e) $\{\mu_\omega | \omega \in \Lambda_\gamma\}$ is the set of ergodic measures on Λ_γ; it is compact and totally disconnected.

Proof: As in the proof of theorem (29.2) we construct a sequence $\overline{\gamma} = \gamma_o,\gamma_1,\gamma_2,\ldots$ of generators for Ψ with $\|\gamma_{i-1},\gamma_i\| < \epsilon_i$, where $\epsilon_i \to 0$ so rapidly that $\gamma = \lim \gamma_i$ is a generator (see (26.10)). We shall not mention these conditions for the ϵ_i any more. We put $M_o = M$, $\overline{M}_o = \overline{M}$, $j_o = 1$ and assume that after i steps we have:

A partition (X_i^j) $(1 \leq j \leq j_i)$ of $\mathfrak{X}$ into Ψ-invariant sets;
a generator γ_i for Ψ;
m.f.t. subshifts $\overline{M}_i^j \subsetneqq M_i^j$ $(1 \leq j \leq j_i)$;
numbers $r_i \in \mathbb{N}$; $\epsilon_i < 2^{-i} \cdot \theta(M)^{-r_i}$ such that for $i \geq 1$

the M_i^j are disjoint $(1 \leq j \leq j_i)$;

$(X_i^j)_{1\leq j\leq j_i} \supset (X_{i-1}^j)_{1\leq j\leq j_{i-1}}$;

$\gamma_i | X_i^j$ is an $\overline{M}_i^j$-generator for $\Psi_{X_i^j}$;

$$\|\gamma_i|X_i^j, \gamma_{i-1}|X_i^j\| < \epsilon_i \quad (\text{and hence } \|\gamma_{i-1},\gamma_i\| < \epsilon_i);$$

$$M_i := \bigcup_j M_i^j \subset M_{i-1};$$

$$h(M_i^j) - h_{inf}(\Psi_{X_i^j}) < 2\epsilon_i$$

and, if $M_{i-1}^{j'}$ is the system from step i-1 which contains M_i^j, then

(**) for all $P \in Bl_{r_i}(M_{i-1}^{j'}) : P \underset{\epsilon_i\text{-reg}}{<} M_i^j$ and $P \underset{d}{<} M_i^j$.

Now we choose $r_{i+1} > r_i$ and $\epsilon_{i+1} < 2^{-(i+1)} \cdot \theta(M)^{-r_{i+1}}$. We apply the lemmas (31.1) and (26.19) on each set X_i^j separately and get

$$X_{i+1}^j,\ M_{i+1}^j,\ \overline{M}_{i+1}^j \quad (1 \leq j \leq j_{i+1})$$

which again have the properties above. Let

$$M' = \bigcap_i M_i, \quad \gamma = \lim \gamma_i.$$

Then $\Lambda_\gamma \subset M'$; for, by construction $\gamma|X_i^j$ is a $\Psi_{X_i^j}$-generator for each (i,j).

Every $\omega \in M'$ is strictly ergodic; for, if $\omega \in \bigcap_i M_i^{j_i(\omega)}$, $(\mathfrak{M}_\sigma(M_i^{j_i(\omega)}))_{i\in\mathbb{N}}$ is a Cauchy filter base in $\mathfrak{M}_\sigma(M)$ by condition (**). Since all sets $\mathfrak{M}_\sigma(M_i^{j_i(\omega)})$ are closed and the sequence decreases, its intersection contains one point which must be μ_ω. Minimality (almost periodicity) of ω also follows from (1).

$\omega \mapsto \mu_\omega$ is continuous on M', because, by the disjointness of the M_i^j, $\mathfrak{M}_\sigma(M_i^{j_i(\omega)} \cap M')$ is a sequence of open and closed neighbourhoods of μ_ω in $\mathfrak{M}_\sigma(M')$ with $\mathfrak{M}_\sigma(M_i^{j_i(\omega)} \cap M') \searrow \mu_\omega$. We also see that $M' = \Lambda_\gamma$; for suppose $\omega \in M'$, $\omega \notin \Lambda_\gamma$. Then $\overline{O(\omega)} \cap \Lambda_\gamma = \emptyset$ ($\overline{O(\omega)}$ is minimal) and μ_ω is not a limit of measures on Λ_γ; on the other hand $\mu_{[\gamma | X_i^{j_i(\omega)}]} \in \mathfrak{M}_\sigma(M_i^{j_i(\omega)})$.

To show d), let $f_i : \omega \mapsto h(M_i^{j_i(\omega)})$, or

$$f_i = \sum_j h(M_i^j) \cdot 1_{M' \cap M_i^j} .$$

f_i is continuous, the sequence decreases and converges uniformly, since for $M_{i-1}^{j'} \supset M_i^j$:

$$h(M_{i-1}^{j'}) > h(M_i^j) \geq h(\Psi_{X_i^j}) \geq h(M_{i-1}^{j'}) - 2\varepsilon_{i-1}.$$

$h(\mu_\omega) \leq f_i(\omega)$ is obvious. For fixed $\omega \in M'$ let ν_i be the measure of maximal entropy on $M_i^{j_i(\omega)}$. Then $\nu_i \longrightarrow \mu_\omega$, and by the upper semi-continuity,

$$\lim f_i(\omega) = \lim h(\nu_i) \leq h(\mu_\omega)$$

and $h(\omega) = \lim f_i(\omega)$ is continuous.

(Corollary (16.12) shows directly that $f_i(\omega) \searrow h(\mu_\omega)$.) □

Let us discuss the continuity conditions c) and d) of the theorem. They imply that for every Ψ-invariant $D \in \Sigma$ with $m(D) > 0$

$$(***)\quad \begin{cases} h_{\sup}(\Psi_D) = \max\,\{h(\mu_\omega) \mid \omega \in \Lambda_{Y|D}\} \\ \\ h_{\inf}(\Psi_D) = \min\,\{h(\mu_\omega \mid \omega \in \Lambda_{Y|D}\}. \end{cases}$$

The purpose of the conditions c), d) is to get an embedding such that the image system does not carry too many ergodic measures. If $(\mathfrak{X},m,\Psi)$ has only finitely many ergodic components, the situation is quite simple: we may apply Jewett's theorem (29.2) on each component separately, and the compact image system has exactly "the same" ergodic measures; it is the disjoint union of finitely many strictly ergodic systems. In the countably infinite case one can do the same; one obtains a locally compact space and compactifies it with the Alexandroff point as a fixed point under the transformation. Now one has still the continuity of $\omega \mapsto \mu_\omega$. But if e.g. all ergodic components have entropy 1, this fixed point is somewhat embarrassing; we would prefer to obtain only new transformations of the same type as those we had before. If we classify ergodic transformations only by their entropy, our conditions are sufficiently restrictive. Thus, we may say that, if for almost no $x \in \mathfrak{X}$ $h_x(\Psi)$ is in the open set $U \subset \mathbb{R}$, then, by (***), for no ω $h(\mu_\omega) \in U$, and the same is valid for all the restrictions Ψ_D ($D \in \Sigma$ invariant, $m(D) > 0$). However, if every $(\mathfrak{X},m_x,\Psi)$ is isomorphic to a Bernoulli shift, c) and d) do not say that every μ_ω has the same property. That better conditions are desirable also follows from the example with the Alexandroff compactification, if we choose all systems (except the fixed point) isomorphic to some nontrivial system with entropy 0. Such examples can also be constructed in the shift space.

Minimal embedding for aperiodic systems

It has been known for a long time (see [142] that minimal topological dynamical systems may carry several ergodic invariant measures. In (19.12) some previous constructions are mentioned which furnish a certain richness of the set of ergodic measures, and in section 27 we constructed a minimal set with exactly two ergodic measures, but without further conditions on the type of these measures. Now we show that any aperiodic m.t. dynamical system (with $h_{sup}(\Psi) < \infty$) is conjugate to an invariant measure on a minimal subshift. We also try to restrict the number of ergodic measures on the minimal set which do not come from the original system. What was said in the last paragraph also applies here: We have restrictions in terms of continuity of the entropy as a function of the measure (of course, in a minimal system we can not have continuous, non-constant, invariant functions of the points), and in terms of the density of the measures coming from the original system. Although this is not quite satisfactory, the situation is somewhat better than in the case of the pointwise strictly ergodic embedding, namely fixed points or periodic points cannot enter because of the minimality.

By the density condition e) of the theorem, the embedding will be strictly ergodic if $(\mathfrak{X},m,\Psi)$ is ergodic. This certainly would not hold without condition e). Note that also in the pointwise strictly ergodic embedding ergodic systems are obviously mapped to strictly ergodic sets.

(31.3) Theorem: Let $(\mathfrak{X},m,\Psi)$ be an aperiodic measure theoretic dynamical system, $\overline{M} \subsetneq M$ m.f.t.-subshifts, $\overline{\gamma}$ an $\overline{M}$-generator for Ψ and $\varepsilon > 0$. Then there exists an M-generator $\hat{\gamma}$ for Ψ such that

a) $\|\overline{\gamma},\hat{\gamma}\| < \varepsilon$

b) $\Lambda_{\hat{\gamma}}$ is minimal

c) The set of ergodic measures on $\Lambda_{\hat{\gamma}}$ is closed and totally disconnected.

d) On $\Lambda_{\hat{\gamma}}$ only ergodic measures have generic points.

e) The convex hull of $\{\mu_{\hat{\gamma}|D} \mid D \in \Sigma$ **is** Ψ**-invariant,** $m(D) > 0\}$ **is dense in** $\mathfrak{M}_\sigma(\Lambda_{\hat{\gamma}})$.

f) The function $\nu \mapsto h(\nu)$ is continuous on $\mathfrak{M}_\sigma(\Lambda_{\hat{\gamma}})$.

Proof: Let d be a metric on $\mathfrak{M}_\sigma(M)$.

As in former cases we construct a sequence $\overline{\gamma} = \gamma_o, \gamma_1, \gamma_2, \ldots$ of generators such that $\|\gamma_{i-1},\gamma_i\| < \varepsilon_i$ where the ε_i are so small that $\hat{\gamma} = \lim \gamma_i$ is a generator (see (26.10)), and the properties of Λ_{γ_i} are improved step by step.

We put $J_o = 1$, $M_o = M$, $M_o^1 = \overline{M}$, $X_o^1 = \mathfrak{X}$ and $\gamma_o = \overline{\gamma}$. Assume now that we have arrived at step i-1 so that we have

$J_{i-1} \in \mathbb{N}$

$M_{i-1}^1, \ldots, M_{i-1}^{J_{i-1}}$ disjoint m.f.t.-subshifts

$M_{i-1} \supsetneq \bigcup_{j=1}^{J_{i-1}} M_{i-1}^j$ an m.f.t.-subshift,

$(X_{i-1}^1, \ldots, X_{i-1}^{J_{i-1}})$ a Ψ-invariant partition of $\mathfrak{X}$,

γ_{i-1} a generator for Ψ such that $\gamma_{i-1}|X_{i-1}^j$ is an M_{i-1}^j-generator for $\Psi_{X_{i-1}^j}$ $\quad (1 \leq j \leq J_{i-1})$.

We determine:

(1) $\tau_i < 2^{-i}$;

$s_i \in \mathbb{N}$ so large that for $1 \le j \le J_{i-1}$

(2) $s_i^{-1} \log \theta_{s_i}(M_{i-1}^j) - h(M_{i-1}^j) < \frac{1}{2}\tau_i$;

(3) $\xi_i < \frac{1}{2} \inf \{d(\mathfrak{M}_\sigma(M_{i-1}^j), \mathfrak{M}_\sigma(\bigcup_{j' \neq j} M_{i-1}^{j'})) \mid 1 \le j \le J_{i-1}\}$

so small that for $\nu \in \mathfrak{M}_\sigma(M)$ with $d(\nu, \mathfrak{M}_\sigma(M_{i-1}^j)) < \xi_i$:

(4) $s_i^{-1} H_\nu(B_{\langle o, s_i - 1\rangle}) < h(M_{i-1}^j) + \tau_i \quad (1 \le j \le J_{i-1})$.

($B_{\langle o,s-1\rangle}$ is the algebra generated by the cylinders ${}_o[P]$ with $P \in Bl_s(M)$), and

(5) $$\begin{cases} d(\nu, \mathfrak{M}_\sigma(M_{i-1}^j)) < \xi_i, \ d(\nu_1, \mathfrak{M}_\sigma(M_{i-1}^j)) < \xi_i, \\ d(\nu_2, \mathfrak{M}_\sigma(\bigcup_{j' \neq j} M_{i-1}^{j'})) < \xi_i \text{ for some } j, \\ \nu = (1-\lambda)\,\nu_1 + \lambda\,\nu_2 \end{cases} \Longrightarrow \lambda < \tau_i\,;$$

(6) $t_i \in \mathbb{N}$ so large and $\epsilon_i > o$ so small that

$t_i > t_{i-1}$, $\epsilon_i < 2^{-i}\,\epsilon_{i-1}$ $(i > 1)$ and

$\nu, \nu' \in \mathfrak{M}_\sigma(M)$, $\forall P \in Bl_{t_i}(M)$: $|\nu(P) - \nu'(P)| < 2\epsilon_i \Rightarrow d(\nu,\nu') < 3^{-i}\xi_i$;

a Ψ-invariant partition $(X_i^1, X_i^2, \ldots, X_i^{J_i})$ of $\mathfrak{X}$ which refines $(X_{i-1}^1, \ldots, X_{i-1}^{J_{i-1}})$ and for which

$$|h_{\sup}(\Psi_{X_i^j}) - h_{\inf}(\Psi_{X_i^j})| < \tau_i \quad (1 \le j \le J_i),$$

$$x, y \in X_i^j \Rightarrow \forall\, P \in Bl_{t_i}(M_{i-1}) : |\mu_{x,\gamma_{i-1}}(P) - \mu_{y,\gamma_{i-1}}(P)| < \epsilon_i.$$

(31.1) gives us disjoint m.f.t.-subshifts M_i^j $(1 \le j \le J_i)$ with $\bigcup_j M_i^j \subsetneqq M_{i-1}$ and a generator γ_i such that for $1 \le j \le J_i$

(7) $\gamma_i | X_i^j$ is an M_i^j-generator for $\Psi_{X_i^j}$;

(8) $\| \gamma_{i-1} | X_i^j, \gamma_i | X_i^j \| < t_i^{-1} \varepsilon_i$

(9) $P \in Bl_{t_i}(M_{i-1}) \Rightarrow P \underset{d}{\prec} M_i^j$ and $P \underset{\varepsilon_i\text{-reg}}{\prec} M_i^j$

(10) $h(M_i^j) < h_{inf}(\Psi_{X_i^j}) + \tau_i$

We remark that (8), (6), (9) imply:

(11) $X_i^j \subset X_{i-1}^{j'}, \ \nu \in \mathfrak{M}_\sigma(M_i^j) \quad \Longrightarrow$

$$d(\nu, \mathfrak{M}_\sigma(M_{i-1}^{j'})) < 3^{-i}\xi_i \quad \text{and} \quad \operatorname{diam}(\mathfrak{M}_\sigma(M_i^j)) < 3^{-i}\xi_i \ .$$

Now we choose $\overline{r}_i \in \mathbb{N}$ so large that for $1 \leq j \leq J_i$

$$P \in Bl_{t_i}(M_{i-1}), \ r' \geq \overline{r}_i, \ Q_1, Q_2 \in Bl_{r'}(M_i^j) \implies$$

$$P \prec Q_1 \quad \text{and} \quad |\mu_{Q_1}(P) - \mu_{Q_2}(P)| < \varepsilon_i,$$

and such that there is a block $C_i \in Bl_{\overline{r}_i}(M_{i-1}) \setminus Bl_{\overline{r}_i}(\bigcup_j M_i^j)$.

We take $L_i \in \mathbb{N}$ so large that for any $P \in Bl_{\overline{r}_i}(M_i^j)$, $P' \in Bl_{\overline{r}_i}(M_i^{j'})$ $(j \neq j')$ there is a transition block $U \in Bl_{L_i}(M_{i-1})$ such that

$$P \cdot U \cdot P' \prec M_{i-1},$$

$C_i \prec U$, but only once, namely at the place $[\frac{1}{2} L_i]$.

We let $r_i > 3\varepsilon_i^{-4}(\overline{r}_i + L_i)$, $S_i^j \in Bl_{r_i}(M_i^j)$ fixed blocks, and

$$\widehat{Bl} = \bigcup_j Bl_{r_i}(M_i^j) \cup \bigcup_{j \neq j'} \{P \in Bl_{r_i}(M_{i-1}) \mid P \prec S_i^j \cdot U \cdot S_i^{j'}\},$$

where the transition block U is always chosen as above. $\widehat{Bl} = Bl_{r_i}(M_i)$ is the defining block system for $M_i \subset M_{i-1}$.

With this definition the induction step is finished.

We now consider $\hat{\gamma} = \lim \gamma_i$ and $\Lambda_{\hat{\gamma}}$.

Since $\mu_{\hat{\gamma}} = \lim \mu_{\gamma_i}$ and μ_{γ_i} is carried by $M_{i'}$ if $i' \leq i$, $\Lambda_{\hat{\gamma}} \subset \bigcap_i M_i$. But since $P \in Bl_{t_{i-1}}(M_{i-1})$, $Q \in Bl_{r_i}(M_i) \Rightarrow P \prec Q$, $\bigcap_i M_i$ is minimal, and hence $\Lambda_{\hat{\gamma}} = \bigcap_i M_i$.

We put $\underline{J} = \{\underline{j} = (j_1, j_2, \ldots) \mid 1 \leq j_i \leq J_i,\ X_{i+1}^{j_{i+1}} \subset X_i^{j_i} \forall i \geq 1\}$. $\underline{J}$ is a compact subset of $\prod_{i \in \mathbb{N}} \{1,2,\ldots,J_i\}$, and

$$\underline{j}, \underline{j}' \in \underline{J},\ j_i = j'_i \Rightarrow \forall\ t \leq i : j_t = j'_t.$$

Let us choose some $\underline{j} \in \underline{J}$ and $\nu_i \in \mathfrak{M}_\sigma(M_i^{j_i})$ $(i \in \mathbb{N})$. By (11)

$$d(\nu_{i-1}, \nu_i) < 2 \cdot 3^{-i+1} \xi_{i-1}$$ i.e. the ν_i converge to a limit which obviously does not depend on the choice of the ν_i and which we therefore call $\nu(\underline{j})$.

The mapping $\underline{j} \mapsto \nu(\underline{j})$ is injective; for let $j_i \neq j'_i$ with $i \geq 2$. Then

$$d(\nu(\underline{j}), \mathfrak{M}_\sigma(M_i^{j_i})) \leq \frac{1}{2} \xi_i$$

$$d(\nu(\underline{j}'), \mathfrak{M}_\sigma(M_i^{j'_i})) \leq \frac{1}{2} \xi_i$$

and by the choice of ξ_i, $\nu(\underline{j}) \neq \nu(\underline{j}')$. The mapping also is continuous; for if $j_i = j_i'$ with $i \geq 2$, we conclude that

$$d(\nu(\underline{j}),\nu(\underline{j}')) \leq \xi_{i+1} + \operatorname{diam}(\mathfrak{M}_\sigma(M_i^{j_i})) \leq \xi_{i+1} + \xi_i.$$

Let $\mathfrak{E} := \{\nu(\underline{j}) \mid \underline{j} \in \underline{J}\}$.

For $\nu \in \mathfrak{E}$, $\nu = \nu(\underline{j})$, we put $\underline{j} = \underline{j}(\nu)$ and $j_i = j_i(\nu)$. Let $\mathfrak{E}_i^j = \{\nu(\underline{j}) \mid \underline{j} \in \underline{J},\ j_i = j\}$ $(1 \leq j \leq J_i)$. These sets form a neighbourhood-base of closed and open sets for the topology on $\mathfrak{E}$ ($\mathfrak{E}$ is compact and totally disconnected by the homeomorphy with $\underline{J}$).

The proof that on $\Lambda_{\hat{Y}}$ only measures $\nu \in \mathfrak{E}$ have generic points is exactly as in (27.2). This implies in particular: ν ergodic $\Rightarrow \nu \in \mathfrak{E}$. To show that all elements of $\mathfrak{E}$ are ergodic, suppose $\nu \in \mathfrak{E}$ and ρ is a probability measure on $\mathfrak{E}$ with $\int \varphi \, d\rho(\varphi) = \nu$.

For $i \in \mathbb{N}$, we put

$$1 - \lambda_i = \rho(\mathfrak{E}_i^{j_i(\nu)})$$

$$\nu_i^1 = (1 - \lambda_i)^{-1} \cdot \int_{\mathfrak{E}_i^{j_i(\nu)}} \varphi \, d\rho(\varphi)$$

$$\nu_i^2 = \lambda_i^{-1} (\nu - \nu_i^1)$$

Then $d(\nu, \mathfrak{M}_\sigma(M_i^{j_i(\nu)})) < \xi_{i+1}$, $d(\nu_i^1, \mathfrak{M}_\sigma(M_i^{j_i(\nu)})) < \xi_{i+1}$ and $d(\nu_i^2, \mathfrak{M}_\sigma(\bigcup_{j \neq j_i(\nu)} M_i^j)) < \xi_{i+1}$, so, by (5) $\rho(\mathfrak{E}_i^{j_i(\nu)}) \geq 1 - \tau_{i+1}$, i.e. ρ is concentrated on ν, and ν is an extremal point of $\mathfrak{M}_\sigma(\Lambda_{\hat{Y}})$.

For the proof of f) it is sufficient to show that h is continuous on the set $\mathfrak{E}$ of ergodic measures.
On $\mathfrak{E}$ let $f_i(\nu) = h(M_i^{j_i(\nu)})$.

Since for $\underline{j} \in \underline{J}$ we have by (10), (11), (4)

$$h(M_{i-1}^{j_{i-1}}) - \tau_{i-1} \leq h(M_i^{j_i}) \leq h(M_{i-1}^{j_{i-1}}) + \tau_i,$$

the sequence (f_i) is a Cauchy sequence of continuous functions; so we have to prove only $h(\nu) = \lim f_i(\nu)$ $(\nu \in \mathfrak{E})$. But $\mu_{M_i^{j_i(\nu)}} \longrightarrow \nu$ together with the upper semi-continuity of h implies $\lim f_i(\nu) \leq h(\nu)$, and since
$d(\nu, \mathfrak{m}_\sigma(M_{i-1}^{j_{i-1}(\nu)}) < \xi_i$,
$h(\nu) < h(M_{i-1}^{j_{i-1}(\nu)}) + \tau_i = f_{i-1}(\nu) + \tau_i$, so $h(\nu) \leq \lim f_i(\nu)$.

□

B I B L I O G R A P H Y

[1] ABRAMOV, L.M., *The entropy of an induced automorphism*. Doklady Akad. Nauk SSSR 128 (1959), 647-650.

[2] ADLER, R.L., A.G. KONHEIM, M.H. McANDREW, *Topological entropy*. Trans. Amer. Math. Soc. 114 (1965), 309-319.

[3] ---, M.H. McANDREW, *The entropy of Chebyshev Polynomials*. Trans. Amer. Math. Soc. 121 (1966), 236-241.

[4] ---, P. SHIELDS, M. SMORODINSKY, *Irreducible Markov shifts*. Annals of Math. Statistics 43 (1972), 1027-1029.

[5] ---, B. WEISS, *Entropy, a complete metric invariant for automorphisms of the torus*. Proc. Natl. Acad. Sci. 57 (1967), 1573-1576.

[6] ---, B. WEISS, *Similarity of automorphisms of the torus*. Memoirs AMS 98 (1970).

[7] ALEKSEEV, V.M., *Quasirandom dynamical systems I*. Math. Sb. 76 (1968), 72-130.

[8] ANOSOV, D.V., *Geodesic flows on closed Riemanian manifolds of negative curvature*. Proc. Steklov Inst. 90 (1967).

[9] AOKI, K., *On symbolic representation*. Proc. Japan Acad. 30 (1954), 160-164.

[10] ---, S. SAIKAWA, *On compact groups which admit expansive automorphisms, to appear.*

[11] ---, H. TOTOKI, *Ergodic automorphisms of T^∞ are Bernoulli transformations*. Publ. RIMS Kyoto Univ., 10 (1975), 535-544.

[12] ARNOLD, V.I, A. AVEZ, *Problèmes ergodiques de la mécanique classique*. Gauthier-Villars, Paris, 1966.

[13] AVEZ, A., *Propriétés ergodiques des endomorphismes dilatants de variétés compactes*. C. R. Acad. Sci, Paris, Sér.A, 266 (1968), 610-612.

[14] AZENCOTT, R., *Diffeomorphismes d'Anosov et schémas de Bernoulli*. C. R. Acad. Sci. Paris Sér.A, 270 (1970), 1105-1107.

[15] BAUER, W., K. SIGMUND, *Topological dynamics of transformations induced on the space of probability measures*. Monatshefte f. Mathematik, 79 (1975), to appear.

[16] BERG, K.R., *Convolution and invariant measures, maximal entropy*. Math. Syst. Theory 3 (1969), 146-150.

[17] BILLINGSLEY, P., *Ergodic theory and information*. John Wiley and Sons, New York 1965.

[18] BOWEN, R., *Topological entropy and axiom A*. Proc. Symp. Pure Math. 14 (1970), 23-41.

[19] ---, *Markov partitions for axiom-A-diffeomorphisms*, Amer. J. Math. 92 (1970), 725-747.

[20] ---, *Markov partitions and minimal sets for axiom-A-diffeomorphisms*. Amer. J. Math. 92 (1970), 907-918.

[21] ---, *Periodic points and measures for axiom-A-diffeomorphisms*. Trans. Americ. Math. Soc. 154 (1971), 377-397.

[22] ---, *Entropy for group endomorphisms and homogeneous spaces*. Trans. Amer. Math. Soc. 153 (1971), 401-413.

[23] ---, *Entropy-expansive maps*, Trans. Amer. Math. Soc. 164 (1972), 323-333.

[24] ---, *Periodic orbits for hyperbolic flows*. Amer. J. Math. 94 (1972), 1-30.

[25] ---, *The equidistribution of closed geodesics*. Amer. J. Math. 94 (1972), 413-423.

[26] ---, *Topological entropy for noncompact sets*. Trans. Amer. Math. Soc. 184 (1973), 125-136.

[27] ---, *Some systems with unique equilibrium states*. Math. Syst. Theory 8 (1974), 193-202.

[28] ---, *Symbolic dynamics for hyperbolic systems*. Amer. J. Math. 95 (1973), 429-459.

[29] ---, *Bernoulli equilibrium states for axiom-A-diffeomorphisms*. Math. Syst. Theory 8 (1975), 289-294.

[30] ---, *ω-limit sets for axiom-A-diffeomorphisms*. J.Diff.Equations 18 (1975), 333-339.

[31] ---, *Equilibrium states and the ergodic theory of Anosov diffeomorphisms*. Lecture Notes in Mathematics, Berlin-Heidelberg-New York, Springer 470 (1975).

[32] ---, D. RUELLE, *The ergodic theory for axiom-A-flows*. Inventiones Math. 29 (1975), 181-202.

[33] ---, P. WALTERS, *Expansive one-parameter flows*. J.Diff.Equations 12 (1972), 180-193.

[34] BRYANT, B.F., *On expansive homeomorphisms*. Pac. J. Math. 10 (1960), 1163-1167.

[35] ---, *Expansive self-homeomorphisms of a compact metric space.* Amer. Math. Monthly 69 (1962), 386-391.

[36] ---, D.B. COLEMAN, *Some expansive homeomorphisms of the reals.* Amerc. Math. Monthly 73 (1966), 370-373.

[37] ---, P. WALTERS, *Asymptotic properties of expansive homeomorphisms.* Math. Syst. Theory 3 (1969), 60-66.

[38] CIGLER, B., *Ein gruppentheoretisches Analogon zum Begriff der normalen Zahl.* J. Reine Angew. Math. 206 (1961), 5-9.

[39] CONZE, J.P., *Points périodiques et entropie topologique.* C. R. Acad. Sci. Paris, Sêr. A 267 (1968), 149-152.

[40] COVEN, E., M. PAUL, *Sofic systems.* Israel J. Math. 20 (1975),no. 2, 165-177.

[41] ---, ---, *Endomorphisms of irreducible subshifts of finite type.* Math. Syst. Theory 8 (1974), 167-175.

[42] COLEBROOK, C., *The Hausdorff dimension of certain sets of non-normal numbers.* Mich. Math. 3, 17 (1970), 103-115.

[43] DENKER, M., *Einige Bemerkungen zu Erzeugersätzen.* Z. Wahrscheinlichkeitstheorie verw. Geb. 29 (1974), 57-64.

[44] ---, *Untersuchungen über eine spezielle Klasse von Zerlegungen eines kompakten, metrischen Raumes.* Thesis, Universität Erlangen-Nürnberg 1972.

[45] ---, *Une démonstration nouvelle du théorème de Goodwyn.* C. R. Acad. Sci. Paris, Sêr. A, 275 (1972), 735-738.

[46] ---, *Measures with maximal entropy.* To appear in: Journées Ergodiques. Lecture Notes in Mathematics.

[47] ---, *Remarques sur la pression pour les transformations continues.* C. R. Acad. Sc. Paris, Sêr. A, 279 (1974), 967-970.

[48] ---, *On strict ergodicity.* Math. Z. 134 (1973), 231-253.

[49] ---, *Finite generators for ergodic, measure-preserving transformations.* Z. Wahrscheinlichkeitstheorie verw. Geb. 29 (1974), 45-55.

[50] ---, E. EBERLEIN, *Ergodic flows are strictly ergodic.* Advances in Math. 13 (1974), 437-473.

[51] ---, M. KEANE, *Generators for almost-topological dynamical systems.* Preprint.

[52] DINABURG, E.I., *An example for the computation of topological entropy.* Usp. Met. Nauk 23 (1968), 249-250.

[53] ---, *The relation between topological entropy and metric entropy.* Dokl. Akad. Nauk SSSR 190 (1970) = Soviet Math. Dokl. 11 Nr.1 (1970), 13-16.

[54] DOWKER, Y.N., *The mean and transitive points of homeomorphisms.* Ann.of Math. (2) 58 (1953), 123-133.

[55] ---, G. LEDERER, *On ergodic measures.* Proc. Amer. Math. Soc. 15 (1964), 65-69.

[56] EISENBERG, M., *Expansive transformation semigroups of endomorphisms.* Fund. Math. 59 (1966), 313-321.

[57] FRIEDMAN, N.A., *Introduction to ergodic theory.* Van Mostrand, 1970.

[58] FURSTENBERG, H., *Strict ergodicity and transformations of the torus.* Amer. J. Math. 83 (1961), 573-601.

[59] ---, *Disjointness in ergodic theory.* Math. Syst. Theory 1 (1967), 1-49.

[60] GOODMAN, T.N.T., *Relating topological entropy with measure theoretic entropy.* Bull. London Math. Soc. 3 (1971), 176-180.

[61] ---, *The relation between topological entropy and measure entropy.* Proc. Symp. on Topological Dynamics and ergodic theory. University of Kentucky (1971), 45-46.

[62] ---, *Maximal measures for expansive homeomorphisms.* J. London Math.Soc. (2) 5 (1972), 439-444.

[63] GOODWYN, L., *Topological entropy and expansive cascades.* Thesis. University of Maryland, 1968.

[64] ---, *Topological entropy bounds measure theoretic entropy.* Proc.Amer. Math.Soc. 23 (1969), 679-688.

[65] ---, *Comparing topological entropy with measure theoretic entropy.* Amer. J.Math. 94 (1972), 366-388.

[66] ---, *A characterization of symbolic cascades in terms of expansiveness and entropy.* Math. Syst. Theory 4 (1970), 157-159.

[67] ---, *The product theorem for topological entropy.* Trans. Amer. Math. Soc. 158 (1971), 445-452.

[68] ---, *Some counter-examples in topological entropy.* Proc.Symp. on Topological Dynamics and Ergodic Theory. Univ. of Kentucky (1971), 47-49.

[69] ---, *Some counter-examples in topological entropy.* Topology 11 (1972), 377-385.

[70] GOTTSCHALK, W.H., G. HEDLUND, *Topological dynamics.* Amer.Math.Coll. Publ. 36 Providence. R.I. (1955).

[71] GRILLENBERGER, Chr., *Zwei kombinatorische Konstruktionen für strikt ergodische Folgen*. Thesis. Universität Erlangen-Nürnberg (1970).

[72] ---, *Constructions of strictly ergodic systems I. Given entropy*. Z. Wahrscheinlichkeitstheorie u. verw. Geb. 25 (1973), 323-334.

[73] ---, *Constructions of strictly ergodic systems II. K-systems*. Z. Wahrscheinlichkeitstheorie u. verw. Geb. 25 (1973), 335-342.

[74] ---, P. SHIELDS, *Construction of strictly ergodic systems. III. A Bernoulli system*. Z. Wahrscheinlichkeitstheorie u. verw. Geb. 33 (1975), 215-217.

[75] GUREVIČ, B.M., *Topological entropy of a countable Markov chain*. Dokl. Akad. Nauk SSSR 187 Nr. 4 (1969) = Soviet Math. Dokl. 10 Nr. 4 (1969), 911-915.

[76] ---, *The invariant measure with maximal entropy for an Anosov diffeomorphism*. Funct. Anal. and its Appl. 4 (1970), 282-289 = Funkt. Anal. i ego Pril. 4 Nr. 4 (1970), 21-30.

[77] ---, *Entropy of the shift and Markov measures in the space of paths of a countable graph*. Dokl. Akad. Nauk SSSR 192 Nr. 5 (1970), 963-965.

[78] HAHN, F., Y. KATZNELSON, *On the entropy of uniquely ergodic transformations*. Trans. Amer. Math. Soc. 126 (1967), 335-360.

[79] ---, W. PARRY, *Minimal dynamical systems with quasi-discrete spectrum*. J. London Math. Soc. 40 (1965), 309-323.

[80] HALMOS, P.R., *Lectures on ergodic theory*. Chelsea, New York 1956.

[81] HAMACHI, T., H. TOTOKI, *A remark on the topological entropy*. Memoirs Fac. Sci. Kyushu Univ. Ser. A. 25 (1971), 300-303.

[82] ---, M. OSIKAWA, *Topological entropy of a non-irreducible intrinsic Markov-shift*. Preprint.

[83] HANSEL, G., *Strict uniformity in ergodic theory*. Math. Z. 135 (1974), 221-248.

[84] ---, J.P. RAOULT, *Ergodicité, uniformité et unique ergodicité*. Indiana Univ. Math. J. 23 (1973), 221-237.

[85] HEDLUND, G.A., *Sturmian minimal sets*. Amer.J. Math.66 (1944), 605-620.

[86] ---, *Endomorphisms and automorphisms of the shift dynamical system*. Math. Syst. Theory 3 (1969), 320-375.

[87] ---, M. MORSE, *Symbolic dynamics*. Amer. J. Math. 60 (1938), 815-866.

[88] ---, ---, *Symbolic dynamics II*. Amer. J. Math. 62 (1940), 1-42.

[89] HEMMINGSEN, E., W. L. REDDY, *Lifting and projecting expansive homeomorphisms*. Math. Syst. Theory 2 (1968), 7-15.

[90] ---, ---, *Expansive homeomorphisms on compact manifolds*. Fund. Math. 64 (1969), 203-207.

[91] HOPF, E., *Ergodentheorie*. Springer, Berlin 1937.

[92] ITO, S., *An estimate from above for the entropy and the topological entropy of a C^1-diffeomorphism*. Proc. Japan Acad. 46 (1970), 226-230.

[93] JACOBS, K., *Neuere Methoden und Ergebnisse der Ergodentheorie*. Springer, Berlin 1960.

[94] ---, *Lecture Notes in ergodic theory*. Aarhus Universitet, 1963.

[95] ---, *Ergodic decomposition of the Kolmogorov-Sinai-invariant*. Proc. Internat. Symp. in Ergodic Theory. Acad. Press, New York (1963), 173-190.

[96] ---, *Lipschitz functions and the prevalence of strict ergodicity for continuous time flows*. Contributions to Ergodic Theory and Probability, Lecture Notes in Mathematics, Berlin-Heidelberg-New York, Springer, 160 (1970), 87-124.

[97] ---, *Systèmes dynamiques Riemanniens*. Czechosl. Math. J. 20 (90) (1970), 628-631.

[98] ---, M.KEANE, *0-1-sequences of Toeplitz type*. Z. f. Wahrscheinlichkeitstheorie verw. Geb. 13 (1969), 123-131.

[99] JACOBSEN, J.F., W.R. UTZ, *The nonexistence of expansive homeomorphisms of a closed 2-cell*. Pac. J. Math. 10 (1960), 1319-1321.

[100] JEWETT, R.I., *The prevalence of uniquely ergodic systems*. J. Math. and Mech. 19 (1970), 717-729.

[101] KAKUTANI, S., *Ergodic theory of shift transformations*. Proc. V Berkeley Symp. 2 (2) (1966), 405-414.

[102] KAMAE, T., *Subsequences of normal sequences*. Israel Z. Math. 16 (1973), 121-149.

[103] KATZNELSON, Y., *Ergodic automorphisms of T^n are Bernoulli shifts*. Israel J. Math. 10 (1971), 186-195.

[104] ---, B. WEISS, *The construction of quasi-invariant measures*. Israel Z. Math. 12 (1970), 1-4.

[105] KEANE, M., *Generalized Morse sequences*. Z. f. Wahrscheinlichkeitstheorie verw. Geb. 10 (1968), 335-353.

[106] ---, *Sur les mesures quasi-invariantes des translations irrationelles*. C.R. Acad. Sci., Paris. Sér. A. 272 (1971), 54-56.

[107] KEYNES, H.B., *Lifting topological entropy*. Proc. Amer. Math. Soc. 24 (1970), 440-445.

[108] ---, *Expansive algebras and expansive extensions*. Proc. of the Symp. on Topol. Dyn. and Ergodic Theory. University of Kentucky (1971).

[109] ---, J.B. ROBERTSON, *Generators for topological entropy and expansiveness*. Math. Syst. Theory 3 (1969), 51-59.

[110] KOLMOGOROV, A.N., *A new metric invariant of transitive dynamical systems and automorphisms of Lebesgue spaces*. Dokl. Akad. Nauk SSSR 119 (1958), 861-864.

[111] ---, V.A. TIKOMIROV, *ε-entropy and ε-capacity of sets in functional spaces*. Uspekhi Mat. Nauk SSSR 14 Nr. 2 (1959), 3-86 = Amer. Math. Soc. Translations (2) 17 (1961), 277-367.

[112] KORNFELD, I.P., *On invariant measures of minimal dynamical systems*. Soviet Math. Dokl. 13 (1972), 87-90.

[113] KRENGEL, U., *On certain analogous difficulties in the investigation of flows in a probability space and of transformations in an infinite measure space*. Functional Analysis - Proc. of a Symp. (1970), Acad. Press, New York.

[114] ---, *Transformations without finite invariant measure have finite strong generators*. Contributions to Ergodic Theory and Probability. Lecture Notes in Mathematics, Berlin-Heidelberg-New York, Springer 160 (1970), 133-157.

[115] ---, *Recent results on generators in ergodic theory*. Transactions of the Sixth Prague Conference on Information Theory, Statistical Decision Functions, Random Processes. Academia, Prague, 1973.

[116] KRIEGER, W., *On entropy and generators of measure preserving transformations*. Trans. Amer. Math. Soc. 149 (1970), 453-464. Erratum: 168 (1972), 519.

[117] ---, *On generators in exhaustive σ-algebras of ergodic measure-preserving transformations*. Z. Wahrscheinlichkeitstheorie verw. Geb. 20 (1971), 75-82.

[118] ---, *On unique ergodicity*. Proc. of the Sixth Berkeley Symp. on Math. Stat. and Probability, Berkeley, Los Angeles, University of California Press, I (1972), 327-346.

[119] ---, *On generators in ergodic theory*. Preprint: Intern. Congress Math. Vancouver 1974.

[120] ---, *On quasi-invariant measures in uniquely ergodic systems*. Invent. Math. 14 (1971), 184-196.

[121] ---, *On the uniqueness of the equilibrium state*. Math. Syst. Theory 8 (1974), 97-104.

[122] KRYZEWSKI , K., *Note on topological entropy*. Bull. Acad. Pol. Sci. 16 (1968), 465-467.

[123] ---, *On expanding mappings*. Bull. Acad. Pol. Sci. 19 (1971), 23-24.

[124] ---, *On connection between expanding mappings and Markov chains*. Bull. Acad. Pol. Sci. 19 (1971), 291-293.

[125] ---, *On a problem of Rohlin*. Bull. Acad. Pol. Sci. 20 (1972), 207-210.

[126] ---, W. SZLENK, *On invariant measures for expanding differentiable mappings*. Studia Mathematica 33 (1969), 83-92.

[127] KRYLOV, N., N. BOGOLIUBOV, *La théorie générale de la mesure dans son application à l'étude des systèmes de la méchanique non linéaire*. Ann. of Math. (2) 38 (1937), 65-113.

[128] KURATOWSKI, K., *Topologie I, II.*, Warszawa 1952.

[129] KUSHNIRENKO, G., *Upper bound of the entropy of a classical dynamical system*. Dokl. Akad. Nauk SSSR 22 (1967), 57-59.

[130] LAM, P.F., *On expansive transformation groups*. Trans. Amer. Math. Soc. 150 (1970), 131-138.

[131] ---, *Homeomorphisms of expansive transformation groups*. Math. Syst. Theory 4 (1970), 249-256.

[132] LIVSIC, A.N., *Periodic trajectories of automorphisms of the torus and approximation of invariant measures*. Izv. Vysŝt. Ucebn. Zaved., Mat. 4 (95) (1970), 64-66 (Russian)

[133] LORENTZ, G., *Metric entropy, widths, and superposition of functions*. Amer. Math. Monthly 69 (1962), 469-485.

[134] MARGULIS, G.A., *On some measures connected with U-flows*. Funkt. Anal. i ego Pril. 4 Nr. 1 (1970), 62-76.

[135] MISIUREWICZ, M., *On expanding maps of compact manifolds and local homeomorphisms of a circle*. Bull. Acad. Pol. Sci. 18 (1970), 725-732.

[136] ---, *On non-continuity of topological entropy*. Bull. Acad. Pol. Sci. 19 (1971), 319-320.

[137] ---, *Diffeomorphisms without any measure with maximal entropy*. Bull. Acad. Pol. Sci. 21 (1973), 903-910.

[138] ---, *Topological conditional entropy*. To appear in Studia Math. 55.

[139] O'BRIEN, T.V., *Expansive homeomorphisms of compact manifolds*. Proc. Amer. Soc. 24 (1970), 769-771.

[140] ---, W.L. REDDY, *Each compact orientable surface of positive genus admits an expansive homeomorphism*. Pac. J. Math. 35 (1970), 737-741.

[141] ORNSTEIN, D., *Ergodic theory, randomness and dynamical systems*. Yale University Press 1974.

[142] OXTOBY, J.C., *Ergodic sets*. Bull. Amer. Math. Soc. 58 (1952), 116-136.

[143] ---, *Stepanoff flows on the torus*. Proc. AMS 4 (1953), 982-987.

[144] ---, *On two theorems of Parthasarathy and Kakutani*. Proc. Int. Symp. on Ergodic Theory, Tulane, New Orleans, Ed. F.B. Wright, Academic Press, M.Y. (1963), 203-215.

[145] ---, S.M. ULAM, *Measure-preserving homeomorphisms and metrical transitivity*. Ann. of Math. (2) 42 (1941), 874-920.

[146] PARRY, W., *Intrinsic Markov chains*. Trans. Amer. Math. Soc. 112 (1964), 55-66.

[147] ---, *Symbolic dynamics and transformations of the unit interval*. Trans. Amer. Math. Soc. 122 (1966), 368-378.

[148] ---, *Zero entropy of distal and related transformations*. Topological Dynamics (Symp. Colorado State Univ. 1967), Benjamin, New York, 1968, 383-389.

[149] ---, *Entropy and generators in ergodic theory*. Math. Lecture Notes Series, Benjamin, New York 1969.

[150] ---, *Topological Markov chains and suspensions*. Preprint. Univ. of Warwick, 1974.

[151] PARTHASARATHY, K.R., *On the category of ergodic measures*. Ill. J. Math. 5 (1961), 648-656.

[152] ---, *Probability measures on metric spaces*. Academic Press, New York 1967.

[153] PETERSEN, K.E., *A topologically strongly mixing symbolic minimal set*. Trans. Amer. Math. Soc. 148 (1970), 603-612.

[154] REDDY, W.L., *The existence of expansive homeomorphisms on manifolds*. Duke Math. J. 32 (1965), 627-632.

[155] ---, *Lifting expansive homeomorphisms to symbolic flows*. Math. Syst. Theory 2 (1968), 91-92.

[156] ---, *On positively expansive maps*. Math. Syst. Theory 6 (1972), 76-81.

[157] ROBBIN, J., *Topological conjugacy and structural stability for discrete dynamical systems*. Bull. Amer. Math. Soc. 78 (1972), 923-952.

[158] ROHLIN , V.A., *On the fundamental ideas of measure theory*. Amer. Math. Soc. Translations. Ser. 1, 10 (1962), 1-54. = Mat. Sb. 25 (1949), 107-150.

[159] ---, *Generators in ergodic theory I, II*. Vestnik Leningrad Univ. Mat. Mech. Astron. (1963), 26-32 and ibid. (1965), 68-72.

[160] ---, *Lectures on the entropy theory of measure-preserving transformations*. Uspekhi Mat. Nauk 22 Nr.5 (1967), 3-56 = Russian Math. Surveys 22 Nr.5 (1967), 1-52.

[161] RUELLE, D., *Statistical mechanics on a compact set with* Z^{ν} *action satisfying expansiveness and specification*. Trans. Amer. Math. Soc. 185 (1973), 237-251.

[162] ---, *A measure associated with axiom-A attractors*. Amer. J. Math. 1975.

[163] SCHMIDT, W., *On normal numbers*. Pac. J. Math. 10 (1960), 661-672.

[164] ---, *Normalität bezüglich Matrizen*. J. Reine Angew. Math. 214/215 (1964), 227-260.

[165] SEARS, M., *Topological models for generators*. Math. Syst. Theory 7 (1973), 32-38.

[166] ---, *Expansiveness on locally compact spaces*. Math. Syst. Theory 7 (1974), 377-382.

[167] SHTILMAN, M.S., *Number of invariant measures with maximal entropy for translations in the space of sequences*. Mt. Zametki 9 (1971), 291-298.

[168] SIGMUND, K., *Generic properties of invariant measures for axiom-A-diffeomorphisms*. Inventiones math. 11 (1970), 99-109.

[169] ---, *On the prevalence of zero entropy*. Israel J. Math. 10 (1971), 281-288.

[170] ---, *Mixing measures for axiom-A diffeomorphisms*. Proc. Amer. Math. Soc. 36 (1972), 497-504.

[171] ---, *On the space of invariant measures for hyperbolic flows*. Amer. J. Math. 94 (1972), 31-37.

[172] ---, *On dynamical systems with the specification property*. Trans. Amer. Math. Soc. 190 (1974), 285-299.

[173] ---, *On normal and quasiregular points for endomorphisms of the torus*. Math. Syst. Theory 8 (1974), 251-255.

[174] ---, *On the time evolution of statistical states for Anosov systems*. Math. Z. 138 (1974), 183-189.

[175] ---, *Invariant measures for continuous transformations*. Global Analysis and its Applications. IAEA 1974 (Vienna), 137-161.

[176] SINAI, Ya.G., *Construction of Markov partitions*. Funct. Anal. and its Appl. 2 (1968), 245-253. = Funkt. Anal. i ego Pril. 2 Nr. 3 (1968), 70-80.

[177] ---, *Markov partitions and C-diffeomorphisms*. Funct. Anal. and its Appl. 2 (1968), 64-89, = Funkt. Anal. i ego Pril. 2 Nr. 1 (1968), 67-89.

[178] ---, *Invariant measures for Anosov's dynamical systems*. Proc. Intern. Congress Math. Nice 1970. 2, Gauthier-Villars, Paris (1971), 929-940.

[179] ---, *Gibbs measures in ergodic theory*. Russ. Math. Surveys 27 Nr.4 (1972), 21-69.

[180] SMALE, S., *Differentiable dynamical systems*. Bull. Amer. Math. Soc. 73 (1967), 747-817.

[181] SMORODINSKY, M., *Ergodic Theory, entropy*. Lecture Notes in Mathematics, Berlin-Heidelberg-New York, Springer, 214, (1971).

[182] SPITZER, F., *A variational characterization of finite Markov chains*. Ann. Math. Stat. 43 (1972), 303-307.

[183] UTZ, W.R., *Unstable homeomorphisms*. Proc. Amer. Math. Soc. 1 (1950), 769-774.

[184] VEECH, W.A., *Strict ergodicity in zero dimensional dynamical systems and the Kronecker-Weyl, theorem mod 2*. Trans. Amer. Math. Soc. 140 (1969), 1-35.

[185] WALTERS, P., *Ergodic Theory - Introductory Lectures*. Lecture Notes in Mathematics, Berlin-Heidelberg-New York, Springer 458 (1975).

[186] ---, *A variational principle for the pressure of continuous transformations*. Mathemat. Institute, University of Warwick, Coventry (1973), 56 pp.

[187] ---, *Some transformations having a unique measure with maximal entropy*. Preprint.

[188] WEISS, B., *Intrinsically ergodic systems*. Bull. Amer. Math. Soc. 76 (1970), 1266-1269.

[189] ---, *Topological transitivity and ergodic measures*. Math. Syst. Theory 5 (1971), 71-75.

[190] ---, *Subshifts fo finite type and sofic systems*. Monatsh. Math. 77 (1973), 462-474.

[191] WEISS, M., *Restricted topological entropy and expanding mappings*. Math. Syst. Theory 7 (1974), 318-322.

[192] WILLIAMS, B., *A note on unstable homeomorphisms*. Proc. Amer. Math. Soc. 6 (1955), 308-309.

[193] ---, *Classification of subshifts of finite type*. Ann. of Math. 98 (1973), 120-153, Errata: Ann. of Math. 99 (1974), 380-381.

[194] BILLINGSLEY, P.: *Convergence of probability measures*. J. Wiley 1968.

[195] CHACON, R., *A geometric construction of measure preserving transformations*. Proc. V Berkeley Symp. Math. Stat. Prob. Univ. of California Press, 1967, Vol. II., part 2, 335-360.

[196] DENKER, M., *Maße für Axiom A Diffeomorphismen*. Diplomarbeit. Universität Erlangen, 1970.

[197] GANTMACHER, F., *Theory of matrices*. New York 1959.

[198] GRILLENBERGER, Chr., *Ensembles minimaux sans mesure d'entropie maximale*. To appear in: Monatshefte für Mathematik.

[199] ---, U. KRENGEL, *On marginal distributions and isomorphisms of stationary processes*. To appear in Mathematische Zeitschrift.

[200] HIRSCH, M.W., C.C. PUGH, *Stable manifolds and hyperbolic sets*. Global Analysis.

[201] KAKUTANI, S., *Examples of ergodic measure-preserving transformations which are weakly mixing but not strongly mixing*. Recent advances in Topological Dynamics. Springer Lecture Notes in Math. 318 (1973).

[202] LEDRAPPIER, F.,*Principe variationnel et systèmes dynamiques symboliques*. Z. Wahrscheinlichkeitstheorie verw. Geb. 30 (1974), 185-202.

[203] SINAI, Ya., G., *Theory of dynamical systems*. Part I: Ergodic Theory. (1970), Aarhus Lecture Notes, Series 23.

[204] HOFBAUER, F., *Examples for the nonuniqueness of the equilibrium state*. Preprint.

[205] SIGMUND, K., *On the distribution of periodic points for β-shifts*. Preprint.

[206] ---, *On minimal centers of attraction*. Preprint.

[207] EBERLEIN, E., *On topological entropy of semigroups of commuting transformations*. Preprint.

[208] MISIUREWICZ, M., *A short proof of the variational principle for a* Z_+^N *action on a compact space*. Preprint.

[209] MESHALKIN, L.D., *A case of isomorphism of Bernoulli schemes*. Dokl. Akad. Nauk SSSR 128 (1959), 41-44. (Russian).

[210] FISCHER, R., *Sofic systems and graphs*. To appear in Monatshefte Math.

[211] ---, *Graphs and symbolic dynamics*. Preprint.

Index

C.

D.

E.

G.

H.

M.

List of Symbols

$\bar{A}$ closure of A (not always!)

int A = interior of A

bd A = boundary of A

card = cardinality

Vol. 399: Functional Analysis and its Applications. Proceedings 1973. Edited by H. G. Garnir, K. R. Unni and J. H. Williamson. II, 584 pages. 1974.

Vol. 400: A Crash Course on Kleinian Groups. Proceedings 1974. Edited by L. Bers and I. Kra. VII, 130 pages. 1974.

Vol. 401: M. F. Atiyah, Elliptic Operators and Compact Groups. V, 93 pages. 1974.

Vol. 402: M. Waldschmidt, Nombres Transcendants. VIII, 277 pages. 1974.

Vol. 403: Combinatorial Mathematics. Proceedings 1972. Edited by D. A. Holton. VIII, 148 pages. 1974.

Vol. 404: Théorie du Potentiel et Analyse Harmonique. Edité par J. Faraut. V, 245 pages. 1974.

Vol. 405: K. J. Devlin and H. Johnsbråten, The Souslin Problem. VIII, 132 pages. 1974.

Vol. 406: Graphs and Combinatorics. Proceedings 1973. Edited by R. A. Bari and F. Harary. VIII, 355 pages. 1974.

Vol. 407: P. Berthelot, Cohomologie Cristalline des Schémas de Caracteristique $p > o$. II, 604 pages. 1974.

Vol. 408: J. Wermer, Potential Theory. VIII, 146 pages. 1974.

Vol. 409: Fonctions de Plusieurs Variables Complexes, Séminaire François Norguet 1970–1973. XIII, 612 pages. 1974.

Vol. 410: Séminaire Pierre Lelong (Analyse) Année 1972–1973. VI, 181 pages. 1974.

Vol. 411: Hypergraph Seminar. Ohio State University, 1972. Edited by C. Berge and D. Ray-Chaudhuri. IX, 287 pages. 1974.

Vol. 412: Classification of Algebraic Varieties and Compact Complex Manifolds. Proceedings 1974. Edited by H. Popp. V, 333 pages. 1974.

Vol. 413: M. Bruneau, Variation Totale d'une Fonction. XIV, 332 pages. 1974.

Vol. 414: T. Kambayashi, M. Miyanishi and M. Takeuchi, Unipotent Algebraic Groups. VI, 165 pages. 1974.

Vol. 415: Ordinary and Partial Differential Equations. Proceedings 1974. XVII, 447 pages. 1974.

Vol. 416: M. E. Taylor, Pseudo Differential Operators. IV, 155 pages. 1974.

Vol. 417: H. H. Keller, Differential Calculus in Locally Convex Spaces. XVI, 131 pages. 1974.

Vol. 418: Localization in Group Theory and Homotopy Theory and Related Topics. Battelle Seattle 1974 Seminar. Edited by P. J. Hilton. VI, 172 pages 1974.

Vol. 419: Topics in Analysis. Proceedings 1970. Edited by O. E. Lehto, I. S. Louhivaara, and R. H. Nevanlinna. XIII, 392 pages. 1974.

Vol. 420: Category Seminar. Proceedings 1972/73. Edited by G. M. Kelly. VI, 375 pages. 1974.

Vol. 421: V. Poénaru, Groupes Discrets. VI, 216 pages. 1974.

Vol. 422: J.-M. Lemaire, Algèbres Connexes et Homologie des Espaces de Lacets. XIV, 133 pages. 1974.

Vol. 423: S. S. Abhyankar and A. M. Sathaye, Geometric Theory of Algebraic Space Curves. XIV, 302 pages. 1974.

Vol. 424: L. Weiss and J. Wolfowitz, Maximum Probability Estimators and Related Topics. V, 106 pages. 1974.

Vol. 425: P. R. Chernoff and J. E. Marsden, Properties of Infinite Dimensional Hamiltonian Systems. IV, 160 pages. 1974.

Vol. 426: M. L. Silverstein, Symmetric Markov Processes. X, 287 pages. 1974.

Vol. 427: H. Omori, Infinite Dimensional Lie Transformation Groups. XII, 149 pages. 1974.

Vol. 428: Algebraic and Geometrical Methods in Topology, Proceedings 1973. Edited by L. F. McAuley. XI, 280 pages. 1974.

Vol. 429: L. Cohn, Analytic Theory of the Harish-Chandra C-Function. III, 154 pages. 1974.

Vol. 430: Constructive and Computational Methods for Differential and Integral Equations. Proceedings 1974. Edited by D. L. Colton and R. P. Gilbert. VII, 476 pages. 1974.

Vol. 431: Séminaire Bourbaki – vol. 1973/74. Exposés 436–452. IV, 347 pages. 1975.

Vol. 432: R. P. Pflug, Holomorphiegebiete, pseudokonvexe Gebiete und das Levi-Problem. VI, 210 Seiten. 1975.

Vol. 433: W. G. Faris, Self-Adjoint Operators. VII, 115 pages. 1975.

Vol. 434: P. Brenner, V. Thomée, and L. B. Wahlbin, Besov Spaces and Applications to Difference Methods for Initial Value Problems. II, 154 pages. 1975.

Vol. 435: C. F. Dunkl and D. E. Ramirez, Representations of Commutative Semitopological Semigroups. VI, 181 pages. 1975.

Vol. 436: L. Auslander and R. Tolimieri, Abelian Harmonic Analysis, Theta Functions and Function Algebras on a Nilmanifold. V, 99 pages. 1975.

Vol. 437: D. W. Masser, Elliptic Functions and Transcendence. XIV, 143 pages. 1975.

Vol. 438: Geometric Topology. Proceedings 1974. Edited by L. C. Glaser and T. B. Rushing. X, 459 pages. 1975.

Vol. 439: K. Ueno, Classification Theory of Algebraic Varieties and Compact Complex Spaces. XIX, 278 pages. 1975

Vol. 440: R. K. Getoor, Markov Processes: Ray Processes and Right Processes. V, 118 pages. 1975.

Vol. 441: N. Jacobson, PI-Algebras. An Introduction. V, 115 pages. 1975.

Vol. 442: C. H. Wilcox, Scattering Theory for the d'Alembert Equation in Exterior Domains. III, 184 pages. 1975.

Vol. 443: M. Lazard, Commutative Formal Groups. II, 236 pages. 1975.

Vol. 444: F. van Oystaeyen, Prime Spectra in Non-Commutative Algebra. V, 128 pages. 1975.

Vol. 445: Model Theory and Topoi. Edited by F. W. Lawvere, C. Maurer, and G. C. Wraith. III, 354 pages. 1975.

Vol. 446: Partial Differential Equations and Related Topics. Proceedings 1974. Edited by J. A. Goldstein. IV, 389 pages. 1975.

Vol. 447: S. Toledo, Tableau Systems for First Order Number Theory and Certain Higher Order Theories. III, 339 pages. 1975.

Vol. 448: Spectral Theory and Differential Equations. Proceedings 1974. Edited by W. N. Everitt. XII, 321 pages. 1975.

Vol. 449: Hyperfunctions and Theoretical Physics. Proceedings 1973. Edited by F. Pham. IV, 218 pages. 1975.

Vol. 450: Algebra and Logic. Proceedings 1974. Edited by J. N. Crossley. VIII, 307 pages. 1975.

Vol. 451: Probabilistic Methods in Differential Equations. Proceedings 1974. Edited by M. A. Pinsky. VII, 162 pages. 1975.

Vol. 452: Combinatorial Mathematics III. Proceedings 1974. Edited by Anne Penfold Street and W. D. Wallis. IX, 233 pages. 1975.

Vol. 453: Logic Colloquium. Symposium on Logic Held at Boston, 1972–73. Edited by R. Parikh. IV, 251 pages. 1975.

Vol. 454: J. Hirschfeld and W. H. Wheeler, Forcing, Arithmetic, Division Rings. VII, 266 pages. 1975.

Vol. 455: H. Kraft, Kommutative algebraische Gruppen und Ringe. III, 163 Seiten. 1975.

Vol. 456: R. M. Fossum, P. A. Griffith, and I. Reiten, Trivial Extensions of Abelian Categories. Homological Algebra of Trivial Extensions of Abelian Categories with Applications to Ring Theory. XI, 122 pages. 1975.

Vol. 457: Fractional Calculus and Its Applications. Proceedings 1974. Edited by B. Ross. VI, 381 pages. 1975.

Vol. 458: P. Walters, Ergodic Theory – Introductory Lectures. VI, 198 pages. 1975.

Vol. 459: Fourier Integral Operators and Partial Differential Equations. Proceedings 1974. Edited by J. Chazarain. VI, 372 pages. 1975.

Vol. 460: O. Loos, Jordan Pairs. XVI, 218 pages. 1975.

Vol. 461: Computational Mechanics. Proceedings 1974. Edited by J. T. Oden. VII, 328 pages. 1975.

Vol. 462: P. Gérardin, Construction de Séries Discrètes p-adiques. »Sur les séries discrètes non ramifiées des groupes réductifs déployés p-adiques«. III, 180 pages. 1975.

Vol. 463: H.-H. Kuo, Gaussian Measures in Banach Spaces. VI, 224 pages. 1975.

Vol. 464: C. Rockland, Hypoellipticity and Eigenvalue Asymptotics. III, 171 pages. 1975.

Vol. 465: Séminaire de Probabilités IX. Proceedings 1973/74. Edité par P. A. Meyer. IV, 589 pages. 1975.

Vol. 466: Non-Commutative Harmonic Analysis. Proceedings 1974. Edited by J. Carmona, J. Dixmier and M. Vergne. VI, 231 pages. 1975.

Vol. 467: M. R. Essén, The Cos $\pi\lambda$ Theorem. With a paper by Christer Borell. VII, 112 pages. 1975.

Vol. 468: Dynamical Systems – Warwick 1974. Proceedings 1973/74. Edited by A. Manning. X, 405 pages. 1975.

Vol. 469: E. Binz, Continuous Convergence on C(X). IX, 140 pages. 1975.

Vol. 470: R. Bowen, Equilibrium States and the Ergodic Theory of Anosov Diffeomorphisms. III, 108 pages. 1975.

Vol. 471: R. S. Hamilton, Harmonic Maps of Manifolds with Boundary. III, 168 pages. 1975.

Vol. 472: Probability-Winter School. Proceedings 1975. Edited by Z. Ciesielski, K. Urbanik, and W. A. Woyczyński. VI, 283 pages. 1975.

Vol. 473: D. Burghelea, R. Lashof, and M. Rothenberg, Groups of Automorphisms of Manifolds. (with an appendix by E. Pedersen) VII, 156 pages. 1975.

Vol. 474: Séminaire Pierre Lelong (Analyse) Année 1973/74. Edité par P. Lelong. VI, 182 pages. 1975.

Vol. 475: Répartition Modulo 1. Actes du Colloque de Marseille-Luminy, 4 au 7 Juin 1974. Edité par G. Rauzy. V, 258 pages. 1975. 1975.

Vol. 476: Modular Functions of One Variable IV. Proceedings 1972. Edited by B. J. Birch and W. Kuyk. V, 151 pages. 1975.

Vol. 477: Optimization and Optimal Control. Proceedings 1974. Edited by R. Bulirsch, W. Oettli, and J. Stoer. VII, 294 pages. 1975.

Vol. 478: G. Schober, Univalent Functions – Selected Topics. V, 200 pages. 1975.

Vol. 479: S. D. Fisher and J. W. Jerome, Minimum Norm Extremals in Function Spaces. With Applications to Classical and Modern Analysis. VIII, 209 pages. 1975.

Vol. 480: X. M. Fernique, J. P. Conze et J. Gani, Ecole d'Eté de Probabilités de Saint-Flour IV-1974. Edité par P.-L. Hennequin. XI, 293 pages. 1975.

Vol. 481: M. de Guzmán, Differentiation of Integrals in R^n. XII, 226 pages. 1975.

Vol. 482: Fonctions de Plusieurs Variables Complexes II. Séminaire François Norguet 1974–1975. IX, 367 pages. 1975.

Vol. 483: R. D. M. Accola, Riemann Surfaces, Theta Functions, and Abelian Automorphisms Groups. III, 105 pages. 1975.

Vol. 484: Differential Topology and Geometry. Proceedings 1974. Edited by G. P. Joubert, R. P. Moussu, and R. H. Roussarie. IX, 287 pages. 1975.

Vol. 485: J. Diestel, Geometry of Banach Spaces – Selected Topics. XI, 282 pages. 1975.

Vol. 486: S. Stratila and D. Voiculescu, Representations of AF-Algebras and of the Group U (∞). IX, 169 pages. 1975.

Vol. 487: H. M. Reimann und T. Rychener, Funktionen beschränkter mittlerer Oszillation. VI, 141 Seiten. 1975.

Vol. 488: Representations of Algebras, Ottawa 1974. Proceedings 1974. Edited by V. Dlab and P. Gabriel. XII, 378 pages. 1975.

Vol. 489: J. Bair and R. Fourneau, Etude Géométrique des Espaces Vectoriels. Une Introduction. VII, 185 pages. 1975.

Vol. 490: The Geometry of Metric and Linear Spaces. Proceedings 1974. Edited by L. M. Kelly. X, 244 pages. 1975.

Vol. 491: K. A. Broughan, Invariants for Real-Generated Uniform Topological and Algebraic Categories. X, 197 pages. 1975.

Vol. 492: Infinitary Logic: In Memoriam Carol Karp. Edited by D. W. Kueker. VI, 206 pages. 1975.

Vol. 493: F. W. Kamber and P. Tondeur, Foliated Bundles and Characteristic Classes. XIII, 208 pages. 1975.

Vol. 494: A Cornea and G. Licea. Order and Potential Resolvent Families of Kernels. IV, 154 pages. 1975.

Vol. 495: A. Kerber, Representations of Permutation Groups II. V, 175 pages. 1975.

Vol. 496: L. H. Hodgkin and V. P. Snaith, Topics in K-Theory. Two Independent Contributions. III, 294 pages. 1975.

Vol. 497: Analyse Harmonique sur les Groupes de Lie. Proceedings 1973–75. Edité par P. Eymard et al. VI, 710 pages. 1975.

Vol. 498: Model Theory and Algebra. A Memorial Tribute to Abraham Robinson. Edited by D. H. Saracino and V. B. Weispfenning. X, 463 pages. 1975.

Vol. 499: Logic Conference, Kiel 1974. Proceedings. Edited by G. H. Müller, A. Oberschelp, and K. Potthoff. V, 651 pages 1975.

Vol. 500: Proof Theory Symposion, Kiel 1974. Proceedings. Edited by J. Diller and G. H. Müller. VIII, 383 pages. 1975.

Vol. 501: Spline Functions, Karlsruhe 1975. Proceedings. Edited by K. Böhmer, G. Meinardus, and W. Schempp. VI, 421 pages. 1976.

Vol. 502: János Galambos, Representations of Real Numbers by Infinite Series. VI, 146 pages. 1976.

Vol. 503: Applications of Methods of Functional Analysis to Problems in Mechanics. Proceedings 1975. Edited by P. Germain and B. Nayroles. XIX, 531 pages. 1976.

Vol. 504: S. Lang and H. F. Trotter, Frobenius Distributions in GL_2-Extensions. III, 274 pages. 1976.

Vol. 505: Advances in Complex Function Theory. Proceedings 1973/74. Edited by W. E. Kirwan and L. Zalcman. VIII, 203 pages. 1976.

Vol. 506: Numerical Analysis, Dundee 1975. Proceedings. Edited by G. A. Watson. X, 201 pages. 1976.

Vol. 507: M. C. Reed, Abstract Non-Linear Wave Equations. VI, 128 pages. 1976.

Vol. 508: E. Seneta, Regularly Varying Functions. V, 112 pages. 1976.

Vol. 509: D. E. Blair, Contact Manifolds in Riemannian Geometry. VI, 146 pages. 1976.

Vol. 510: V. Poènaru, Singularités C^∞ en Présence de Symétrie. V, 174 pages. 1976.

Vol. 511: Séminaire de Probabilités X. Proceedings 1974/75. Edité par P. A. Meyer. VI, 593 pages. 1976.

Vol. 512: Spaces of Analytic Functions, Kristiansand, Norway 1975. Proceedings. Edited by O. B. Bekken, B. K. Øksendal, and A. Stray. VIII, 204 pages. 1976.

Vol. 513: R. B. Warfield, Jr. Nilpotent Groups. VIII, 115 pages. 1976.

Vol. 514: Séminaire Bourbaki vol. 1974/75. Exposés 453 – 470. IV, 276 pages. 1976.

Vol. 515: Bäcklund Transformations. Nashville, Tennessee 1974. Proceedings. Edited by R. M. Miura. VIII, 295 pages. 1976.